Eurasia: Sustainable Development, Security, Cooperation 2019

SHS Web of Conferences Volume 71 (2019)

Samara, Russia
25 - 26 October 2019

Editor:

Valentina Mantulenko

ISBN: 978-1-5108-9866-0

Printed from e-media with permission by:

Curran Associates, Inc.
57 Morehouse Lane
Red Hook, NY 12571

CURRAN ASSOCIATES INC. proceedings.com

Some format issues inherent in the e-media version may also appear in this print version.

For additional information, please contact EDP Sciences – Web of Conferences at the address below.

EDP Sciences – Web of Conferences
17, Avenue du Hoggar
Parc d'Activité de Courtabœuf
BP 112
F-91944 Les Ulis Cedex A
France

Phone: +33 (0) 1 69 18 75 75
Fax: +33 (0) 1 69 28 84 91

contact-edps@webofconferences.org

Additional copies of this publication are available from:

Curran Associates, Inc.
57 Morehouse Lane
Red Hook, NY 12571 USA
Phone: 845-758-0400
Fax: 845-758-2633
Email: curran@proceedings.com
Web: www.proceedings.com

TABLE OF CONTENTS

Preface

Modern Eurasian integration processes have been developing dynamically in the last three or four years, although the real mechanisms and tools are far from being fully developed. Currently, the achievements of international cooperation are most noticeable in the system of higher education and science: the unification of educational programs in the framework of the Bologna process, international research grants, international cooperation between universities in educational projects (exchange programs, double degree programs, joint master's programs), etc.

At the same time, integration processes on the Eurasian continent differ significantly from the European Union experience. Their peculiarity is rooted in the different historical, economic, cultural, political development of countries and peoples united in the Shanghai Cooperation Organization (SCO), the Eurasian Economic Community (EAEU) and other international organizations. This creates additional, sometimes insurmountable, difficulties for intercultural integration.

The project 574060-EPP-1-2016-1-KZ-EPPKA2-CBHE-SP "Enhancement of higher education and corporate sectors integration in accordance with new social environment" (ENINEDU), which lays in the basis of these conference intends to enhance business-HEI cooperation in Euroasian area by changing the paradigm of education in line with Bologna process requirements. Improvement of such cooperation would result in gradual effectiveness increase of national development strategy, priority fulfillment, as well as change the graduate entrepreneurship and employment situation.

The idea of the project came from a composition of partner university experience. People's Friendship University and Al-Farabi Kazakh National University (KazNU) have had long cooperation with major international, Kazakhstan and Russian companies. L.N. Gumilyov Eurasian National University (ENU) has experience in both local and European projects and specific business-HEI cooperation practice in their education area. Samara State University of Economics and Orel State University have expertise of operation and cooperation with employers in different regions with various economic conditions. The inter-university discussion of graduate practical training problems have long been conducted at various intercollegiate conferences in the countries of these project consortium.

While performing the ENINEDU project the participants have come to conclusion that all cultural, educational or economic processes of Euroasian integration in long -term trend should result in higher possibilities of investment and competition expenditure, leading to hastier development of market economy.

Involvement of HEI's plays a crucial part of the project. Most of the published outcomes are aimed at changing the educational process at universities of the Euroasian area in order to promote the development of Eurasian economic integration, as well as expanding and strengthening international contacts of the Eurasian continent. This includes the business-HEI cooperation in EU overview, greater connections with businesses and increased number of joint projects with them. Regionally and nationally this project outcome would contribute to the creation of education-business clusters, thus increasing the education effectiveness. The latter will further lead to wider attractiveness of HEI's for potential students and increased stability of HEI's development.

The state regulators themselves would be interested in the recommendations from such a wide range of stakeholders, including EU partners, partner countries HEI's and business community. Locally, regionally and nationally the project outcome implementation is decreasing the municipal and national expenses for state unemployment subsidies, and at the same time increasing the income through taxation of both graduates and business activity.

The International Conference "Eurasia: Sustainable Development, Security, Cooperation" – 2019, being an initiative of three universities within this project, was aimed at analyzing geopolitical and economic situation in Central Eurasia which is quickly changing. The conference organizers tried to bring together researches and practioners from political, economic, legal, cultural and educational areas to discuss the above mentioned changes and challenges from different perspectives.

Svetlana Ashmarina, *Valentina* Mantulenko & *Anna* Zotova
EURASIA-2019 Organizing Committee,
Samara, Russia

edp sciences

Web of Conferences

Statement of Peer review

In submitting conference proceedings to *Web of Conferences*, the editors of the proceedings certify to the Publisher that

1. They adhere to its **Policy on Publishing Integrity** in order to safeguard good scientific practice in publishing.
2. All articles have been subjected to peer review administered by the proceedings editors.
3. Reviews have been conducted by expert referees, who have been requested to provide unbiased and constructive comments aimed, whenever possible, at improving the work.
4. Proceedings editors have taken all reasonable steps to ensure the quality of the materials they publish and their decision to accept or reject a paper for publication has been based only on the merits of the work and the relevance to the journal.

Title, date and place of the conference

"EURASIA SUSTAINABLE DEVELOPMENT, SECURITY, COOPERATION", 25th-26th October, 2019, Samara, Russia

Proceedings editor(s):

Valentina Mantulenko, PhD, Assistant Prof. Samara State University of Economics, Samara, Russia

Date and editor's signature

Oct. 23, 2019

The Relevance and Effectiveness of Adaptation Measures in Russian and Foreign Companies

A.I. Tikhonov[1], and *V.G.* Konovalova[2*]
*Corresponding author: vg_konovalova@guu.ru.

[1] Moscow Aviation Institute (National Research University), Moscow, Russia
[2] State University of Management, Moscow, Russia

Abstract. This article discusses current issues related to the adaptation of new employees in enterprises. Previously most companies did not attach serious importance to this, but over the past 20 years the use of adaptation programs as an effective personnel management tool has grown markedly in the world and Russian markets. We consider the main problems of creating and using new adaptation programs in the practice of Russian and foreign companies, because competent adaptation of personnel allows us to simplify the comfortable entry of new employees into the team and to ensure increased labor productivity. As a result of creating favorable psychological and social conditions, the goals of an individual person and the entire professional group come closer together.

Keywords: HR management, employee adaptation, adaptation program, talent management.

1 Introduction

Since the 2000s the process of adapting new employees has become an important part of working with staff in many large companies. By 2008 among large companies in North America, UK and the Asia-Pacific region – one of the most economically developed regions in the world – 62% had a developed system for adapting newcomers, and in 2009 this share was estimated at 77% [1]. In Russia the high penetration rate of adaptation programs on the market was achieved in the next decade: in 2017, 76% of companies used a formal adaptation program in one form or another among Russian companies. Among large companies (with a staff of more than 1 thousand people), the share of firms using adaptation programs in the Russian market is even higher at 86% [2].

2 Problem Statement

To date the process of adaptation of beginners is one of the basic in working with staff. At the same time the adaptation programs themselves can vary greatly from company to company, and by no means do they always correspond to the needs of business or applicants. These problems as well as general aspects of building adaptation programs are considered in the article.

Investments in adaptation programs are beneficial for employers for several reasons. According to statistics, in the United States, workers aged 18 to 37 change on average 10 jobs each, and a significant proportion of staff turnover falls on the employees working in the company for the first year [1]. The departure of a newly hired employee not only forces the company to spend money on finding and hiring a new specialist, but also reduces the productivity of existing staff. After hiring another candidate, employees again have to spend time getting to know and helping a novice, sacrificing the performance of work duties. The presence of adaptation programs directly affects staff turnover in the company – according to a study conducted in Poland, the inclusion of employees in a well-thought-out adaptation program increases the likelihood that they will remain in the company for the next three years by 69% [3].

A study conducted in 2000 by the Harvard University showed that up to 80% of staff turnover in companies can be explained by errors in hiring and in the process of adapting new employees [1]. At the same time, the adaptation process is not obliged to be limited only to the introduction of the employee into the work team at the beginning of his work. Many researchers in the field of personnel management see in adaptation programs a tool whose final purpose is to achieve the maximum level of productivity from the employee in the shortest possible time elapsed since hiring [1]. That is why the duration of adaptation programs can stretch for up to 90 days and go far beyond interacting with a new employee of the HR department, including his colleagues and the leadership in the active work with newcomers. According to a survey conducted by a researcher at Irish National College on the issue of employee adaptation in large international companies with offices in Dublin, 63% of HR managers in large companies refuse to consider adaptation programs to be exclusively an HR tool, calling it rather a general business tool. At the same time, all 100% of the HR managers surveyed agreed that this tool is important for the company [1].

Adaptation programs may include not only the technical part of acquainting an employee with new information, but also immersion in the corporate environment - familiarization with the corporate culture and values of the company, getting to know the team in an informal setting (up to participating in joint informal activities), constant feedback with leadership on progress in fulfilling work tasks (the rule is widely known according to which a new employee needs to be given feedback on his success through not share, month and three months of work). A structured and balanced process of adaptation makes employees happier and more involved in the work process. A hired job seeker not only feels suitable for a new job, but also feels a sense of belonging to the goals and strategy of the employer, which affects his productivity [2, 3, 4]. Thus the process of adaptation of newcomers directly affects not only the company's expenses on personnel, but also on its income by increasing the productivity of the team.

3 Research Questions

Among the companies that do not have an adaptation program (as mentioned above, there are 24% of those on the Russian market), the most frequent reason for refusing to use them is the lack of corresponding need - this is the opinion of 33% of those companies that do not have adaptation programs at all. The same number of companies consider the main reason the lack of the necessary human resources for the implementation of the program (that is, there is no person in the team who could take on this responsibility) [2].

19% of HR managers surveyed admitted that they lacked a budget or that their management was opposed. 14% have no understanding of how to organize the process of adaptation of beginners. Thus, the resources for introducing adaptation programs on the Russian market can hardly be called completely exhausted, and in the future the prevalence of using this tool may increase [5].

4 Purpose of the Study

The purpose of this study is a comparative analysis of the effectiveness of various adaptation measures. We consider the main problems of creating and using new adaptation programs in the practice of Russian and foreign companies. The choice of the purpose of the study is determined by the fact that competent adaptation of the personnel makes it possible to simplify the comfortable entry of new employees into the team and to ensure increased labor productivity.

5 Research Methods

Not all employees of a certain company can go through adaptation programs. In a survey conducted by Head Hunter among applicants and employers in the Russian labor market, respondents were asked to choose the options for implementing adaptation programs that were present in their work practice. The survey results showed that on average, employers chose more options than job seekers - this was reflected in the higher popularity rates of each of the answer options in the employer rating compared to the job applicant rating. For example, only 30% of applicants said they had experienced an introductory briefing on corporate rules, while among companies, 91% said that they had such a briefing (of those applicants and employers in whose practice, in principle, adaptation programs were present). This difference can be explained not only by the selectivity in using adaptation programs for different categories of personnel, but also by the difference in the perception of adaptation measures - a conversation with a manager about corporate rules can be perceived by an employee as informal, but at the same time be part of the manager's real responsibilities [5].

In foreign companies, the attitude of employees towards adaptation programs on average is more positive than in Russian [6, 7]. According to a survey conducted by an employee of the Irish National College among employees of large financial international companies, 26% of employees consider their employer's effective adaptation program to be 64% more effective [1]. Thus 80% of specialists recognize the effectiveness of the existing adaptation program, which is more than an indicator of the effectiveness of any of the adaptation measures in the opinion of Russian applicants [5].

Russian employers and applicants recognize mentoring as the most effective adaptation measure (99% and 75% approving, respectively) [5]. The practice of mentoring can have a different meaning: the most widespread practices are referred to in English as buddy-programs, that is, formalized programs to help an employee painlessly join the team from a specially selected colleague.

According to a survey by the Irish National College, 75% of companies have corresponding elements in their adaptation programs (68% of respondents among applicants have encountered them). Elements of mentoring and coaching in foreign companies are much less common - they are present in only 12% of those companies that have adaptation programs. But among the applicants, 53% came across elements of mentoring - this suggests that the transfer of directly working knowledge is also a common practice in the process of mentoring, although it may be less obligatory and formalized. Coaching turned out to be a rare occurrence in the experience of applicants as well - only 15% experienced it [1].

In Russian realities, the practice of helping an employee with adaptation in the team is accompanied by training in working skills even more often. So 76% of applicants and 89% of employers (of those who have come across adaptation programs) explain mentoring as the transfer of practical professional skills related to work responsibilities. At the same time 65% of applicants and 85% of employers see in mentoring the practice of explaining the general corporate processes of work related to internal systems of interaction, the use of special programs, and so on. And only 55% of applicants and 72% of employers recognize the mentor as the duty of general support for a novice - notification of the unspoken rules of the team, assistance in resolving everyday issues, and so on [5]. Thus the teaching aspect is expressed in Russian mentoring programs more than social or psychological.

6 Findings

The importance of adaptation programs is obvious to the vast majority of Russian job seekers and employers. Adaptation programs turn out to be widespread in the global and Russian markets, but the quality of programs in Russian companies far from always meets the requirements of applicants. One way to solve the problem of discrepancies between the requests of applicants and the wishes of employers is to build a feedback system with employees. In foreign companies, 62% of companies have a feedback system with newcomers (from those who, in principle, practice the use of adaptation programs) [1]. It is the presence of feedback that makes it possible to better adapt adaptation programs to the needs of applicants and to avoid the repetition of errors during adaptation.

Employers should pay more attention to the socialization practices of new employees – this is the aspect that is the main one in many large companies from technological sectors that practice adaptation programs [8, 9]. Practices such as introducing new tools for newcomers to communication (from digital chats or websites to premises allocated inside the office), conducting joint events (joint athletic competitions or obligatory distribution of coffee with donuts by newly arrived employees are real cases of US companies [1]), the extension of the adaptation program for up to three months: all this contributes to the retention of staff and the speedy bringing of the productivity of new employees to the maximum level of productivity.

7 Conclusion

In a situation where the labor market is increasingly dependent on job seekers rather than employers – and this is precisely the direction in which the labor market in Russia has been moving in recent decades – companies have to allocate additional funds in order to retain staff, search for and hire talented employees. The importance of adaptation measures in organizations that face these problems most often (for example, in such technology-intensive industries as IT or banking) should encourage employers to pay more attention to progressive and team-based working methods, including with new employees.

References

1. J. O'Brien, *Employee onboarding: An analysis of best practice in employee onboarding and their implementation in the financial services industry in Ireland* (National College of Ireland, Dublin, 2013). URL: http://trap.ncirl.ie/868/1/johnobrien.pdf. Accessed: 31.08.19..
2. J. Kaji, B. Hurley, N. Gangopadhyay, R. Bhat, A. Khan, *Leading the social enterprise: Reinvent with a human focus. 2019 Deloitte Global Human Capital Trends* (2019). URL: https://www2.deloitte.com/content/dam/Deloitte/cz/Documents/human-capital/cz-hc-trends-reinvent-with-human-focus.pdf. Accessed: 31.08.19.
3. K.L. Pike, New employee onboarding programs and person organization fit: An examination of socialization tactics. *Seminar Research Paper Series*, Paper 24 (2014). URL: https://digitalcommons.uri.edu/cgi/viewcontent.cgi?referer=https://www.google.com/&httpsredir=1&article=1043&context=lrc_paper_series. Accessed: 31.08.19.
4. J. Nekoranec, L. Nagyová, Adaptation of employees in the organization and its importance in terms of human resource management. *Revista Academiei Forțelor Terestre,* **1**(73), 114-120 (2014).
5. Head Hunter, *Almanac HeadHunter 2017* (2017). URL: https://hh.ru/article/21200. Accessed: 31.08.19. [in Rus.].
6. A. Zalesna, The impact of the process of adaptation and knowledge sharing on the assessment of suitability of a new employee in the company: Case studies. *Torun Business Review*, **14**(1), 67-80 (2016).
7. M. Tomcikova, P. Zivcak, The proceedings of adaptation process in the company. *Applied Computer Science*, **1**(8), 58-63 (2012).
8. M. Nation, C. Crusto, A. Wandersman, K.L. Kumpfer, D. Seybolt, E. Morrissey-Kane, K. Davino, What works in prevention: Principles of effective prevention programs. *American Psychologist*, **58**(6-7), 449-456 (2003).
9. F.G. Castro, M. Barrera Jr., C.R. Martinez Jr., The cultural adaptation of prevention interventions: Resolving tensions between fidelity and fit. *Prevention Science*, **5**(1), 41-45 (2004).

Enhancing Export Diversification in Resource-Rich Economies – Policies and Evidence from Kazakhstan

P. Zhelev[1,2*]
* Corresponding author: pzhelev@unwe.bg.

[1]University of National and World Economy, Sofia, Bulgaria
[2]Korkyt Ata Kyzylorda State University, Kyzylorda, Kazakhstan

Abstract. Kazakhstan has managed to benefit from its considerable endowments of natural resources during the commodity price boom in the 2000s. However, the global crisis of 2008-09 has demonstrated that relying on a single sector prone to high price volatility cannot be a viable strategy for development. The government of Kazakhstan well realizes that and for already 2 decades has been pursuing various industrial policy initiatives to build a more diversified and competitive economy. The paper aims to examine how successful were those policies by applying various indicators of export diversification. The results show that progress has been very limited as the country's export basket is highly concentrated and still dominated by oil. This suggests that Kazakhstan should ensure better implementation, coordination, monitoring and evaluation of its diversification initiatives.

Keywords: export, diversification, policy, natural resources.

1 Introduction

Kazakhstan, the biggest economy in Central Asia, is a resource-rich country endowed with considerable amounts of natural resources including coal, oil, gas, uranium, copper, iron, zinc. At the end of 2018 Kazakhstan possessed 2.4% (10th position in the world) of the total world proved coal resources, 0.5% (24th position) of natural gas and 1.7% (12th position) of oil [1]. Capitalizing on its extensive natural resources and favourable world conjuncture with exceptionally high commodity prices (driven mainly by the booming demand for natural resources of China) it has registered impressive growth rates since the early 2000s. That made possible building infrastructure, fighting poverty, increasing GDP per capita and already in 2006 entering the upper-middle-income group of countries.

However, this remarkable economic performance and the related increase in living standards have been largely due to the country's extractive industries and especially oil, accounting for more than half of exports and a large percentage of total government revenue [2]. This concentration on a single sector that moreover is notorious for being characterized by high price volatility makes Kazakhstan extremely vulnerable to external shocks. The Great Recession and the end of the commodity price boom with the subsequent slump in the economic growth in resource-rich Central Asian countries have demonstrated that such a growth model based on commodities export cannot be sustained in the long-term.

While Kazakhstan's proven oil reserves are estimated at 30 billion barrels and based on current production levels (about 1.67 million barrels per day as of 2018 according to BP data [1]) it will take more than 42 years to full depletion of its proven oil resources, the country will be affected much sooner by the forthcoming developments in the global market for fossil fuels. In a recent study the EBRD (2018) expects significant changes in fossil fuel industries that due to technological changes and fulfilment of climate change commitments embodied in the UN's SDGs and the Paris Agreement by major countries will lead to a transition to a green global economy which will keep prices of oil subdued. That will pose significant economic risks to Kazakhstan and requires resolute policy measures to encourage growth of the non-oil sectors and promote structural transformation of the economy [3]. In various country reports the OECD has also identified economic diversification as one of the major challenges for Kazakhstan [4].

The Kazakh government well realizes that it is imperative for the national interest to develop a more diversified and competitive economy and in line with that it has enacted various industrial policy initiatives since the end of the 1990s.

In this context, the main objective of the paper is to examine how successful were the industrial policy initiatives pursued by Kazakhstan in the last decade by analysing the export diversification of its economy.

2 Methods

Various studies discussed economic diversification in Kazakhstan. Based on the Herfindahl-Hirschman Index for the period 2005-2015, Bayramov and Orujova (2017) find out that the exports of Kazakhstan were highly concentrated that raises concerns regarding the sustainability of long-term economic growth in the economy [5].

A report of the Asian Bank for Development (2018) used the number of products exported with revealed comparative advantage (RCA) to measure the diversification of the Kazakh economy. It covered data for the period 1995-2015 and found out that overall diversification deteriorated, as the total number of products with comparative advantage fell from 82 in 1995 to 45 to 2010, and picked up only slightly to 60 in 2015 [2]. Discussing Kazakhstan's diversification initiatives Howie (2018) also contends that they have not resulted in increasing the range of the country's exports [6].

Since previous studies cover data up to 2015 it is instructive to see whether the continuous implementation and updating of the economic diversification policy of Kazakhstan has managed to produce better results in the following years by using more recent data.

Export diversification could be measured by various trade indicators. Since any single indicator has its limitations we are going to apply a broader set of indicators:

- share of manufactured goods vs. primary commodities in total merchandise exports;
- share of the top three goods exports in total exports of the country;
- number of exported products.

In addition, we are going to use several more complex trade indicators computed by the UNCTAD, presented in table 1.

Table 1. Selected trade indicators used for measuring export diversification.

Trade indicator	Formula	Meaning
Concentration index (Herfindahl-Hirschmann Index – HHI)	$H_j = \frac{\sqrt{\sum_{i=1}^{n}\left(\frac{x_{ij}}{X_j}\right)^2} - \sqrt{1/n}}{1-\sqrt{1/n}}$ where xij = value of export for country j and product i $X_j = \sum_{i=1}^{n} x_{ij}$ n = number of products (SITC Revision 3 at 3-digit group level)	HHI index value closer to 1 indicates a country's exports are highly concentrated on a few products. On the contrary, values closer to 0 reflect exports are more homogeneously distributed among a series of products.
Diversification index	$S_j = \frac{\sum_i \lvert h_{ij} - h_i \rvert}{2}$ where: hij = share of product i in total exports of country j hi = share of product i in total world exports	The index is calculated by measuring the absolute deviation of the export structure of a country from the world structure. It takes value between o and 1 and a value closer to 1 indicates greater divergence from the world pattern.
RCA index	$RCA_{ij} = \frac{x_{ij}}{x_{it}} \div \frac{x_{wj}}{x_{wt}}$ where Xij and Xwj, are values of country i's and world exports of product j, respectively; Xit and Xwt are values of country i's total exports and world total exports, respectively	RCA index measures the existence of comparative advantage by comparing the country of interests' export profile with the world average. Country i has a CA in a product j when its ratio of exports of product j to its total good exports exceeds the same ratio for the world as a whole, i.e when RCA>1.

Source: compiled by the author based on UNCTAD [7].

3 Policies for export diversification in Kazakhstan

Kazakhstan has been trying to diversify its economy since 1997 when it introduced for the first time a long-term development strategy Kazakhstan-2030 which laid the ground for further strategic documents and policies for economic diversification. Accordingly, a Strategy for Industrial and Innovative Development 2003-2015 (SIID 2003-2015) was adopted which aimed at supporting modernization and diversification of the Kazakh economy. While over the period the dependence on extractive industries continued, one of the major outcomes of the programme was the establishment of technology parks, industrial zones and institutes of development [6].

The institutions that were established to carry out the diversification strategy included: the Development Bank of Kazakhstan, the Investment Fund of Kazakhstan, the National Innovation Fund, and the Export Insurance Corporation. With a view to improve efficiency and management in 2006 Kazyna Sustainable Development Fund was created and it absorbed all of the development agencies. In 2008 it merged with the Samruk State Holding Co. to form Sovereign Wealth Fund Samruk-Kazyna JSC where the sole shareholder is the government of Kazakhstan. Samruk-Kazyna JSC Group includes enterprises operating in several industries – oil, transport and logistics, chemical and nuclear, mining and smelting, energy, mechanical engineering and real estate and as of 2018 the Fund and its portfolio companies share in the economy amounted to KZT 4 trillion [8].

After the Global Financial Crisis, the SIID 2003-2015 was revised and the National Program for Forced Industrial and Innovative Development (2010–2014) (NPFIID) was launched. Its objective is to "guarantee stable and well-balanced economic growth by means of diversification and improvement of its competitive ability." To achieve concentration of resources of government and businesses priority sectors for development were identified: *traditional* (oil and gas sector, ore mining and smelting complex, atomic and chemical industry); *based on national demand* (machinery, building industry, pharmaceutics); *export-oriented sectors* (agro-industrial complex, light industry, tourism); *sectors "of economy of the future"* (ICTs, biotechnologies, alternative energetics, space activities) [9].

In 2015 the Ministry of Economy evaluated the NPFIID and found out that out of the 191 evaluation criteria 147 were achieved (44 not) and despite some positive impact on the economy the major goal of diversification was not attained. The factors behind that included: large coverage of economic sectors, projects, and indicators considered by the programme; deterioration of external economic factors; and underfunding of the NPFIID. Furthermore, there was little involvement of the private sector and the regions in the implementation of the programme, inefficient coordination between state bodies, and a poor methodology for monitoring and evaluation [2].

Based on the experience with NPFIID and as its logical continuation and following the long-term priorities of the Strategy "Kazakhstan-2050" for the implementation of key direction "Accelerating economic diversification", the State program on industrial-innovative development of Kazakhstan for 2015-2019 (SPIID) was designed [10]. The SPIID has a bigger budget (KZT 1.7 billion), less priorities (ferrous and nonferrous metallurgy, oil processing, petro-chemistry and agro-chemistry, food production, car manufacturing, and electro-technical machine building) and improved coordination, monitoring and control. Furthermore, the program aims at fostering the development of regional industrial clusters, improvement of the business climate, stimulating entrepreneurship and SMEs and attracting foreign direct investment by providing fiscal and non-fiscal stimuli. A special focus is put on internationalization and a major criterion for providing state support is export-orientation of the project.

In the next section we will explore the results achieved by the industrial policy in Kazakhstan in terms of diversifying the economy away from extractive industries.

4 Export diversification of Kazakhstan during the last decade

One of the major goals of diversification is to generate exports outside the resource sector in order to lower export volatility and vulnerability to external shocks. Looking at the data for the structure of exports of Kazakhstan on Fig.1 we see that primary commodities continue to dominate and even increase their significance in the country's export basket. In 2008 they accounted for 85.4% of the exports while in 2018 their share was already 87.6%. At the same time manufactured goods have decreased their share from 13.8% to 12.4%.

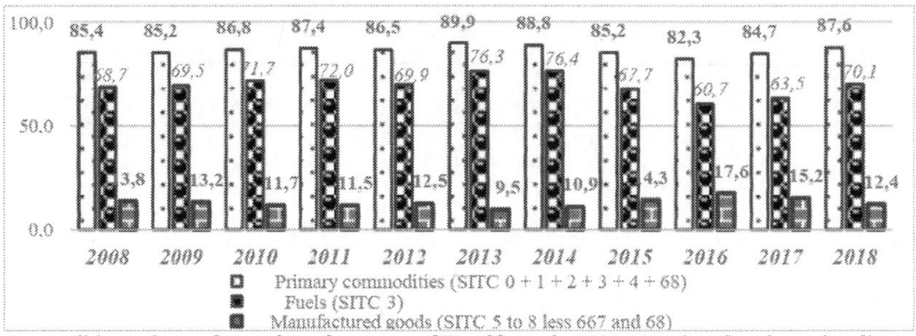

Fig. 1. Primary commodities and manufactured goods exports of Kazakhstan in 2008-2018 (% of total merchandise exports), Source: compiled by the author based on UNCTAD data.

Another perspective is given by looking at more detailed export data. It shows that the share of the top 3 products in Kazakhstan's exports has been hovering between 67% and 71% and in 2018 it stood at 69.3%. Moreover, just one product (defined at the 6-digit HS code) – crude petroleum oil – accounted for the staggering 62% of all merchandise exports in 2018. This represents extremely high concentration and dependence on just a single product. The other two top export products in the Kazakh export list – refined copper (3.7%) and natural gas (3.6%) are also highly resource-intensive and expand the resource dependence of the country (Fig. 2).

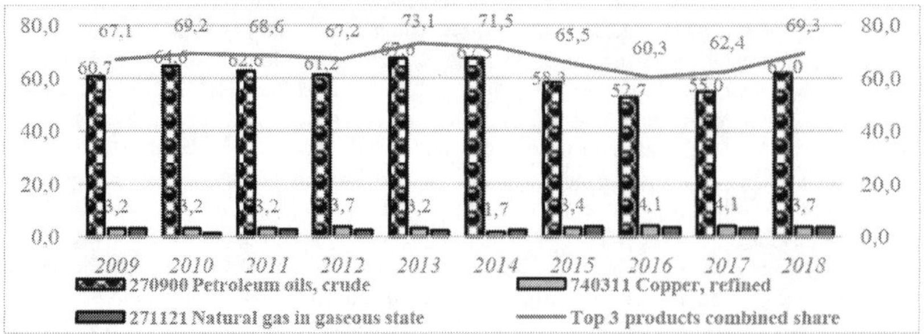

Fig. 2. Share of top three 6-digit HS products in total merchandise exports of Kazakhstan in 2009-2018 (%), Source: compiled by the author based on International Trade Centre data [11].

The high export concentration is confirmed by both the concentration and diversification indices. During the last 10 years they have increased their values in the case of Kazakhstan which means that concentration of the country's exports and its divergence from the world export pattern have become more pronounced (Table 2). This is in contrast to the trend in 3 groups of countries to which Kazakhstan belongs (upper-middle-income economies, exporters of petroleum and emerging markets) which for the period 2008-2018 have managed to improve their performance in terms of export diversification.

Table 2. Selected indicators for export diversification of Kazakhstan and selected group of countries.

	Number of exported products		**Concentration index (HHI index)**		**Diversification index**	
	2008	***2018***	***2008***	***2018***	***2008***	***2018***
Kazakhstan (KZ)	210	235	0.59	0.60	0.74	0.79
Upper-middle-income economies	260	260	0.15	0.09	0.23	0.20
Selected exporters of petroleum	258	260	0.60	0.46	0.66	0.64
Emerging markets: Europe and KZ	258	257	0.17	0.13	0.31	0.29

Source: compiled by the author based on UNCTAD data.

On the positive side, Kazakhstan has managed to increase the number of exported products – from 210 in 2008 to 235 in 2018 (Table 3). However, the comparison with the countries from the same income group and region and those specialized in exports of petroleum again shows significant lagging behind.

Table 3. Product groups with gained and lost comparative advantage of Kazakhstan over 2008-2018.

Product groups with gained CA			Product groups with lost CA		
Product (3-digit SITC)	RCAi in 2008	RCAi in 2018	Product (3-digit SITC)	RCAi in 2008	RCAi in 2018
272 Crude fertilizers	0.9	4.9	971 Gold	1.1	0
344 Petroleum gases	0.8	3.4	325 Coke of coal	1.1	0.1
343 Natural gas	0.8	2.6	523 Metallic salts	1.3	0.5
047 Other cereal meals and flour	0.3	1.6	335 Residual petroleum product	2.6	0.6
684 Aluminium	0.5.	1.5	268 Wool	2.8	0.4
351 Electric current	0.2	1.3	611 Leather	4.5	0.2
524 Other inorg. chemicals	0.95	1.2			
288 Waste and scrap	0.3	1.2			
062 Sugar confectionery	0.1	1.1			
091 Margarine	0.3	1.1			
122 Tobacco, manufactured	0.3	1.0			

Source: compiled by the author based on UNCTAD data.

Finally, we analyse the product groups in which Kazakhstan has managed to gain comparative advantage and the ones where the economy has lost it during the last decade. Table 2 indicates that over the period 2008-2018 diversification in the country increased as the products where competitiveness was achieved outnumbers by 5 products the ones where it was lost. Although this is an encouraging sign, the largest increase in products with positive RCA came from the primary commodities group. However, there are two product groups (electric current and inorganic chemicals) among the ones with gained comparative advantage which could be classified as technology-intensive and contributed not just to diversification but also the sophistication of the Kazakh export basket.

5 Conclusion

The results from the study show that reaching export diversification turns out to be a difficult long-term process. According to the most indicators applied Kazakhstan has achieved very little progress in diversifying its exports into non-extractive sectors during the review period even though it has been pursuing active industrial policies for 2 decades. This suggests that Kazakhstan should ensure better implementation, coordination, monitoring and evaluation of its diversification initiatives. That requires maintaining good governance and institutions and development of human capital including in the public administration. Furthermore, the success of such strategies rests on the efficient engagement of all stakeholders (businesses, NGOs, academia, the regions).

References

1. British Petroleum, *Statistical review of world energy* (2019). URL: https://www.bp.com/en/global/corporate/energy-economics/statistical-review-of-world-energy.html. Accessed: 05.09.2019.
2. K. Anderson, G. Capannelli, E. Ginting, K. Taniguchi (Eds.), *Kazakhstan: Accelerating economic diversification* (ADB, Mandaluyong City, Philippines, 2018). URL: https://www.adb.org/publications/kazakhstan-economic-diversification. Accessed: 05.09.2019.
3. European Bank for Reconstruction and Development, *The fiscal implications for Kazakhstan of worldwide transition to a greener global economy* (2018). URL: https://www.ebrd.com/news/publications/special-reports/the-fiscal-implications-for-kazakhstan-of-worldwide-transition-to-a-greener-global-economy.html. Accessed: 05.09.2019.
4. The Organisation for Economic Co-operation and Development (OECD), *Enhancing competitiveness in Central Asia* (2018). URL: https://www.oecd.org/about/sge/enhancing-competitiveness-in-central-asia-9789264288133-en.htm. Accessed: 05.09.2019.
5. V. Bayramov, L. Orujova, Volatility, diversification and oil shock in resource-rich Turkic countries: Avenues for recovery. *Bilig*, **83**, 303-329 (2017).
6. P. Howie, Policy transfer and diversification in resource-dependent economies: Lessons for Kazakhstan from Alberta. *Politics & Policy*, **46**(1), 110-140 (2018).
7. UNCTAD, URL: https://unctadstat.unctad.org/wds/ReportFolders/reportFolders.aspx. Accessed: 05.09.2019.
8. Sovereign wealth fund «Samruk-Kazyna» JSC website. URL: https://sk.kz/about-fund/?lang=en. Accessed: 05.09.2019.
9. National Program of forced industrial and innovative development of the Republic of Kazakhstan 2010-2014, "Kazakhstanskaya pravda" March 31, 2010, # 74 (26135).
10. M.M. Idrisov, Program of industrial and innovative development of the Republic of Kazakhstan for 2015-2019. *Economy, Entrepreneurship and Law*, **5**(4), 253-272 (2015). DOI: 10.18334/epp.5.4.2111.
11. International Trade Centre, *International trade statistics 2001-2019* (2019). URL: http://www.intracen.org/itc/market-info-tools/trade-statistics/. Accessed: 05.09.2019.

Using Artificial Neural Networks for Equalizing Time Series Considering Seasonal Fluctuations

J. Vrbka[1], *J.* Horák[1*], and *V.*Machová[1]
* Corresponding author: 8913@mail.vstecb.cz.

[1]Institute of Technology and Business, School of Expertness and Valuation, České Budějovice, Czech Republic

Abstract. The objective of this contribution is to prepare a methodology of using artificial neural networks for equalizing time series when considering seasonal fluctuations on the example of the Czech Republic import from the People's Republic of China. If we focus on the relation of neural networks and time series, it is possible to state that both the purpose of time series themselves and the nature of all the data are what matters. The purpose of neural networks is to record the process of time series and to forecast individual data points in the best possible way. From the discussion part it follows that adding other variables significantly improves the quality of the equalized time series. Not only the performance of the networks is very high, but the individual MLP networks are also able to capture the seasonal fluctuations in the development of the monitored variable, which is the CR import from the PRC.

Keywords: artificial neural networks, seasonal fluctuations, import, prediction, time series.

1 Introduction

Time series forecasting is currently one of the most significant fields of statistics in terms of development [1]. The main reason for the growing influence of this discipline is mainly the fact that it is capable of dealing successfully with the description of dynamic systems, which are often faced in practice [2]. The application of time series analysis occurs in every field of human activity, e.g. physics, technology, medicine, social sciences, and economics as well. Using time series is one of the most significant quantitative methods used in data analysis in economics [3].

Analysis of time series is important especially for the planning and forecasting crucial decisions in specific fields. Moreover, he claims that specifically, it is the field of operative management (forecasting the quantity of production), finance (forecasting the investment income), industry, economy (forecasting significant economic indicators) [4].

It is possible to forecast the indicators of foreign trade, i.e. export and import, based on time series as well [5]. Considering the subject of this paper, i.e. forecasting the Czech Republic import from the People's Republic of China, it is appropriate to examine the terms of the import and foreign trade of both countries more closely.

Import is defined by the value of goods which has been received from abroad or crossed the frontier for limited or permanent stay in the country. It means that it involves the goods for consumption, cultivation, repair or re-export [6]. Both the Czech Republic and the People's Republic of China are open economies [7]. China is the biggest exporter and the second biggest importer for the Czech Republic, the ratios are 12.4% and 10.3% [8]. Chinese import into the Czech Republic is of larger volume, the structure of which is concentrated on investment goods and on such products that are further used for subsequent production [9]. Were recognize the groups of SITC 6 – market products, SITC 7 – machines and transport equipment and SITC 8 – industrial products as parts of the prevailing tradable groups [10].

It is typical for our times that information is recorded in the form of chronologically structured data, i.e. time series [2]. Line of time series was characterized as a sequence of substantially and spatially observations, which are timely structured from the past to the present. As mentioned above, time series is used in various disciplines, such as economics, medicine, physics or even meteorology [1]. There is an effort to define the development of monitored indicators in the past, find out the causes of variability and subsequently forecast the future in such disciplines [3].

If a seasonal component is the content of time series, it is necessary to clear it because it can obscure the dynamics of monitored phenomena and prohibit an objective comparison within the given year. The term of seasonal clearance can be understood as a removal of the seasonal component from the line of time series. In a model approach, the seasonal component represents a seasonal influence which is defined as a set of direct and indirect causes that reoccur annually. Concerning the analysis of seasonality, the first task is to define seasonal fluctuations in terms of their statistical significance. If the existence of seasonal component in the line of time series is proved, it is followed by the quantification of seasonal fluctuations [11].

If the data are seasonally cleared, it is possible to apply them for the construction of econometric models. There is a risk in the form of the reduced or distorted estimates of parameters even in this way. It is also very difficult to identify what the influence of seasonal clearing is on the accuracy of predictions. There is a general problem the essence of which is whether the line of time series is correctly seasonally cleared. It is possible to come across several criteria in practice which evaluate the quality of seasonal clearing. However, all of them are imperfect [12]. Prediction is a very demanding task in practice and there have been developed various methods for it. The methods on the basis of which it is possible to carry out the analysis and the prediction of time series and which take into account seasonal oscillations are for instance the decomposition of time series, seasonal differentiation, HP filter, a spectral analysis or Box-Jenkinson methodology [13].

Here has been an increased number of references to artificial neural networks (ANN) related to the analysis of time series and forecasting [14]. Artificial neural networks have been employed for prediction in the economy for decades [15]. Artificial neural networks are nowadays applied mainly for solutions to possible future problems [16]. The use of artificial neural networks is appropriate for complicated operations which we are not able to analytically identify. This is the reason why they are mostly used for the modelling of very demanding strategic decisions [17]. Neural networks are applied for classification, regression, an approximation of functions, understanding, identification of texts, generation of languages, etc. apart from dealing with time series [18-19]. Artificial neural networks differ from classical methods of by the ability to cope with a large quantity of data and the accuracy of results. Further advantages are their relatively easy use in predictions and complex problems [20]. Neural networks are very flexible in their use [21]. One of their disadvantages, is a requirement for the large quantity of data relating to a sample [22]. Consequently, it is necessary to carry out a lot of test observations, which is not very comfortable for users.

If we focus on the relation of neural networks and time series, it is possible to state that both the purpose of time series themselves and the nature of all the data are what matters. The purpose of neural networks is to record the process of time series and to forecast individual data points in the best possible way. It is necessary to direct neural networks in the right way so that they could deal with time series well [23].

2 Data and methods

The subject of analysis is the data on the Czech Republic import from the People´s Republic of China. The data is available on the websites of the World Bank, etc. For the analysis, the data on monthly balance starting from January 2000 and ending with July 2018 (in EUR) will be used, which is 223 input data. Table 1 shows descriptive characteristics of the input data set.

Table 1. Characteristics of input data set

Samples	Month (Input variable)	Imports (Output (target)
Minimum (Training)	36526.00	4.421794E+07
Maximum (Training)	43252.00	1.216031E+09
Median (Training)	39916.52	4.384395E+08
Standard deviation (Training)	1949.82	2.799478E+08
Minimum (Testing)	36586.00	6.038031E+07
Maximum (Testing)	43282.00	1.152282E+09
Median (Testing)	39702.30	4.362519E+08
Standard deviation (Testing)	2174.13	3.301845E+08
Minimum (Validation)	36951.00	8.455607E+07
Maximum (Validation)	43040.00	1.289245E+09
Median (Validation)	40047.88	4.726722E+08
Standard deviation	3096.16	3.810380E+08
Minimum (Overall)	36526.00	4.421794E+07
Maximum (Overall)	43282.00	1.289245E+09
Median (Overall)	39904.26	4.431816E+08
Standard deviation (Overall)	1963.77	2.918713E+08

Source: Own processing.

An important phenomenon is the development of the Czech Republic import from the People´s Republic of China over time. The input data histogram and other selected statistical characteristics are presented in Figure 1 in graphical form.

The histogram corresponds to normal distribution. The data will be processed using DELL´s Statistica software, version 12. Regression using artificial neural networks will be carried out, when multilayer perceptron networks (MLP) and radial basis function (RBF) will be generated. There will be two sets of artificial neural networks generated:

Dependent variable = the Czech Republic import from the People´s Republic of China, independent variable = time.

Dependent variable = the Czech Republic import from the People´s Republic of China, continuous independent variable = time, categorical variable in the form of the month in which the value was measured = seasonal fluctuation.

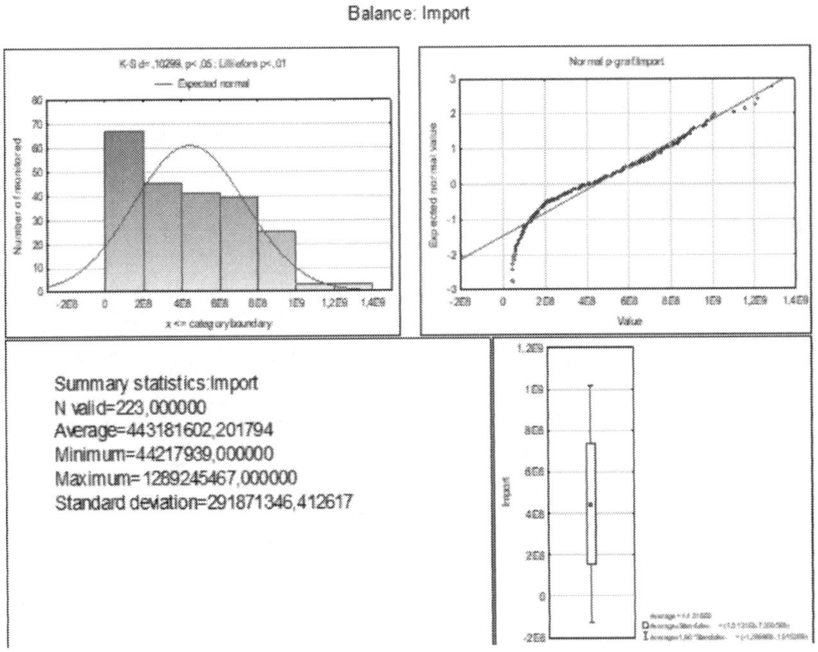

Fig 1. Basic statistical characteristics (Source: Own processing)

The process of work will be the same for both data sets. Time series will be divided into three sets - training, testing and validation data sets. Based on the training data set, which will contain 70 % of the input data, neural networks will be generated. Testing and validation data sets, both containing 15 % of the input data, will be used to verify the reliability of the neural network. A total of 10,000 neural networks will be generated, out of which 5 with the best characteristics will be retained. To determine the reliability of the neural network, the least squares method will be used. Generating of neural networks will be finished when there is no reduction of the value of the sum of the squares.

On the neural networks, whose sum of squares to the actual development of the Czech Republic import from the People´s Republic of China will be zero in ideal case, or at least as low as possible. The hidden layer of the MLP networks will contain at least 2, and no more than 50 neurons. The RBF network hidden layer will contain at least 21, but no more than 30 neurons. The delay of the time series will be 1. For the hidden and output layers, 5 distribution functions will be considered (linear, logistic, hyperbolic tangent, exponential, and sinus). Other settings will remain default. Finally, the results of both retained neural networks will be compared, and the findings from the research will be formulated.

3 Results

3.1 Neural structures 1

Based on the methodology used, 10,000 neural networks were generated, out of which 5 with the best parameters were retained (see Table 2).

Table 2. Neural structures 1 – Overview of retained neural networks

	Network	Train. perform.	Test. perform.	Valid. perform.	Train. error	Testing error	Validation error	Train. algorit	Error function	Activation of hidden layer	Output activation function
1	RBF1-23-1	0.952560	0.945596	0.966572	3.565842E+15	6.094306E+15	4.010215E+15	RBFT	Sum of squares	Gaussian	Identity
2	RBF1-29-1	0.965412	0.961986	0.965336	2.611062E+15	4.393356E+15	4.113568E+15	RBFT	Sum of squares	Gaussian	Identity
3	RBF1-23-1	0.957046	0.968388	0.964778	3.228416E+15	4.177702E+15	4.563611E+15	RBFT	Sum of squares	Gaussian	Identity
4	RBF1-23-1	0.944243	0.939097	0.968911	4.165960E+15	6.720191E+15	3.653697E+15	RBFT	Sum of squares	Gaussian	Identity
5	RBF1-29-1	0.958260	0.947902	0.964785	3.138918E+15	5.813955E+15	4.361519E+15	RBFT	Sum of squares	Gaussian	Identity

Source: authors.

It clearly results from the table that all 5 retained networks are RBF networks. The input layer contains one variable only – time. The output layer also contains only one variable, which is the Czech Republic import from the People´s Republic of China. The hidden layers contain 23 neurons (3 of the retained networks) and 29 neurons (2 out of the five retained networks). All networks used the RBFT training algorithm. The activation function for the hidden layer was Gaussian function, while the output activation function was Identity for all networks. Training, testing and validation performance of all retained networks is very interesting. Generally, it is necessary to find a network whose performance is ideally the same in all data sets. The error should be as small as possible (here, it shall be noted that the distribution of the data into the data sets was random).

Table 3 shows the correlation coefficients of the training, testing and validation data sets, that is, the values of the performance of networks in these data sets.

Table 3. Neural structures 1 – Correlation coefficients of individual data sets

	Imports (Training)	**Imports (Testing)**	**Imports**
1.RBF 1-23-1	0.952560	0.945596	0.966572
2.RBF 1-29-1	0.965412	0.961986	0.965336
3.RBF 1-23-1	0.957046	0.968388	0.964778
4.RBF 1-23-1	0.944243	0.939097	0.968911
5.RBF 1-29-1	0.958260	0.947902	0.964785

Source: Own processing.

As it can be see, the performance of all retained neural networks is relative high and almost identical. The small differences detected do not have any impact on the performance of the retained neural networks. The values of the correlation coefficients for the training data set are above 0.94, where the highest value of the correlation coefficient is detected in the case of the 2nd retained RBF 1-29-1 network. In the case of the testing data set, the correlation coefficients show similar values (0.94-0.96), where the highest value was detected in the case of the 3rd retained RBF 1-23-1 network. The values of the correlation coefficient in the case of the validation data set are about 0.96, where the 4th retained network, RBF 1-23-1, shows the highest value. At first sight (in terms of correlation coefficient), the 2nd retained network, RBF 1-29-1, may appear to be the most suitable. However, it is necessary to carry out a more detailed analysis of the results obtained, and to give the basic statistical characteristics of the individual data sets for all retained neural networks (see Table 4).

Table 4. Neural structures 1 – Statistics of individual data sets by retained neural structures

Statistics	**1.RBF 1-23-1**	**2.RBF 1-29-1**	**3.RBF 1-23-1**	**4.RBF 1-23-1**	**5.RBF 1-29-1**
Minimal forecast (Training)	5.501300E+07	3.874122E+07	54127502	68736086	45177864
Maximal forecast (Training)	1.046923E+09	1.027789E+09	910686099	980087664	933641653
Minimal forecast (Testing)	6.673260E+07	5.661960E+07	56468857	104458273	78575258
Maximal forecast (Testing)	8.796578E+08	9.238689E+08	885524267	911563076	884408431
Minimal forecast (Validation)	6.454159E+07	6.859920E+07	87361281	115492148	100278886
Maximal forecast (Validation)	9.329360E+08	9.529149E+08	944079041	979827258	929389373
Minimal residuals (Training)	-	-	-	-	-
Maximal residuals (Training)	3.623555E+08	3.277480E+08	426881585	372478096	344316239
Minimal residuals (Testing)	-	-	-	-	-
Maximal residua (Testing)	3.437825E+08	2.653424E+08	270803176	451092385	303367708
Minimal residuals (Validation)	-	-	-	-	-
Maximal residuals (Validation)	3.563094E+08	3.363306E+08	345166426	309418209	359856094
Minimal standard residua	-	-	-3	-4	-4
Maximal standard residuals	6.068116E+00	6.414038E+00	8	6	6
Minimal standard residuals	-	-	-3	-3	-3
Maximal standard residuals	4.403740E+00	4.003211E+00	4	6	4
Minimal standard residuals	-	-	-2	-3	-2
Maximal standard residuals	5.626567E+00	5.243932E+00	5	5	5

Source: authors.

In ideal case, individual statistics of the neural networks shall be identical in all data sets, both in terms of minimal and maximal forecasts, minimal and maximal residuals, and maximal standard residuals. In the case of equalized time series, the differences are minimal; however, it is not possible to identify which of the retained structures shows the best results. It is necessary to create a graphical representation of the actual development of the Czech Republic import from the People´s Republic of China, as well as the development of the forecasts of the individual retained networks (see Figure 2).

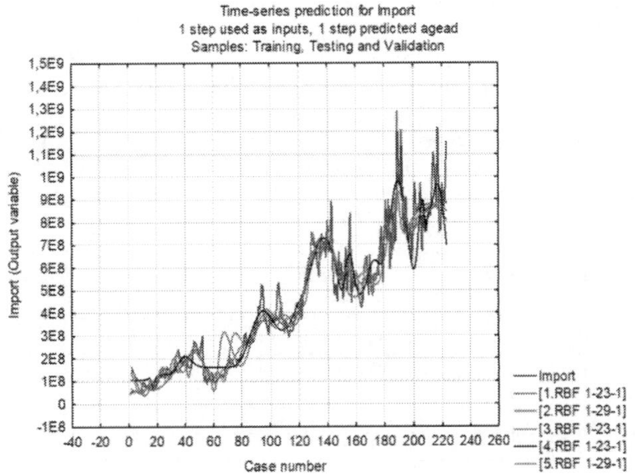

Fig. 2. Neural structures 1 – Development of the European Union import from the People´s Republic of China forecast by neural structures and compared with the actual import in the monitored period (Source: authors)

The figure shows that all five retained networks´ forecasts of the development of the Czech Republic import from the People´s Republic of China are very similar. The shape of the balanced time series is also very similar to the actual import development; all networks thus appear to be very interesting. All five networks forecast the basic direction of the development of the Czech Republic import from the People´s Republic very well. These networks are also able to forecast local minimum and maximum. In terms of practical applicability, all networks appear to be applicable. The figure also confirms that the 2nd retained network (RBF 1-29-1) shows the best performance (however, the differences between the retained networks are minimal).

3.2 Neural structures 2

In the second case, there were generated 10,000 neural networks. Again, five of them with the best parameters were retained. The overview of the retained networks can be seen in Table 5.

Table 5. Neural structures 2 – Retained neural networks

	Network	Train. perform.	Test. perform.	Valid. perform.	Train. error	Test. error	Valid. error	Train. algorithm	Error function	Activation of hidden layer	Output activation function
1	MLP 13-3-1	0.903128	0.952125	0.914862	7.382464E+15	5.065919E+15	8.457511E+15	BFGS (Quasi-Newton) 5	Sum of squares	Identity	Exponential
2	MLP 13-4-1	0.908948	0.957133	0.917074	7.868172E+15	4.504405E+15	7.801584E+15	BFGS (Quasi-Newton) 5	Sum of squares	Identity	Logistic
3	MLP 13-6-1	0.905782	0.943762	0.915119	7.503161E+15	6.411903E+15	8.631853E+15	BFGS (Quasi-Newton) 27	Sum of squares	Sinus	Logistic
4	MLP 13-9-1	0.906607	0.961278	0.917146	8.794756E+15	4.659435E+15	7.853076E+15	BFGS (Quasi-Newton) 17	Sum of squares	Sinus	Identity
5	MLP 13-5-1	0.911396	0.951877	0.916795	7.325275E+15	4.984896E+15	8.179707E+15	BFGS (Quasi-Newton) 5	Sum of squares	Identity	Logistic

Source: authors.

It results from the table that the retained networks are only multilayer perceptron networks. The radial basis function networks did not show characteristics suitable for being retained for the best performance. The input layer is characterized by two variables, one of them being time as a continuous variable, characterized by one neuron in the input layer. The second variable is the month of the measuring, which is categorical variable characterized by twelve neurons in the input layer. The hidden layers of the retained networks contain 3 – 9 neurons. The output layer is logically characterized by one neuron and one output variable, which is the import (from the People´s Republic of China to the Czech Republic). The training algorithm used for all generated and retained networks was the Quasi-Newton algorithm, but used in three alternatives. The error function for all networks was the sum of the least squares. For three retained structures, the activation of the hidden layer was used by means of the Identity function, for two networks, it was the sinus function. The output activation function was the logistic function for three of the networks, in the case of the first retained network the exponential function was used, while for the fourth retained network, it was the Identity function. The correlation coefficients of the training, testing and validation data sets, that is, the performance values of the networks in these data sets are shown in Table 6.

Table 6. Neural structures 2 – Networks performances for individual data sets

	Imports (Training)	**Imports (Testing)**	**Imports (Validation)**
1.MLP 13-16-1	0.903128	0.952125	0.914862
2.MLP 13-6-1	0.908948	0.957133	0.917074
3.MLP 13-8-1	0.905782	0.943762	0.915119
4.MLP 13-9-1	0.906607	0.961278	0.917146
5.MLP 13-13-1	0.911396	0.951877	0.916795

Source: authors.

It shows that the performance values of all retained neural networks are quite high and also approximately the same. The small differences detected do not have any significant impact on the performance of the retained networks. The correlation coefficients for the training data set are above 0.90, while the 5th retained network, MLP 13-13-1, showed the highest value. For the testing data set, the correlation coefficients values are even higher, between 0.95-0.96, where the highest coefficient value is in the case of the 4th retained network, MLP 13-9-1.

The retained networks in the case of validation data set show the value of the correlation coefficient of 0.91-0.92, where the highest value is shown by the 4th retained network, MLP 13-9-1. At first sight, this network might appear to be the most suitable one. However, it is necessary to carry out a more detailed analysis of the results obtained, and to give basic statistical characteristics of the individual data sets for all retained neural networks (see Table 7).

Table 7. Neural structures 2 – Statistics of individual data sets by retained neural structures

Statistics	**1.MLP 13-3-1**	**2.MLP 13-4-1**	**3.MLP 13-6-1**	**4.MLP 13-9-1**	**5.MLP 13-5-1**
Minimal forecasts (Training)	1.345832E+08	9.022528E+07	1.112790E+08	-	8.606204E+07
Maximal forecasts (Training)	1.247507E+09	1.051503E+09	1.038887E+09	1.217395E+09	1.029910E+09
Minimal forecasts (Testing)	1.477843E+08	1.011464E+08	1.088374E+08	3.231568E+07	1.037890E+08
Maximal forecasts (Testing)	1.136650E+09	1.081451E+09	9.643342E+08	1.160234E+09	1.027815E+09
Minimal forecasts	1.614028E+08	1.221337E+08	1.320116E+08	4.810283E+07	1.190544E+08
Maximal forecasts	1.084525E+09	9.944837E+08	9.317799E+08	1.073948E+09	9.507107E+08
Minimal residuals (Training)	-	-	-	-	-
Maximal residuals (Training)	4.086713E+08	3.291968E+08	3.137593E+08	4.431121E+08	3.939937E+08
Minimal residuals (Testing)	-	-	-	-	-
Maximal residuals (Testing)	2.998577E+08	2.312087E+08	3.787620E+08	3.451835E+08	3.003628E+08
Minimal residuals	-	-	-	-	-
Maximal residuals	3.914654E+08	4.929653E+08	4.290046E+08	3.795827E+08	4.433647E+08
Min. standard residuals	-	-	-	-	-
Max. standard residuals	4.756347E+00	3.711237E+00	3.622217E+00	4.724999E+00	4.603384E+00
Min. standard residuals	-	-	-	-	-
Max. standard residuals	4.212949E+00	3.444970E+00	4.730128E+00	5.056889E+00	4.254202E+00
Min. standard residuals	-	-	-	-	-
Max. standard residuals	4.256691E+00	5.581166E+00	4.617533E+00	4.283379E+00	4.902214E+00

Source: authors.

In ideal case, the individual statistics of the neural networks shall be horizontally the sane in all data sets in terms of minimal and maximal forecasts, minimal and maximal residuals, and minimal and maximal standard residuals. In the case of equalized time series, the differences are minimal, but it is not possible to determine unequivocally, which of the retained networks show the best characteristics. However, the 4th retained network MLP 13-9-1 appear to show very good results.

It appears to be necessary to make a graphical representation of the actual development of the Czech Republic import from the People´s Republic of China, as well as the development of forecasts of the individual retained networks (see Figure 3). The figure indicates that all retained neural networks´ forecasts of the development of the Czech Republic import from the People´s Republic of China are very similar. The curve of the equalized time series is also very similar to the actual development of the import; therefore, all retained networks appear to be very interesting. All 5 networks forecast the basic development trend of the Czech Republic import from the People´s Republic of China very well. Moreover, these networks are able to make good predictions of local minimum and maximum. In terms of practical applicability, all networks appear to be applicable. The figure also confirms that the 4th retained network, MLP 13-9-1, show the best results (nevertheless, the differences between the retained networks are minimal).

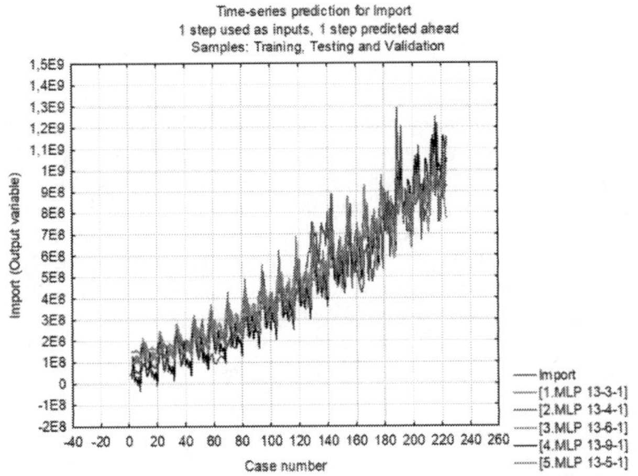

Fig. 3. Neural structures 2 – Development of the European Union import from the People´s Republic of China forecast by neural networks compared with actual import in the monitored period (Source: authors)

3.3 Comparison of A and B results

All generated and retained neural networks show very high performance in equalizing time series. The correlation coefficients of both RBF and MLP networks are higher than 0.9, which indicates a very high performance. However, assessing the performance only according to the correlation coefficients, it could be stated that the RBF networks show significantly better results than the MLP networks, where the sensitivity to seasonal fluctuations has been tested. It could be stated that the difference of max. 0.05 from the correlation coefficient value is not big, and does not play any role in the interpretation of the results. This is true to certain extent. However, comparing the graphical representation of the equalized time series with the actual course of the CR import from the PRC, it is clear at first sight that in particular the MLP networks are able to capture the seasonal fluctuations in the export, and the equalized time series successfully follow the past course of the monitored variable.

However, the correlation coefficients of training and validation data sets achieve lower values than those of the RBF networks. This can be mainly due to improper division of the data set into training, testing and validation data sets. This was done separately for each part of the experiment, that is, separately for the generation of the A and B networks results. The data were selected by means of the function randomsample. Although it is less likely, it is possible that an error occurred and the subsets are not equalized and cannot adequately represent the original data set. Another possible reason can be the calculation of the correlation coefficient, which works with mean values. These mean values can be better in the case of the RBF (although the results are not better in general). The position of the equalized time series compared to the actual course of the variable can thus be "advantageous" but the result can be inaccurate. Therefore, the conclusion should be that the time series equalized by means of MLP show much better results, with the 4th retained network, MLP 13-9-1 shows the best performance in general.

4 Conclusion

The objective of the contribution was to prepare a methodology of using artificial neural networks in equalizing time series and considering seasonal fluctuations on the example of the Czech Republic import from the People´s Republic of China. From the discussion part it follows that adding other variables significantly improves the quality of the equalized time series. Not only the performance of the networks is very high, but the individual MLP networks are also able to capture the seasonal fluctuations in the development of the monitored variable, which is the CR import from the PRC. Time series are relatively accurate in their course, and there are only slight fluctuations. This indicates possible applicability of MLP for equalizing time series affected with seasonal fluctuations.

This was achieved by adding categorical variable, which was intended to detect possible seasonal fluctuation. Moreover, it can be generalized that the seasonal fluctuations can be monitored in multiannual cycles, in a one-year cycle (monthly, quarterly), monthly cycles, weekly cycles. It always depends on what variable is being monitored. In the case of the CR import from the PRC, mainly the fluctuations in the individual months in a year were monitored. It was proved that the variable "month" enabled to achieve better results. There was also generated a neural network which is able to forecast the monitored period better than the other networks. It was the 4th retained network, MLP 13-9-1. As for the forecast itself, it is necessary to carry out an experiment with the time series delay that could bring interesting results in this case. The objective of the contribution was achieved.

References

1. M.K. Rafsanjani, M. Samareh, Chaotic time series prediction by artificial neural network. *Journal of Computational Methods in Sciences and Engineering*, **16**(3), 599-615 (2016).
2. X. Wang, M. Hang, Improved extreme learning machine for multivariate time series online sequential prediction. *Engineering Applications of Artificial Inteligence*, **40**, 28-36 (2015).
3. F. Fernandez-Navarro, M. A. de la Cruz, P.A. Gutierrez, A. Castano, C. Hervas-Martinez, Time series forecasting by recurrent product unit neural network. *Neural Computing & Applications*, **29**(3), 779-791 (2018).
4. R. Chandra, Competition and collaboration in cooperative coevolution of Elman recurrent neural networks for time-series prediction. *IEEE Transactions on Neural Networks and Learning Systems*, **26**(12), 3123-3136 (2015).
5. J. Vrbka, Z. Rowland, P. Šuleř, Comparison of neural networks and regression time series in estimating the Czech Republic and PRC trade balance. In J. Horák (Ed.), *Innovative Economic Symposium 2018 – Milestones and Trends of World Economy (IES2018), SHS Web of Conferences*, **61**, 01031. Beijing, China (2019).
6. Czech Statistical Office, *Když se řekne zahraniční obchod [When you say foreign trade]* [online], Available at: https://www.czso.cz (2018).
7. J. Gourdon, S. Monjon, S. Poncet, Trade policy and industrial policy in China: What motivates public authorities to apply restrictions on exports? *China Economic Review*, **40**, 105-120 (2016).
8. V. Stehel, P. Šuleř, Foreign trade between China and the Czech Republic. *Littera Scripta*, **9**(3), 84-95 (2016).
9. T. De Castro, Z. Stuchlíková, China-V4 trade relations – A Czech perspective. In S. Mráz, K. Brocková (Eds.), *Proccedings of International Scientific Conference on Current Trends and Perspectives in Development of China - V4 Trade and Investment*, 12-14 March 2014, Bratislava (pp. 46-60). Bratislava: Vydavateľstvo EKONÓM (2014).
10. P. Higgins, T. Zha, W. Zhong, Forecasting China's economic growth and inflation. *China Economic Review*, **41**(C), 46-61 (2016).
11. F. Rodrigues, I. Markou, F.C. Pereira, Combining time-series and textual data for taxi demand prediction in event areas: A deep learning approach. *Information Fusion*, **49**, 120-129 (2019).
12. P. Rostan, A. Rostan, The versatility of spectrum analysis for forecasting financial time series. *Journal of Forecasting*, **37**(3), 327-339 (2018).
13. T. Klieštik, M. Mišánková, K. Valášková, L. Švábová, Bankruptcy prevention: New effort to reflect on legal and social changes. *Science and Engineering Ethics*, **24**(2), 791-803 (2018).
14. M. Vochozka, J. Horák, P. Šuleř, Equalizing seasonal time series using artificial neural networks in predicting the Euro-Yuan Exchange rate. *Journal of Risk and Financial Management*, **12**(2), 76 (2019). DOI: 10.3390/jrfm12020076.
15. M. Tkáč, R. Verner, Artificial neural networks in business: Two decades of research. *Applied Soft Computing*, **38**, 788-804 (2016)
16. H.T. Pao, A comparison of neural network and multiple regression analysis in modelling capital structure. *Expert Systems with Applications*, **35**(3), 720-727 (2008).
17. E. Guresen, G. Kayakutlu, Definition of artificial neural networks with comparison to other networks. *Procedia Computer Science*, **3**, 426-433 (2011).
18. H. Altun, A. Bilgil, B.C. Fidan, Treatment of multi-dimensional data to enhance neural network estimators in regression problems. *Expert Systems with Applications*, **32**(2), 599-605 (2007).
19. V. Boguslauskas, R. Mileris, Estimation of credit risk by artificial neural networks models. *Engineering Economics*, **64**(4), 7-14 (2009).
20. Z. Rowland, J. Vrbka, Using artificial neural networks for prediction of key indicators of a company in global world. In T. Kliestik (Ed.), *Proceedings of 16th International Scientific Conference on Globalization and its Socio-Economic Consequences*, Rajecké Teplice, Slovakia (pp. 1896-1903). Zilina: GEORG (2016).
21. D. Santin, On the approximation of production functions: A comparison of artificial neural networks frontiers and efficiency techniques. *Applied Economics Letters*, **15**(8), 597-600 (2008).
22. V. Stehel, J. Vrbka, Z. Rowland, Using neural networks for determining creditworthiness for the purpose of providing bank loan on the example of construction companies in South Region of Czech Republic. *Ekonomicko-Manažerské Spektrum*, **2016**(2), 62-73 (2016).
23. Y. H. Hu, J.N. Hwang, *Handbook of neural network signal processing* (CRC Press, Boca Raton, 2001).

Private Partnership as a Global Trend of Digital Government

E.L. Sidorenko[1,*], and *Z.I.* Khisamova[2]
*Corresponding author: 12011979@list.ru.

[1]Moscow State Institute of International Relations (University) of the Ministry of Foreign Affairs, Moscow, Russia
[2] Krasnodar University of the Ministry of internal affairs of Russia, Krasnodar, Russia

Abstract. Currently, the concept of creating a digital government is firmly entrenched in the long-term development strategies of most countries of the world. However, the implementation of digital state projects requires capital investments to ensure the operation and maintenance of systems and technologies in a sustainable manner. Faced with excessive financial constraints, the governments of most countries were not able to fully provide open financial support for digital projects. In the context of budgetary savings and demand for improving the quality of public services, the search for innovative methods of financing projects has gained particular importance. This led to a steady global trend of public-private partnership in the provision of digital public services to the population. The purpose of this study was to identify the main trends in digitalization of the public administration system based on public-private partnerships, as well as to reveal the features and effectiveness of the implemented models. This study reveals the features of public-private partnership mechanisms, identifies the risks of such interaction, and presents examples of successful global practices to minimize these risks.

Keywords: digital public administration, digital trends, public-private partnerships, digital transformation.

1 Introduction

For over 18 years, the UN has been studying the pace of digital government in the world. At the same time, the 2018 Review noted not only a global increase in the level of digitalization of public administration, but also the presence of a steady trend in enhancing public-private cooperation in this area [1]. Such a partnership ensures the implementation of the integrated approach that involves the introduction of advanced technologies, and the provision of affordable, reliable and fast services to a wide range of users. The idea of public-private partnership is not new. But in the context of the digital economy, it acquired a new sound and posed a number of urgent issues related to determining the areas of responsibility of the state and private structures in implementing digital state programs, as well as establishing the ultimate beneficiary of digital management models.

2 Problem Statement

The digital lifestyle, caused by the introduction of the latest technologies, such as blockchain, big data, Internet of things, artificial intelligence, predetermined the qualitative transformation of management processes [2, 3]. The analysis of international surveys allows us to highlight a number of key trends in the digitalization of public administration. The most commonly used digital public services in 2018 were utility bills, tax filing, fines and fees; filing applications for birth and marriage certificates; registration of vehicles; applying for a driver's license and identity cards and registering new enterprises [4]. In Russia, the situation regarding the popularity of services is somewhat different. The most popular was the service of obtaining information on the amount of pension savings (5.5 million requests), then the service of registering vehicles (1 million applications), filing applications for a new passport (750 thousand applications) and passing exams for driving (670 thousand applications). The top 5 most demanded services also included the service "mobile voter" (550 thousand applications), which allows you to choose the most convenient polling station for elections [5].

Another trend inherent in all countries involved in the active introduction of IT services in public administration is the active transition to mobile applications [6]. In Russia, this trend is also obvious. The Public Services mobile application was downloaded more than 17 million times last year, which is 2 times higher than the previous year. The average monthly audience of the Public Services mobile application has grown from 5.6 million users to 9.3 million, and the number of visits per month - from 45 million to 72 million. Active expansion of mobile applications has contributed to the growth of payments to pay tax and court debts. So, the total volume of mobile payments exceeded 20 billion rubles [5].

But as practice shows, with all the obvious advantages of state digital programs, they cannot always fully reflect the needs of the population in terms of establishing feedback with the provider of public services. As our survey showed, 40% of people using digital public services note an overly formal approach to the solution. This is largely due to the objective limited participation of public and business associations in the development of such digital programs.

It is no coincidence that one of the modern global trends in digitalization is the expansion of public-private partnership in digital social services.

Public-private partnerships are types of partnerships, which include all stages of state interaction in the form of individual executive authorities and business representatives in implementing projects, from operating facilities and providing services on behalf of the state, to flexible methods of financing these services. Moreover, cooperation is based on the exchange of risks and benefits [7].

Among the main advantages of public-private partnership can be identified the following:

- Significant improvement in the quality of public services provided through the introduction of advanced processes and innovative technologies of the private sector;
- Minimization of financial and other risks of the project due to the appropriate distribution of risks throughout the project implementation period;
- Stimulation of private socially-oriented initiatives and projects;
- Active use of alternative sources of financing of state infrastructure and public services.

There are already examples in the world when corporations partially or fully take on administrative and government functions. For example, Microsoft provides data storage about medical records in a cloud accessible to clients when changing a medical institution. Such an initiative allows corporations to obtain the necessary data set used for testing and expanding the capabilities of developed services, and for state and citizens to get free access to various information services [8].

Humanitarian projects are of particular importance in the framework of public-private partnerships.

The Féitel project in Peru is an early and successful example of a global public-private partnership that pioneered an innovative approach to broadband access in rural areas, which is now widely used around the world: the auction with the smallest subsidies. This is an effective mechanism to minimize the subsidies required by commercial telecommunications companies to expand the network in non-profit areas by providing a contract to a bidder seeking the lowest subsidy. Despite its shortcomings, this innovative program brought a number of social benefits, and since then the initiative has expanded from public telephone communications to Internet access [9].

The Rural Knowledge Network is a multi-stakeholder alliance standing for the use of digital technology to empower rural communities through the creation of rural knowledge centers. The project was launched as part of the initiative of the M Swaminathan Research Foundation (MSSRF) in South India in 1998, during which the concept of rural knowledge centers was tested. Subsequently, the project evolved into the "Mission 2007: Every Village is a Knowledge Center" campaign to extend the benefits of ICT access to 600,000 villages in India. The initiative grew from a small pilot project to a mass movement that influenced similar initiatives in Asia and Africa and mobilized high-level support from public, private and public organizations [10].

The bulk of humanitarian projects are unfolding in areas with a humanitarian catastrophe. So, in Mozambique, a health information network (MHIN) has been created. The Collaboration Project between SPA AED-Satellife and Mozambique Ministry of Health include the application of an innovative and affordable technological solution, including mobile networks and the use of Personal Digital Assistants by people unfamiliar with the use of digital technologies, in particular, medical workers, who are often elderly and therefore more resistant to the introduction of new technologies. Thanks to the project, it was possible to reduce costs and increase productivity in terms of collecting data from areas, including monitoring paper and digital data collection systems [10].

An example is the mobile service for identifying counterfeit medical devices through their instant verification using a unique code sent to SMS. Such a drug verification system is free and implemented in Ghana, Kenya and Nigeria.

An important stage in the development of the public-private partnership program was the concept of "lean government". This platform connects developers, content providers, government agencies, service providers and citizens on one platform, allowing them to interact with each other [11].

Such a partnership is successfully used in the UK. It should be noted that the UK was the first country in the world to develop a public-private partnership concept for public health services projects. For almost 3 decades of such partnerships, 130 projects have been implemented in the country totaling more than £ 12 billion. Currently, the concept of public-private partnership is also being implemented in the framework of providing digital public services to the population. This ensures the provision of efficient, cost-effective and measurable public services, while minimizing financial risks. Today, through a public-private partnership, the country plans to reduce ICT spending by 50% through new approaches, while increasing the pace of implementation and the scale of digital government [12]. The experience of Great Britain is also successfully implemented in the provision of services in Canada, Ireland, Portugal, Australia, Japan and Sweden [13].

3 Research Questions

A review of the best practices of implementing digital projects in the framework of public-private partnerships raises the question of what the economic and legal prospects for expanding this activity may be, and how appropriate it is to transfer certain public functions to commercial structures or public organizations. The issues of sharing responsibility in the event of an error or unlawful actions, as well as criteria for assessing the quality of the services provided are also important. Unfortunately, the current level of development of state and social digital projects does not allow talking about resolving the raised issues. Rather, on the contrary, as the digital economy penetrates public administration and public monitoring system and enforcement problems become more acute and chronic.

4 Purpose of the Study

The purpose of the study is to identify the main trends in digitalization of the public administration system on the basis of public-private partnerships, as well as to reveal the features and effectiveness of the introduced models in the context of adaptability of law.

5 Research Methods

The methodological basis of the study is a systematic approach to modern practice of assessing the effectiveness of public-private partnership based on analysis of its legal and economic aspects. In the processing of factual material, such traditional scientific methods as dialectical, logical, scientific generalization, content analysis, comparative analysis, synthesis, SWOT analysis, etc. were used. The application of these methods allows ensuring the validity of the study, theoretical and practical conclusions, and developing proposal. The authors also use a set of statistical methods: summaries and groupings, average and relative values, index method, analysis of the series of dynamics, selective method, and others.

6 Results

The expansion of public-private partnership creates the risks of getting a huge amount of data, including personal information, into the hands of third parties, which is fraught with negative consequences [14]. To prevent such phenomena, it is necessary not only to create reliable data storage and processing systems that prevent unauthorized access, but also to build an effective legislative framework that clearly defines the degree of responsibility of all participants in such relations, as well as the negative consequences of possible leaks.

Switzerland is illustrative in this aspect, which allowed minimizing these risks. So, in Geneva, an initiative to attract private partners for the provision of digital public services is being implemented [15]. So, a private partner produces and technically provides a service to an eligible end user individual. In this case, the legal responsibility for the provision of relevant public services and full control over confidential information remains with the state. The contract between public and private partners establishes the terms of cooperation, remuneration and fines (in case the service is not provided or abuse is committed). The advantage of the scheme is that the end user reduces time costs until the state agency fully completes the manufacturing and control processes before receiving the service. Thus, the contractor is responsible to the state for the functionality of digital services, but the state is responsible to the end user for the provision of digital services and compliance with the rules [16].

7 Conclusion

The active development of public-private partnerships is a steady trend in the development of the digital economy in general and the digitalization of the public administration system in particular. Being closely associated with the digital state program, it is subject to systemic distortions and shortcomings associated with determining the most significant areas for joint efforts.

Within the framework of international cooperation, it seems promising to increase investment in the development of human capital and telecommunications infrastructure, expanding public-private partnerships in the field of healthcare, education and environmental protection.

To coordinate the activities of public servants, it is necessary to create a common concept of public-private partnership, which provides thoughtful management of human resources and comprehensive disclosure procedures based on compliance with information security requirements and the preservation of personal data. New forms of institutional framework for effective coordination, cooperation and accountability should be created at the state level, and in collaboration with relevant non-state actors.

References

1. United Nations, *United Nations e-government survey 2018. Gearing e-government to support transformation towards sustainable and resilient societies 2018* (2019). URL: https://publicadministration.un.org/egovkb/Portals/egovkb/Documents/un/2018-Survey/E-Government%20Survey%202018_FINAL%20for%20web.pdf. Accessed: 16.09.2019.
2. A.J. Meijer, M.P.R. Bolivar, J.R. Gill-Garcia, From e-government to digital era governance and beyond: Lessons from 15 years of research into information and communications technology in the public sector. *Journal of Public Administration Research and Theory*, Virtual Issue, 1-6 (2018). URL: https://static.primary.prod.gcms.the-

infra.com/static/site/ipart/document/EGov-Introduction.pdf?node=691b14abd3cfe516876e&version=9612:e229f05e70f9d491413c. Accessed: 16.09.2019.
3. C. Santiso, Will blockchain disrupt government corruption? *Stanford Social Innovation Review* (2018). URL: https://ssir.org/articles/entry/will_blockchain_disrupt_government_corruption Accessed: 16.09.2019.
4. O.R. Ashaye, Z. Irani, The role of stakeholders in the effective use of e-government resources in public services. *International Journal of Information Management*, **49,** 253-270 (2019). DOI: 10.1016/j.ijinfomgt.2019.05.016.
5. Ministry of Digital Development, Telecommunications and Mass Media of the Russian Federation, *In 2018, the audience of the state services mobile application almost doubled* (2019). URL: https://digital.gov.ru/ru/events/38786/ Accessed: 21.09.2019. [in Rus.].
6. E. Falco, R. Kleinhans, Beyond technology: Identifying local government challenges for using digital platforms for citizen engagement. *International Journal of Information Management*, **40,** 17-20 (2018). DOI: 10.1016/j.ijinfomgt.2018.01.007.
7. S. Sharma, Exploring best practices in public–private partnership (PPP) in e-government through select Asian case studies. *The International Information & Library Review*, **39**(3-4), 203-210 (2007).
8. Microsoft, *HealthVault* (2019).URL: https://www.healthvault.com. Accessed: 12.09.2019.
9. World Bank Group, *Case studies on Pro-Poor PPPs* (2018). URL: https://ppp.worldbank.org/public-private-partnership/case-studies-pro-poor-ppps. Accessed: 16.09.2019.
10. Association for progressive communications, *Pro-poor ICT access toolkit* (2010). URL: https://www.apc.org/en/pubs/policy/pro-poor-ict-access-toolkit. Accessed: 21.09.2019.
11. L.A. Ivanovskaya, The implementation of public-private partnerships in the e-government projects based on the principle of "platform management". *Gosudarstvenno-Chastnoye Partnerstvo,* **3**(4), 291-302 (2016). DOI: 10.18334/ppp.3.4.37089. [in Rus.].
12. Government of UK, *Healthcare: Public Private Partnerships* (2013). URL: https://www.gov.uk/government/publications/public-private-partnerships/public-private-partnerships. Accessed: 21.09.2019.
13. J. Martins, L. Veiga, Innovations in digital government as business facilitators: Implications for Portugal. *GEE Papers*, **97** (2018). URL: http://www.gee.gov.pt/RePEc/WorkingPapers/GEE_PAPERS_97.pdf. Accessed: 15.09.2019.
14. J.F. Yuan, A.P.C. Chan, W. Xiong, M.J. Skibniewski, Q.M. Li, Perception of residual value risk in public private partnership projects: Critical review. *Journal of Management in Engineering*, **31**(3), 15 (2015).
15. N. Veeramootoo, R. Nunkoo, Y.K. Dwivedi, What determines success of an e-government service? Validation of an integrative model of e-filing continuance usage. *Government Information Quarterly*, **35**(2), 161-174 (2018). DOI: 10.1016/j.giq.2018.03.004.
16. A. Sandoz, J.-R. Eudes, R. Prévot, Public-private partnership in e-government: A case implementation. In P. Kropf, M. Benyoucef, H. Mili (Eds.), *International MCETECH Conference on e-Technologies* (pp. 203-211). IEEE: Los Alamitos (2008).

Macroeconomic Relationships for Russian Economic Regions

A.N. Bereznyatskiy[1], and *B.E.* Brodsky[1,*]
[*] Corresponding author: bbrodsky@yandex.ru.

[1]Central Economics and Mathematics Institute of the Russian Academy of Sciences, Moscow, Russia

Abstract. This paper deals with macroeconomic relationships for Russian economic regions. An economic region is the main item of the regional statistics of the Russian Federation. Nowadays there are near 90 of regions and according to classification these can be industrial, agro- or subsidized regions A study is based upon the disaggregated macromodel for the Russian economic region. Then classification of the Russian regions by types of economic activity is made. Finally, we construct macroeconomic relationships for Russian economic regions specific by their type of economic activity This paper can be considered as the first attempt to create the model of the Russian regional level.

Keywords: economy of Russia, structural modeling, disaggregated macroeconomic model, regions of Russia, applied econometric analysis.

1 Introduction

We can mention three main approaches to creation of applied macroeconomic models: DSGE (Dynamic Stochastic General Equilibrium modeling), VAR, and the non-equilibrium structural method (Cowles commission approach). Nowadays DSGE methods are severely criticized by modern economists [1-8].

VAR model are not often selected as methodology for creation of applied macroeconomic models. "A curse of dimensionality" is the main idea which limits the application of VAR models to the analysis of the Russian economy.

Therefore at the macro level the third approach to applied macroeconomic modeling (Cowles commission) is more often used. Below we dwell upon the basic principles of structural macroeconomic model of Russian regions.

2 Methodology

In the early 1990s, after the liberalization of prices and foreign trade, three macro-sectors of economic activity emerged in Russia's real production sphere. Different in terms of competition for domestic and foreign markets these sectors usually include: Export-Oriented Sector (EOS); Domestic-Oriented Sector (DOS), Natural Monopolies (NM).

So, for the proposed model the following system of assumptions is accepted: three-sector structure of the production sector; the different price's priorities in the EOS, DOS, NM; exogeneity of prices. Further, the model assumes that the products of each sector are parametricized by: p – price level (base index), Y – real production output, Inc – aggregated income level, D – total subsidies, investment into a sector.

In addition, the following symbols in the model are used: δ – the direct cost ratio, ν – the average nominal wage, ω – the world (export) price, Ex, Im – real exports and imports, respectively, π – the rate of inflation, β – elasticity of the output with respect to the labor, e – exchange rate dollar/rouble.

These variables accompanied by subscript e,d,m indicating the belonging sector of EOS, DOS, NM, respectively. For example, Y_e is a real output of EOS, p_d the price level of DOS. Double indices are often used, for example Y_{em} – the real volume of intermediate products supplied from the EOS sector (the first index – from where?) for the NM sector (second index – where?).

Basic equations of the disaggregated 3 –sector model of the Russian economy are presented in the above papers of [9-11]. Here we point out what it means for understanding the macrodynamics of the Russian economic regions. Final equations of this model are as follows:

$$(Y_e)_{t+1}\beta_e\alpha_{e1}\left(1+(\pi_e)_{t+1}\right)=l_e\left(\frac{e\omega_e}{p_e}Ex(\omega_e)+Y_e\alpha_{e2}+Y_d\alpha_{e3}\right)_t+l_e(D_e)_t \qquad (1)$$

where α_{e1}, α_{e2}, α_{e3} – coefficients depending on the prices ratio: p_d/p_e, p_m/p_e.

$$(Y_d)_{t+1}\beta_d\alpha_{d1}(1+(\pi_d)_{t+1})=l_e\left(\frac{e\omega_d}{p_d}Ex(\omega_d)+Y_d\alpha_{d2}+Y_e\alpha_{d3}\right)_t+l_d(D_d)_t \qquad (2)$$

where α_{d1}, α_{d2}, α_{d3} – coefficients depending on the prices ratio: $(p_e/p_d)_t$, $(p_m/p_d)_t$.

The resulting system of equations allows us to predict the dynamics of variables.

In particular:

any increase in the world price for crude oil and energy resources prices ω_e has a positive impact on the output of the EOS and DOS sectors and, as a result, on aggregate output Y;

the increase in world prices for exported products of the DOS sector ω_d (agricultural products, weapons, high technologies) leads to an increase in the output of the DOS and EOS sectors;

the growth of the real exchange rate of the ruble p_d/e leads to a decrease in the output of the EOS and DOS sectors;

the growth of deflated prices and tariffs for the products of the NM sector p_m/p_d leads to a decrease in the output of the EOS And DOS;

the growth of the total amount of subsidies D/p_d deflated by basic price level p_d leads to the growth of output of truly subsidized regions;

the growth of real investment in fixed assets Inv/p_d leads to an increase in the output of the EOS and DOS sectors. However, excessive credit resources rates α_e, α_d will lead, on the contrary, to an increase in the number of bankruptcies of real sector enterprises and to a decrease in the output of EOS and DOS sectors.

With a small share of exports of the DOS sector output Y_d will also be small in comparison with Y_e, and when predicting the dynamics of aggregate output, it is possible to be limited in dynamics prediction by indicator Y_e. However, when the Y_d increases its role, the model becomes a system of two difference equations describing the dynamics of Y_e and Y_e indicators.

3 Classification of the Russian regions by types of economic activity

Classification of Russian regions by types of economic activity was done to achieve more accurate calibration of the equations of the model. A fairly simple heuristic algorithm of clustering is proposed in this report. The basic idea is: at the input of the algorithm is the GRP (gross regional product) structure vector with the normalized (by aggregate Russian values) components, the number of clusters in general is equivalent to the number of economic activity types (see [12]). We refer a region to a particular cluster by the maximum value of the ratio of the output of the corresponding type of economic activity of the region to the aggregate Russian output of the corresponding type of economic activity.

4 Macroeconomic relationships for the Russian economic regions

As a result, classification of the Russian regions into truly subsidized, agricultural, and industrially developed regions is proposed in the report. Then we constructed macroeconomic relationships for each type of a region. In particular, for truly subsidized regions of Russia the following relationship holds

$$\log GRP_t = \alpha_0 + \alpha_1 \log SUBSIDIES_t + \varepsilon_t \quad (3)$$

where

GRP_t – gross regional product (deflated for base index),

$SUBSIDIES_t$ – volume of subsidies to the region,

ε_t – residuals which assumed to be stationary.

For Republic of Dagestan, from the sample 2001-2016 (yearly data) the following econometric coefficients were obtained: α_0 = -28.075 ***(0.931), α_1 = 1.831*** (0.055) (the error level 1%). In the above constructed system we used correction of the nominal value of *SUBSIDIES* for the basis price index on the consumer market of the region. After that we used the test of Dickey-Fuller ([13]), [14]) for testing stationarity (non-stationarity) of time series. This test tells us that all initial variables have the order $I(1)$. This result gives us reasons for searching cointegration relationship.

Residuals in the constructed relationship are stationary (test of Davidson-MacKinnon [15]). So, the constructed macroeconometric relationship is of true cointegration type. $R^2 = 0.98$. The analogous results were obtained for other types of regions.

5 Conclusions

The development of macro-model of the Russian economic region is considered in the article. The core idea (methodology and equations) of the approach is illustrated with the disaggregated macroeconomic model of Russia. The main conclusions of the analytical model are statistically tested with real data and generally confirmed at the regional level.

Classification of Russian regions by types of economic activity was done to achieve more accurate calibration of the equations of the model. It was essential to distinguish between many regions steadily (truly) subsidized, for which the influence of macro-factors could be overwhelmed by regularly received subsidies. The analysis of regions in terms of subsidies revealed the stability of the group of subsidized regions over time, as well as the stability of the form of the empirical curve of the distribution of regions by the index value – the ratio of accumulated subsidies of the region to the accumulated GRP.

The practical importance of the results follows from the basic idea to construct macroeconomic relationships of the Russian macroeconomic regions and to make the first attempt to create the macroeconomic model of the Russian regional level.

Acknowledgements

The authors are grateful for the support by the grant of Russian Science Foundation (project 17-18-01080).

References

1. T. Cooley, Calibrated models, *Oxford Review of Economic Policy*, **13**(3), 55-69 (1997).
2. D.C. Colander, Testimony presented to U.S. House of Representatives Science and Technology Committee Hearing entitled "Building a Science of Economics for the Real World", Serial No. 111-106, July 20 (2010).
3. R.M. Edge, R.S. Gurkaynak, How useful are estimated DSGE model forecasts for Central Banks? *Brookings Papers on Economic Activity, Economic Studies Program, the Brookings Institution*, **41**(2(Fall)), 209-259 (2010).
4. J. Fernandez-Villaverde, The econometrics of DSGE models, *Journal of the Spanish Economic Association*, **1**(1), 3-49 (2010).
5. J.J. Heim, *An econometric model of the US economy* (Palgrave-MacMillan, London, 2017).
6. R.C. Fair, *Estimating how the macroeconomy works* (Cambridge Press, Harvard University, 2004).
7. N.G. Mankiw, The macroeconomist as scientist and engineer. *Journal of Economic Perspectives*, **20**(4), 29-46 (2006).
8. R.M. Solow, Testimony presented to U.S. House of Representatives Science and Technology Committee Hearing entitled "Building a Science of Economics for the Real World", Serial No. 111-106, July 20, 2010 (2010).
9. S. Aivazian, A. Bereznyatskiy, B. Brodsky, Macroeconomic modeling of the Russian economy. *Applied Econometrics*, **47**, 5-27 (2017). [in Rus.].
10. S. Aivazian, A. Bereznyatskiy, B. Brodsky, Modeling socio-economic dynamics of the Russian administrative branch. *Bulletin of CEMI*, **4** (2018). DOI: 10.33276/S0000186-7-1. [in Rus.].
11. B.E. Brodsky, S.A. Aivazian, A.N. Bereznyatskiy, Non-equilibrium structural models of the real sector of the Russian economy. *Economics and Mathematical Methods*, **55**(2), 65-80 (2019). [in Rus.].
12. S. Aivazian, M. Afanasiev, A. Kudrov, Clustering methodology of the Russian Federation regions with account of sectoral structure of GRP. *Applied Econometrics*, **41**, 24-46 (2016). [in Rus.].
13. D.A. Dickey, W.A. Fuller, Distributions of the estimators for the autoregressive time series with a unit root. *Journal of the American Statistical Association*, **74** (366), 427-431 (1979).
14. D.A. Dickey, *Estimation and hypothesis testing for nonstationary time series*. Retrospective Theses and Dissertations. 6267. Iowa State University (1976). URL: https://lib.dr.iastate.edu/rtd/6267. Accessed: 04.09.2019.
15. R. Davidson, J.G. MacKinnon, *Estimation and inference in econometrics* (Oxford University Press, New York, N.Y., 1993).

Ukrainian Experience of State Regulation of Stock Exchange Activity

N. Reznik[1], *G.* Abuselidze[2,*], *A.*Ostapchuk[1], and *V.* Tkachuk[1]

[*] Corresponding author: george.abuselidze@bsu.edu.ge.

[1]National University of Life and Environmental Science of Ukraine, Kiev, Ukraine
[2]Batumi Shota Rustaveli State University, Batumi, Georgia

Abstract. Theme of stock market is on the level of development in some countries. It faces a number of problems for the first time and there are no mechanisms for its overcoming. There are offers how to overcome the most popular problems of countries where the stock market just start to develop. The main part is belonging to state intervention and formation of legislative bases. Due to methods of analysis and generalization the article offers measures and mechanism of action for countries which have the same problems like Ukrainian. This research can be applied as it is or become the basis for development of new recommendations for the regulation of stock market activity.

Keywords: stock market, world trade, foreign investments, basic assets, stock exchange activity.

1 Introduction

The development of stock markets in the world is one of the most urgent themes of recent years. In some countries, this area is at the level of development, it faces a number of problems for the first time and there are no mechanisms for its overcoming. One of these countries is Ukraine, which is currently in a difficult position and can be a good example for highlighting and demonstration the difficulties that may arise in the processes of development and formation of the stock market. Thus, the article analyzes and uses the state of the stock market of Ukraine to offer a mechanism of effective state regulation of stock market activity [1]. This mechanism for regulating of stock market activity can also be used for other countries that, in their own development, have similar problems at a certain level. This research can be applied as it is or become the basis for developing new recommendations for the regulation of stock market activity. In recent years, bills and stock exchange development programs have been elaborated, but it have not yet found a place in the legislative field, and the economic system in Ukraine today cannot guarantee producers and buyers full protection against inflation and financial losses in the future. Unlike Ukraine, in other countries of the world, there are modern fixed-term markets and markets for forward and futures stock exchange contracts, which provide the maximum protection of all participators of trade. These and other factors of influence hinder the process of developing a «healthy» stock market by slowing down of market economy development.

2 Methods

In order to achieve the goal, general scientific and special research methods were used: methods of analysis and synthesis, systematization and generalization, a classification-analytical method for characterizing principles at stock exchange processes, as well as changes in the financial sphere that take place. In the new economy, correlation-regression analysis to determine the relationship between of the functioning and development of stock trading in Ukraine and factors affecting its level.

3 Internal level of state policy

To format of management principles for stock exchange processes it is expedient to allocate the levels of influence of state regulation on the development of stock trading, depending on which the system of coordination and regulation of wholesale trade processes should be formed, which significantly simplifies the transition from simpler management methods to more highly effective and progressive methodological bases for the introduction of modern management principles. It is expedient to implement the state policy in the field of stock exchange trading on four levels: local, regional, national and international [2].

Table 1. Levels of state policy in the field of stock exchange trade.

Level of introduction of state policy in the field of stock exchange trading	Function of level
Local	Formation of state policy within stock exchanges
Regional	Regulation of stock exchange relations within the territorial stock exchange markets
National	Improvement of the system of stock exchange relations within the country
International	Conducting the state stock exchange policy at the interstate level

Source: Authors.

At the local level, it is necessary to improve a number of positions that form a system of stock exchange relations and will be able to synchronize the interaction between the state and the self-regulated organizations of stock exchanges. The organizational structure of the commodity exchange was formed during the evolution of stock exchange institutions and today it is an organizational form, which includes normative and executive components. Based on the foregoing, we can form the key principles for improving the functions performed by the bodies of the internal management of the stock exchange and other organizational units.

On the part of state regulation, it is possible to introduce a system of standardization of stock exchanges, to introduce world trade technologies, to create a favorable investment climate, to improve the legislative and legal framework and to promote the training of specialists. In the process of self-regulation, it is possible to improve the functioning of the arbitration mechanism, to reorganize and optimize the work of the stock exchange bodies, to create a mechanism for accelerating the conclusion of agreements and minimize risks [3]. All these measures on both sides will lead to the process of certification and quality control, to monitoring, to provide clearing and financial guarantees and to strengthen the information function. Unfortunately, despite all advantages of self-regulating stock exchange organizations, it is not necessary to expect positive results at the final stage, because a considerable role is assigned to the state administration in Ukraine, so the most of functions of stock trading cannot be realized without support and intervention of the state bodies. Although should not forget that excessive pressure on the part of the state to internal work of stock exchanges may have also negative influence on the efficiency of its work due to contradictions.

The next level of implementation of the state policy on stock trading is regional. At this level, as in others, there is a set of mistakes on the part of the state bodies and the lack of a clear mechanism of policy shaping for developing stock exchange relations. Significant is the participation of local bodies and local governments in stock exchange activities at the regional level. Of course, such participation involves consistency of action both at the state level and directly at the level of the governing bodies of stock exchanges.

Increasingly, there is a problem of participation of small business entities in stock trading, since the stock market involves the wholesale purchase and sale of small quantities of goods or raw materials, which effectively prevents the entry of small commodity producers into the legalized wholesale market. At the same time, functioning of small and medium-sized businesses is a priority direction of the economy of many leading countries of the world. The rapid development of a small business will positively influence the activation of the market economy, give it flexibility, stimulate the resources of the population, and mobilize the anti-monopoly potential. Since the stock exchange trade is a legalized form of wholesale trade, there is a need to attract small and medium-sized businesses to trading processes on the stock exchanges by enabling business entities to sell or buy at stock exchanges [4]. The specifics of the commodity exchanges' activities are the sale and purchase of a large number of goods, which is why a certain problem for small businesses to distribute relatively small quantities of products through the stock trading network. Most small businesses do not get guarantees on the product market, and the risks that arise at all stages of its activity often lead to negative consequences.

The first step should be reforming the legislative framework for expanding the opportunities and rights for small businesses, as well as reforming the general legal and regulatory framework for the functioning of stock trading in Ukraine. The next steps will be creation of voluntary associations of small business entities to satisfy the common needs of product sales on stock exchanges. Since small enterprises that produce certain type of goods or services can be located in different regions, creation of so-called unions (clusters) or groups of small commodity producers is possible through a single network by creating a web-portal, which will contain the necessary information [5].

It is important to form unified system of standardization and certification of products, which becomes the object of trade at stock exchanges. Product certification has to be state-owned and sustainable, as close as possible to world standards. To increase the interest of small commodity producers is possible with creating conditions for granting certain advantages for agricultural producers. An interest of entity, which receives opportunity to grow one or another crop, is to make organic or ecologic farming, if authorized government agencies monitoring it. When the final product is released to the stock market, the commodity producer has an appropriate certificate, which is evident that the product meets a certain standard [6]. Eventually, such process provides a number of benefits to all parties involved in stock trade.

The national level implies conducting of state policy in a broader sense within the country. It cannot be said that now there is no general government policy in the field of stock exchange, but it is conducted at a rather low level and needs deep reforms. Global trends in economic, industrial and social development form a clear idea of how to act on the state level. At the same time, it is impossible to say certainty that global stock exchange experience of trade can be completely designed in Ukraine, because the existing economic system here is at a so low level that needs a radical adjustment. To stabilize the processes of the stock exchange development, it is necessary to form a clear mechanism of action on the part of the state. Today, the state pays lack of attention to the scientific component of stock exchange activity. Scientists in the field of economics and stock exchange practitioners have lack of financial and investment foundation to analyze and develop forecasts. Data that used to conduct research is difficult to find. In line with this, there is a problem of developing innovations in the field of research. State policy in the field of innovation is a separate important component of the whole system of state regulation. The priority directions of its development in Ukraine should be: support of basic scientific research in the field of economics and stock activities; development and support of industries and increase of enterprise' competitiveness; financial support; improvement of the support system for commodity producers; involving potential investors [7].

4 External level of state policy

The next level of state policy is international. In the field of stock trading, this concerns, first of all, the entry into the international wholesale market of domestic commodities and the attraction of foreign investment. To form effective methodological principles, it is necessary to put the following goals before the government and the state bodies: develop a methodology and system of state regulation of stock trading, taking into account the peculiarities of economic and social development of the country; to determine the main economic, social and scientific directions of the development of stock exchange relations in the future, both in general for the country and for certain regions; to identify and specify ways to achieve goals; to form the organizational and legal basis for processes of transition to modern market relations; to monitor the implementation of the objectives, programs and other elements by the relevant authorities and institutions [8]. The only scheme of action at the four levels of management of stock exchanges is shown in Table 2.

Table 2. System of measures of state regulation of stock exchange trade in Ukraine.

Level	Measures of influence
Local	• reorganization of the structure of stock exchanges • minimization of risks in conclusion of transactions • increasing the level of marketing services • formation of a modern informational environment
Regional	• introduction of regional programs for the support and development of small enterprises • increase participation of local and district state institutions in the development of stock exchange activities • creation of training programs at the state expense • creation of web-portals and Internet-services for the association of small business entities • providing for the possibility of concluding stock exchange agreements on the terms of SPOT, forward and futures • certification of elevators and warehouses
National	• support of basic scientific research in the field of economics and stock activities • development and support of industries and increase of enterprise' competitiveness • financial support • improvement of the support system for commodity producers • involving potential investors
International	• develop a methodology and system of state regulation of stock trading, taking into account the peculiarities of economic and social development of the country • to determine the main economic, social and scientific directions of the development of stock exchange relations in the future, both in general for the country and for certain regions • to identify and specify ways to achieve goals • to form the organizational and legal basis for processes of transition to modern market relations • to monitor the implementation of the objectives, programs and other elements by the relevant authorities and institutions

Source: Authors.

The transparency of the stock market will became the result of the formation of methodological principles for the functioning of the state stock exchange trading policy, as well as the acceleration of international capital turnover and the creation of conditions for macroeconomic planning at the national and international levels. Activating the development of the informational space and increasing the use of Internet resources will ensure the involvement of interested stock market participants; expand the marketing opportunities of stock exchanges [9]. The formation of an effective legal framework will enable stock exchange participants to obtain guarantees for the execution of stock exchange transactions and the ability to insure price risks. The stock market will in fact become a place for free pricing in these conditions, which will promote highly effective economic relations. Since the main place of concentration of goods and services is the stock exchange, it is necessary to pay attention to the problems of functioning of stock exchanges dependence of its quantity on the efficiency of activity. The number of registered and operating stock exchanges in Ukraine is quite large. It is necessary to reorganize commodity stock exchanges, leaving or combining small and medium-sized stock exchanges into more powerful units, and to increase the number of brokerage offices. For example, leave regional commodity stock exchanges or those stock exchanges whose activities are justified by results. From the side of state bodies and a special newly created body in the field of regulation of stock exchange activities, which will conduct the procedure of licensing stock activities, it is necessary to consider the rationality of the territorial location of stock exchanges. It is also possible to achieve optimization in the process of reducing the number of stock exchanges by monitoring its activities and controlling the reporting time about activities of stock exchange institutions. This implies the establishment by a law of a certain period during which the registered stock exchange unit has to submit a report of the established form regarding the results of its activities.

Today in the system of economic relations of the world considerable attention is paid to marketing activity, which covers the formation of demand for stock services and satisfaction of stock exchange interests of clients. One of the key goals of stock marketing is constant involvement of new clients. Significant role is taken by the risks as among the factors influencing on functioning of the stock market. Another task of stock marketing is to identify the sources of risks and minimize its. It is worth considering that the right choice of marketing strategies depends on a number of factors, such as the position of product groups on the stock market, the level of marketing costs, a set of marketing measures. Marketing events on the stock exchanges include such stages as studying potential customers, studying the client's motives who entering into a contract, analyzing the stock market, studying the types of stock services and possible operations, studying the components of the promotion of stock commodities to the end point, that is, the client, research of the competitive environment and potential competitors and implementation of promotional measures. The significant role of self-regulatory bodies in stock exchange activities should be re-emphasized. These bodies are actively involved in the functioning of the stock market and trade.

According to the legislation of Ukraine, self-regulatory bodies are defined as commercial non-profit organizations which are not part of system of state bodies' management. Generally, self-regulatory organizations were created as regulators of financial services markets, and subsequently it began to extend its functions and today play an important role in many processes that are related to stock exchange activities.

It is appropriate to attribute to the powers and functions of self-regulatory organizations: development of rules and standards for stock exchange activities; protection of the participants rights of stock exchange relations; monitoring and forecasting of the stock market; introduction of proposals for the improvement of the legislation in the field of stock exchange activities; organization of professional training; formation and maintenance of information base; delegation of representatives of self-regulatory organizations to state bodies. However, the organizational and other aspects of self-regulatory organizations are not sufficiently researched and require attention from the legal, personnel and regulatory framework. Formation of common mechanism for regulating stock exchange activities needs for closer interaction between self-regulatory organizations and the newly established regulatory body. This is possible with division of powers between all bodies that are related to the system of stock exchange activities [10]. The main goal of the self-regulatory bodies' activity is to increase the efficiency of the commodity stock market through the economic, technical and organizational measures, the implementation of which should take place with the participation of a common supervisory body with self-regulatory organizations.

For the effective functioning of stock exchanges and stock exchange relations, it is offered:

- to create a department for the development of stock exchange activities, which will be able to regulate issues that are involved with the state bodies;
- to develop a structure, provisions, concept and a state program that will fully outline the field of influence and activity of the department for the development of stock exchange activities [11].

5 Conclusions

The mechanism of the functioning and development of stock trading in Ukraine can be created through the establishment of clear standards of activity, taking into account modern requirements. Implementation of the state policy in the area of stock exchange activity at the local level will ensure its development within the stock exchange units, at the regional level – the regulation of stock exchange relations within the territorial stock exchange markets, at the national level – the improvement of stock exchange relations at the state level, at the international level – the conduct of the external state stock exchange policy. There is a need to attract small and medium-sized businesses to trading processes on stock exchanges by enabling business entities to sell or buy products or services at stock exchanges, where an important component is the formation of a unified system of standardization and certification of products that becomes the object of trading on stock exchanges.

Strengthening the function of marketing activities in the system of stock exchange relations involves the formation of demand for stock services and customer satisfaction. One of the key goals of stock exchange marketing is the continuous attraction of new clients, which should take place taking into account the factors affecting the functioning of the stock market, where risk plays a significant role, and the task of stock marketing is to identify the sources of risk and minimize it. The implementation of effective marketing activities includes the study of potential customers, customer motivation, stock market analysis, stock market research, the study of the components of the stock products promotion, the study of the competitive environment, the implementation of promotional activities. Reorganization of commodity stock exchanges should take place by uniting small and medium-sized stock exchanges into more powerful, increasing number of brokerage offices.

The transparency of the stock market will be result of the introduction of a mechanism for the direct participation of the state into the stock exchange activities. The acceleration of international capital turnover and the creation of conditions for macroeconomic planning at the national and global levels will be too. Activating the development of the information space and increasing the use of Internet-resources will ensure attraction of new participants into stock trading, expand the marketing opportunities of stock exchanges, and the formation of an effective legal framework will enable stock exchange participants to obtain guarantees for the implementation of stock exchange transactions and minimize risks.

References

1. A. Agapova, Conventional mutual index funds versus exchange-traded funds. *Journal of Financial Markets* **14**(2), 232-343 (2011).
2. T.G. Andersen, D. Dobrev, E. Schaumburg, Integrated quarticity estimation: Theory and practical implementation. *Nonlinear and Financial Econometrics Conference: A Tribute to A. Ronald Gallant* (Toulouse, France, May 19-21, 2011) (2011). URL: http://idei.fr/sites/default/files/medias/doc/conf/finet/annee_2011/quarticity_rev_1105_v8%5b2%5d.pdf. Accessed: 15.07.2019.
3. V. Cheung, R. Ingram, K. Patel, Next generation risk management. *Market Voice Magazine,* 25th September 2018. (2018). URL: https://marketvoice.fia.org/articles/next-generation-risk-management. Accessed: 15.07.2019.
4. L.K. Grant, Trading risk: Enhanced profitability through risk control (John Wiley & Sons Hoboken, USA, New Jersey, 2004).
5. J.J. Murphy, Technical analysis of the financial markets: A comprehensive guide to trading methods and applications (Prentice Hall Press, New Jersey, 1999).
6. S. Mikhailov, U. Noegel, Heston's stochastic volatility model implementation, calibration and some extensions. *Wilmott Magazine,* **1,** 74-79 (2005).
7. P.D. Leedy, J.E. Omrod, Practical research: Planning and design (8th ed.) (Prentice Hall Press, New Jersey, 2005).
8. N.P. Reznik, Entrepreneurial firm: Organizational aspect (Kyiv International University, Ukraine, Kyiv, 2016).
9. N.P. Reznik, International practice of investing in the agrarian sector. *Culture of the People of the Black Sea*, **147**, 77-80 (2008).
10. W.F. Sharpe, G.J. Alexander, J.W. Bailey, Investments (6th ed.) (Prentice Hall Press, New Jersey, 1998).
11. W.F. Sharpe, Investors and markets: Portfolio choices, asset prices and investment advice (Princeton University Press, Princeton, NJ, 2006).

Development Model for Program-Target Budgeting in Constituent Entities of the Russian Federation

L.M. Mulendeeva[1], *A.G.* Glukhova[1*], and *G.V.* Glukhov[1]
[*] Corresponding author: hecate1@mail.ru.

[1] Samara State University of Economics, Samara, Russia

Abstract. Over the past few years, the introduction of program-target method, that could improve the efficiency of planning public expenditures, has been the priority of the budget policy of the Russian Federation. The aim of this paper is to construct a model for the development of program-target budgeting. As the result of the study the development model for program-target budgeting in the constituent entities of the Russian Federation was constructed. Financial, organizational, methodological and regulatory factors were taken into account in the process of model construction. The application of this model will contribute to the growth of program-target budgeting viability and will increase the efficiency of budget expenditures.

Keywords: budget policy, expenditure planning, program-target budgeting, public finances, efficiency.

1 Introduction

The public finance management system of the Russian Federation has faced with the task of improving the budget expenditures efficiency and their optimization. This task can be achieved through program-target budgeting. Program-target budgeting may be defined as the budget planning process on the corresponding level of the budget system. That process involves both the formation and distribution of budget expenditures for financial support of the implementation of state programs formed on the basis of the priority goals of the socio-economic development of the territory, and monitoring the budget spending effectiveness. Providing the proper organization of program-target budgeting, the quality of budget funds management improves, the processes of formation, implementation and monitoring of state programs are significantly simplified, and the effectiveness of public spending is increased.

Up to now, there has been no detailed investigation of the stages that constitute the whole process of program-targeted budgeting. In addition, there is still very little scientific understanding of the factors affecting program-targeted budgeting at the regional level. Thus, there is an urgent need to construct a model for the development of target budgeting in the constituent entities of the Russian Federation.

This paper attempts to consider the following tasks:

- to determine the content of the main stages of targeted budgeting at the regional level;
- to identify and systematize the main factors affecting the development of program-targeted budgeting in the constituent entities of the Russian Federation;;
- to construct a model for the development of targeted budgeting in the constituent entities of the Russian Federation.

The specific objective of this study is to construct a model for the development of targeted budgeting in the constituent entities of the Russian Federation. To achieve this objective the main factors affecting targeted budgeting in the regions of Russia need to be analyzed. The model will increase the efficiency of spending state budget funds.

2 Methods

This study uses the following methods: theoretical (dialectical logic, methods of scientific abstraction); diagnostic (analysis, modeling); empirical (comparison and generalization, grouping); experimental (observation and practical calculations); methods of mathematical statistics and graphic representation of the results.

Experimental data for this study were collected using the state programs of the Samara region.

3 Results

Nowadays, many changes have been implemented because adapting new public management and financial methods to public institutions and organizations are differed. Especially a new trend which supports improving performance has been a matter of public administration, so performance indicators, performance agreements and strategic plans have been preferred in the public sector context in different countries [1].

Performance-based budgeting aims to improve the efficiency and effectiveness of public expenditure by linking the funding of public sector organizations to the results they deliver, making systematic use of performance information [2].

According to Nataliya Melnichuk, the quality of the budgeting performing and the level of socio-economic development of the country depends on the correct definition of budget targets [3].

It is argued in Robert K. Goertz that target-setting process is very important in the light of suggestions about how the budget process can be adapted to promote government [4].

Today, when implementing targeted budgeting, authorities experience a number of problems. For example, it is not uncommon for developers of government programs to adjust target and final indicators to the allocated amounts of funding, due to budgetary opportunities, which negatively affects the effectiveness of program implementation [5]. Many state programs in Russia are developed in several versions: "a program plan", which represents the need for budgetary allocations for the implementation of program activities, or "stipulated in the draft budget", which represents the amount of budget provisions allocated from the budget of the RF constituent entity for the implementation of program activities [6]. This strategy is unacceptable and discredits the program approach itself.

In addition, there are no provisions in the RF regulatory acts that articulate the need for the state programs to contain the goals and objectives specified in strategic planning documents. [7]. At the level of the constituent entities of the Russian Federation there is an inconsistency in the adoption of state programs, the simultaneous implementation of an unreasonably large number of small programs that are little related to each other [8].

The conclusion to emerge from the literature analyzed is that unified conceptual methodological provisions for the development and implementation of state programs have not been formulated yet [9]. And as it is argued in R.R. Yarullin in Russia targeted budgeting is mainly focused on ensuring compliance with formal procedures. There are no real incentives and effective tools for increasing the efficiency, accountability and transparency of the use of budgetary resources, determined by the objectives of public policy [10].

Five major stages were revealed in the process of program-target budgeting in the constituent entities of the Russian Federation.

1. Listing state programs in accordance with the priorities of the socio-economic development of the RF constituent entity and approving the main parameters of the programs.

2. Defining the goals and objectives, a set of measures, their dates and stages of implementation, as well as participants, resource support, indicators of implementation of state programs.

3. Determining the financial capabilities of the budget of the constituent entity of the Russian Federation, forecasting indicators of state financial support for the period of their operation.

4. Approving the volume of budget allocations to provide financial support for the implementation of state programs in a constituent entity of the Russian Federation for the next financial year and for the planning period.

5. Evaluation of spending budget funds efficiency, making decisions on the need for further implementation of state programs.

Analysis of the development of budget planning at the level of constituent entities of the Russian Federation revealed the factors contributing to the development of program-targeted budgeting in the regions. These factors were divided into three groups: financial, organizational and methodological, and regulatory.

1. Financial factors:

- an objective justification of the need for financial support to implement state programs;
- budget funds sufficiency to financially support the implementation of state programs;
- higher budgets appropriations sufficiency to ensure financing of state programs jointly administered by federal and regional authorities;
- attraction of extrabudgetary sources of financing.

2. Organizational and methodological factors:

- coordination of state programs with strategic planning documents;
- the availability of uniform organizational and methodological foundations of targeted budgeting, including methods for assessing its effectiveness;
- interagency coordination of actions in the process of state programs development and implementation;
- effective examination of government programs for resource support and target indicators determining, etc .;
- the availability of a risk management system in the course of the state programs implementation.

3. Regulatory factors reflect the appropriate regulatory framework for state programs development and implementation at the regional level.

Each stage of program-targeted budgeting is implemented under the influence of relevant factors, which makes it possible to develop a model for the development of program-targeted budgeting in the constituent entities of the Russian Federation (Figure 1).

The model reflects three areas of development of targeted budgeting at the regional level.

A. The logical structure of programs. This direction involves the achievement of the following tasks.

1. Coordination of budget and strategic planning of the territory.
2. Setting a specific, unambiguous goal of the state program and defining the corresponding tasks.
3. Setting a specific goal and responsibility for each ministry participating in the state program.
5. The correlation of goals, objectives, activities, and target indicators of programs.
6. Promoting interconnection with related state programs.

B. Financial support of the programs.

1. Development of programs based on the resources of budgets.
2. Determining the optimal size of budget allocations for the implementation of state programs.

3. Attraction of extrabudgetary financial resources.
4. Change in the program indicators values depending on fluctuations in funding levels.
5. Availability of alternative financing options for programs.

C. Monitoring the effectiveness of budget spending.

1. The presence of an effective methodology for assessing the effectiveness of budget expenditures.
2. Auditing the program effectiveness.
3. Expanding the powers of the financial management regional ministry in the implementation of control over budget expenditures effectiveness.

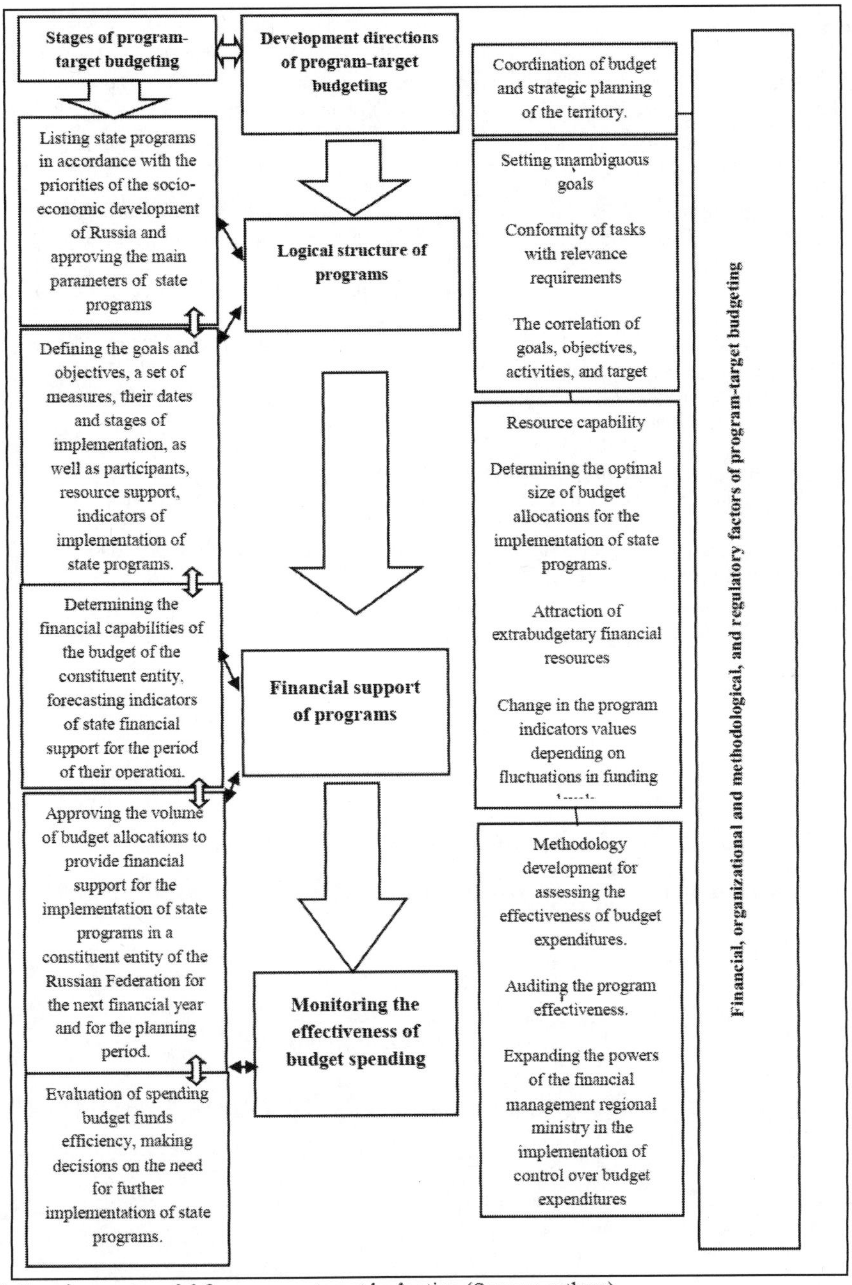

Figure 1. Development model for program-target budgeting (Source: authors)

4 Conclusion

Thus, this study has determined the content of the major stages of program-targeted budgeting at the regional level. The second major finding was that the main factors affecting the development of program-targeted budgeting have been systematized. This allowed constructing a model for the development of targeted budgeting in the constituent entities of the Russian Federation. The findings of this study have practical implications. The results of this research should help to improve the efficiency of state budget funds spending.

References

1. M.E. Altundemir, G.G. Gonca, Performance-based budgeting on strategic planning: The case study in Turkish higher education system. *New Trends and Issues Proceedings on Humanities and Social Sciences*, **3**, 263-270 (2016).
2. M. Robinson, D. Last, *A basic model of performance-based budgeting* (International Monetary Fund, Washington, D.C., 2009).
3. N. Melnichuk, Budget planning in financial management. *Baltic Journal of Economic Studies*, **1**(2), 95-100 (2015).
4. R.K. Goertz, Target-based budgeting and adaptations to fiscal uncertainty. *Public Productivity & Management Review*, **16**(4), 425-429 (1993).
5. O.V. Andreeva, A.A. Sukhoveyeva, Program-targeted budgeting in the system of social protection of the population in the region. *Economic Sciences*, **7**(92), 105-109 (2012). [in Rus.].
6. N.N. Shash, A.I. Borodin, A.A. Tatuev, Problems of transition to the program budget: New challenges of the budget policy of Russia. *Finance and Credit*, **14**(590), 2-9. (2014). [in Rus.].
7. V.F. Sharov, *Improving the efficiency of budget expenditures of the budget system of the Russian Federation* (Unity-Dana, Moscow, 2018). [in Rus.].
8. P.S. Zvyagintsev, Program-targeted planning method as the basis for creating a new industrialization of Russia. *Questions of Economics and Law*, **9**(63), 41-47 (2013). [in Rus.].
9. A.A. Tushov, Targeted programs in the system of state regulation of the economy. *Economic Sciences*, ***3***(88), 151-154 (2012). [in Rus.].
10. R.R. Yarullin, Implementation of program-objective forming of budgets of the Russian federation subjects, *Actual Problems of Economics and Law*, **3**, 223-230 (2014). [in Rus.].

Statistical Research and Modeling of Retail Turnover in the Russian Federation

E.I. Sukhanova[1,*], *S.Y.* Shirnaeva[1], and *E.G.* Repina[1]
*Corresponding author: elsu5463@gmail.com.

[1]Samara State University of Economics, Samara, Russia

Abstract. Retail trade turnover represents one of the most fundamental socio-economic indicators. Constant monitoring of such indicators has a pivotal role in modernization of the Russian economy. Retail trade turnover reflects countries' economic capacity and standard of living. This paper proposes a statistically significant econometric model of the retail trade turnover dependence with respect to different factors. Such factors as consumer prices index, unemployment level and average monthly designed salary were taken as the explanatory variables. The variables were selected based on Granger causality test and time series analysis of several socio-economic indicators. For each explanatory variable, a statistically significant model *ARIMA* (1, 1, 0) was constructed, with the help of which the predicted values of the explanatory variables for October, November and December 2019 were calculated, which were used to forecast the retail turnover for these months.

Keywords: retail trade turnover, time series, econometric model, forecast.

1 Introduction

In Russia today retail trade is one of the fast-developing industries with good prospects. Currently the Russian retail trade sector is characterized by the emergence of new efficient sales techniques, the increase in retail areas, and rapid growth in Internet sales. Big chain stores emerge and intensively develop. The growth in retail turnover gives rise to the profits of commercial organizations, including small business firms. That, in turn, promotes the growth in population wealth and provides new jobs. In 2018 retail and wholesale trade share comprised 12,8% of GDP [1]. That outnumbered the contribution from all other sectors. Standard of living and economic capacity of the country are determined by such indicators as volumes, dynamics and structure of retail trade [2].

Previous research has analyzed the issue of retail trade volumes. In several studies attention has been focused on the modeling and forecasting of retail trade. Those studies differ in methods of modeling, data type, set of variables, and sample size. In [3] the model of retail turnover is examined in relation to consumer confidence indicator, while the authors of [4] suggest taking into account calendar effects. In its turn, [5] provides mathematical statistical analysis of retail trade turnover in the RF considering financial and economic factors. To forecast the retail trade turnover, in [6] it is proposed to use the theory of cointegration, and in [7] – neural networks.

The results of statistical analysis and econometric modeling for retail turnover in the RF are shown in this study. The period represents a decade from 1999 to 2019. The authors of this paper analyze retail turnover in a monthly dynamics for the last 20 years. It allows tracing the dynamics of this indicator. It also provides a sufficient sample that enhances the chances to construct statistically significant models and promotes the quality of forecast. The constructed econometric model takes into account the influence that other socio-economic indicators have on retail trade turnover.

There are several primary aims of this study:

1. To analyze the dynamics and structure of retail turnover.
2. To identify factors (economic and social indicators), that have a significant impact on retail turnover.
3. To build an econometric model of the retail turnover dependence on selected indicators.
4. To make a forecast of retail turnover for the next time periods.

2 Methods

The methodology used in this study was a combination of statistical and econometric methods. For econometric modeling of retail turnover, various indicators were considered. The indicators reflected the state of the construction sector, industry, trade, and employment. The retail turnover was chosen as a performance indicator.

The time series of the retail trade turnover and other indicators in the monthly dynamics for the period from January 1999 to July 2019 were analyzed. The data base for the study was formed on short-term economic indicators from the RF official statistics [8]. The volume of the sample totaled 247 observations (for each time series). Gretl, Excel and EViews programs were used for calculations and analysis.

3 Findings

3.1 Modeling

The retail turnover comprises food turnover (food products, including drinks and tobacco products) and non-food goods. Over the past 20 years, the structure of retail trade turnover has not changed significantly. The share of non-food products was 51-53%, for food products it was 46-48%. A slight increase in the share of the turnover of non-food products, for example, in the crisis years of 2008 and 2014 (53.4% and 53% respectively) was due to the depreciation of the ruble and the following "boom" in purchases of durable goods.

The dynamics of changes in retail trade turnover for 1999-2019 is shown in Fig. 1.The monthly dynamics of the time series for the indicator under study was analyzed and it was suggested that there is a linear trend and a multiplicative seasonal constituent with a period of 12. For the purpose of the study, the seasonal constituent was eliminated.

Generated time series is first-order integrated. This conclusion was made based on the augmented Dickey and Fuller test results (*ADF*-test) [9] with constant for the initial levels (observed value of test statistics $t = -2{,}88$, p-value = 0,1695) and for the first differences (observed value $t = -13{,}34$, p-value = 0,0000).

One purpose of this study was to construct an econometric modeling of retail turnover (Y, billion rubles). To form the set of explanatory variables for the model different socio-economic indicators were taken into account. Those indicators were selected with augmented *ADF*-test (to determine the order of time series integration of variables under study) and Granger causality test (to reveal cause and effect correlation between dependable variable Y and the variables under study). The following indicators were selected:

X_1 – consumer price index (% of the previous month);

X_4 – unemployment level (%);

X_7 – average monthly designed salary (rubles).

The following multifactor linear regression model was chosen for modeling:

$$Y_t = b_0 + b_1X_{1,t} + b_4X_{4,t} + b_7X_{7,t} + \varepsilon \quad (1)$$

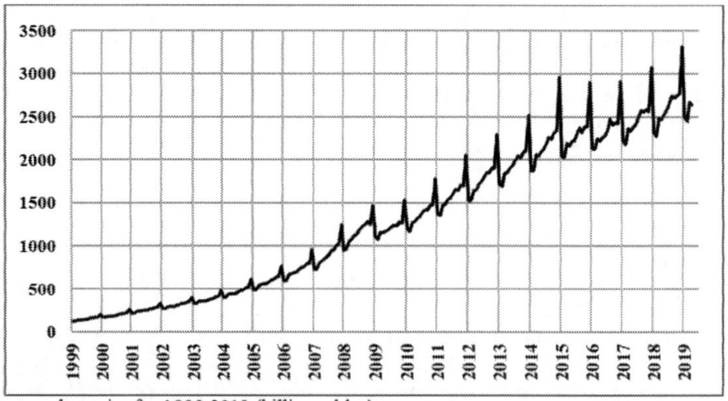

Fig. 1. Retail trade turnover dynamics for 1999-2019 (billion rubles).
Source: Authors' calculations based on [8].

Ordinary least squares method was chosen to evaluate the parameters of the model (1). Table 1 illustrates the results of econometric modeling and the quality indicators of the model that has been built (determination coefficient R^2 and observed value of F-statistics).

Table 1. Results of retail turnover modeling (Y, bln rubles).

Variable	Parameter estimate	Standard error	t-statistics	p-value
const	–476,27	135,09	–3,53	0,0223
X_1	7,09	3,08	2,30	0,0401
X_4	–18,02	–3,16	5,70	0,0183
X_7	0,06	0,001	60,00	<0,0001
R^2 = 0,9862; F(3, 243) = 341,13; p-value (F) = 0,000.				

Source: Authors.

Thus, the retail turnover model is written as:

$$\hat{Y}_t = -476{,}27 + 7{,}09X_{1,t} - 18{,}02X_{4,t} + 0{,}06X_{7,t} \quad (2)$$

Our calculations demonstrate the statistical significance of both the model (2) with 5% significance level and parameters estimates. Determination coefficient $R^2 = 0{,}9862$ validates the quality of the constructed model. The results imply that if consumer price index increases for 1%, retail turnover will grow for 7,09 billion of rubles. The growth in unemployment rate of 1% will provoke the fall for 18,02 billion of rubles in retail trade. The growth of average monthly designed salary of 1 ruble could be followed by retail turnover growth for 0,06 billion of rubles on average.

To study of the residues of the model (2), the following statistical tests were used [10]:

1) augmented *ADF*-test (the test resulted in a hypothesis on time invariance of the model residues observed value $t = -14{,}32$, *p*-value= 0,0001);

2) *Q*-test by Ljung and Box, which demonstrated the lack of autocorrelation for the model residues (observed value $Q = 17{,}30$, *p*-value = 0,3471);

3) chi-squared test which resulted in hypothesis about normal distribution of the residues (observed value $\chi^2 = 3{,}75$, *p*-value = 0,2298);

4) White test for heteroscedasticity of the residues (the results allow hypothesizing on the absence of heteroscedasticity on 5% level of significance: observed value $\chi^2 = 5{,}68$, *p*-value = 0,7291).

The residues of the model obey Gauss-Markov theorem for multiple linear dependence. Thus the model could be used for forecasting.

3.2 Forecasting

The stage of forecasting retail turnover by the econometric model (2) was preceded by finding the forecast values of the explanatory variables (X_1, X_4, X_7). For time series every variable was provided with the selected model *ARIMA*(*p*, *d*, *q*) [11], which was the best to describe it. The model *ARIMA*(1, 1, 0) was chosen for the explanatory variables (the same for the majority of the RF short-term economic indicators [12]). Using this model the forecast values for explanatory variables were calculated for October, November and December 2019. Table 2 demonstrates forecast calculation results for retail turnover and forecast values of explanatory variables for the same periods. The results are modified taking into account a seasonal component.

Table 2. Forecast of retail turnover.

Forecast period	X_1 (%)	X_4 (%)	X_7 (rubles)	Y (bln rubles)
October 2019	100,1	4,5	47745,3	2990,4
November 2019	100,1	4,4	47754,1	3000,3
December 2019	100,0	4,4	48172,6	3643,2

Source: Authors.

The results show that providing the same economic conditions and official statistics the trend towards retail turnover growth will remain till the end of 2019.

4 Conclusion

This study has identified that retail turnover for the previous 20 years has been demonstrating the stable trend towards growth. There were some other major results of the research, namely:

- the retail turnover model, which is statistically significant, with good forecasting properties was built in the course of this research;
- consumer price index, unemployment rate and average monthly designed salary were included in the econometric model as explanatory variables (based on the analysis of different socio-economic indicators Granger causality test);
- for every selected variable a statistically significant model *ARIMA*(1, 1, 0) was constructed, that was used to calculate forecast values of explicative variables for October, November and December 2019, which were used to forecast the retail turnover for these months.

The technique presented in this work can be used to modeling and forecasting various socio-economic indicators presented as time series.

References

1. V. Pavlyushina, V. Brilliantova, N. Kulaeva, *Bulletin of current trends in Russian economy. Personal consumption of goods*, **47** (2019). URL: http://ac.gov.ru/files/publication/a/21585.pdf. Accessed: 12.07.2019.
2. N.I. Merkushova, E.N. Bobrova, Statistical research of retail growth factors. *Bulletin of Samara State University of Economics*, **9**(143), 73-76 (2016). [in Rus.].
3. M. Čižmešija, Z. Orlović, Consumer confidence indicator as a leading indicator of changes in retail trade turnover. *Ekonomski Pregled*, **69**(1), 3-19 (2018). DOI: 10.32910/ep.69.1.1.
4. J. Arnerić, A. Čeh Časni, M. Čular, Adjusting for calendar effects of real retail trade turnover time series. *Croatian Operational Research Review*, **6**(2), 305-319 (2015). DOI: 10.17535/crorr.2015.0024.
5. A.I. Nikitenko, N.A.Reent, Mathematical and statistical analysis of the retail trade turnover in Russia. *Math Application to Economic and Technical Research*, **1**(6), 116-123 (2016). [in Rus.].
6. A.A. Sheveleva, Analysis and forecasting of retail trade volumes. *Proceedings of young scientists of Altai State University*, **15**, 173-176 (2018). [in Rus.].
7. S.V. Konikhin, Application of neural network modeling to forecast retain turnover in the process of investment decision making. *Modern Economy Success*, **4**, 42-46 (2017). [in Rus.].
8. GKS, *Short-term economic indicators of the Russian Federation*. (2019). URL: http://www.gks.ru/wps/wcm/connect/rosstat_main/rosstat/ru/statistics/publications/catalog/doc_1140080765391. Accessed: 30.08.2019. [in Rus.].
9. E.I. Sukhanova, S.Y. Shirnaeva, Statistic-econometric approach towards the research of macroeconomic indicators dynamics. In: S.I. Ashmarina (Ed.), *Proceedings of the 13th International Scientific and Practical Conference "Problems of Enterprise Development: Theory and Practice"* (pp. 248-252), Samara: Samara State University of Economics (2014). [in Rus.].
10. E.I. Sukhanova, S.Y. Shirnaeva, N.A. Zaychikova, Modeling and forecasting financial performance of business: Statistical and econometric approach. In: V.V. Mantulenko (Ed.), *International Scientific Conference "Global Challenges and Prospects of the Modern Economic Development" (GCPMED-2019)* (pp. 487-496). *European Proceedings of Social and Behavioural Sciences,* **57** (2019). London: Future Academy. DOI: 10.15405/epsbs.2019.03.48.
11. E.I. Sukhanova, S.Y. Shirnaeva, Different approaches to macroeconomic processes simulation and forecasting. *Fundamental Research*, **12-2**, 406-411 (2015). [in Rus.].
12. E.I. Sukhanova, S.Y. Shirnaeva, A.G. Mokronosov, Econometric models for forecasting of macroeconomic indices. *International Journal of Environmental and Science Education*, **11(16)**, 9191-9205 (2016).

Public-Private Partnership in the Implementation of the Industrial Policy of Russia

A.D. Kasatov[1,*]
*Corresponding author: kasatov47@mail.ru.

[1]Samara State University of Economics, Samara, Russia

Abstract. The article views the creation of public-private partnership as an important direction in the activation of the innovative type of the development of the economy of Russia. The need to combine the efforts of the public and private business in solving this problem is connected, firstly, with the economic sanctions of the West and not only financial sanctions but also with the refusal to supply high-tech equipment to Russia and, secondly, even more threatening prospects of the dynamics of the innovative development of the manufacturing sector. That is why the main attention is paid to the impact of public-private partnership on the development of integrated business in the context of the implementation of the strategic task of the innovative way of the Russian economy development. It is shown that public-private partnerships are one of the most common instruments for attracting private investments in infrastructure projects both in Russia and abroad. But the format of public-private partnerships for the implementation of Russia's industrial policy in the field of innovation should be used not only for the infrastructure development but also for the development of Russian industrial enterprises and the implementation of national projects.

Keywords: public-private partnership, industrial policy, innovative development way, national projects.

1 Introduction

In modern conditions the relevant problem is the implementation of the Federal Law of the Russian Federation "On Industrial Policy in the Russian Federation" and, above all, in the formation of a high-tech, competitive industry that ensures the transition of the state economy from the export-raw materials type of development to the innovative type of development [1]. This is due to the preservation of a critical degree of moral and physical depreciation of fixed assets of the industry of the Russian Federation. During the period of market reforms, the creation of an organizational and economic mechanism of providing stable reproduction of fixed assets of such quality which is necessary for the production of competitive industrial products. On the background of the preservation of a significant share of uncompetitive fixed assets, there is a hypertrophied increase in financial investments in various types of commercial transactions that are not directly related to solving of the key investment problems of the innovative development of industry.

Overcoming of the technical and technological weakness of the industrial enterprises, bringing them into line with Western analogues is associated with the increase in production efficiency, product quality and competitiveness. In this regard, the main task is to attract investments in the innovative development of the industrial companies. According to the Federal State Statistics Service, the share of Russian organizations engaged in technological innovation decreased from 7.9% in 2010 to 7.5% in 2017, and the proportion of innovative goods, work, services in the total volume of the shipped goods, work performed and services amounted to: 10.4% in 2000, 4.8% in 2010, 8.4% in 2015, 8.5% in 2016 and 7.2% in 2017 [2].

In the innovative respect, the development of industries is seriously aggravated, firstly, by the economic sanctions of the West, and not only financial sanctions but also by the refusal to supply high-tech equipment to Russia; secondly, the dynamics of the innovative development of the manufacturing sector is even more threatening in the future. Both of these positions in their unity once again lead to the need for active interaction between the state and the private business.

That is why the creation of IBS on the basis of public-private partnership is an urgent problem for the modern Russian economy. Due to their financial, economic and production potential, IBS is the most appropriate form of public-private partnership, which allows implementing of investment projects of national importance. The formation of IBS with a mixed form of ownership makes it possible, on the one hand, to ensure the priority of state interests and, on the other hand, to attract private investments on the basis of private investors' participation in the capital of IBS. The most common form of public-private partnership is a holding-type corporation.

2 Methods

Methods of strategic and innovative analysis, methods and means of economic, statistical and logical analysis were used in the research. The information base was compiled by the data of Rosstat (Federal Service of State Statistics), the materials from financial and economic publications of Russia, the Internet, publications from periodicals, scientific publications of domestic and foreign researchers, presented in the form of articles, monographs and theses as well as the materials of international scientific conferences.

3 Results

In solving of the burning problems of ensuring of the investment growth of fixed capital and the increasing of the industrial competitiveness, an important role is played by the large domestic business. In integrated business structures such as public-private partnerships there are created objective preconditions for the concentration of financial resources for the solving of priority investment problems associated with the modernization of the economy on an innovative basis. The main directions of the modern trends in the development of domestic integrated business in Russia, which should include integrated industrial groups, strategic alliances, clusters, etc., are widely covered in the economic literature [3, 4].

In the Russian economy, state activity is noticeable at the level of strategic integration processes, especially in the most important types of the economic activity. State bodies either act as the organizers of the integration, or do not impede the inclusion of enterprises with state participation in the share capital formed at the initiative of joint-stock companies in the corporate structures [5].

Throughout the world the trend of implementing of partnership projects between the state and business is gaining strength. The basis of public-private partnership (PPP) is the concept that achieving more favorable price / quality ratio of services that have traditionally been presented by the public sector is connected with the ability to use the private sector managerial potential (including qualifications, management and finance), as well as the distribution of the project risks among the participants.

With the adoption of the Federal Law "On Public-Private Partnerships, Municipal-Private Partnerships in the Russian Federation and Amendment of Certain Legislative Acts of the Russian Federation" [6] this interaction received a legal basis, the purpose of which is to attract private investments in the Russian economy and to improve the quality of goods and services, the provision of which to the customer is managed by the government bodies and local authorities.

Public-private partnership is one of the most common instruments for attracting private investments in infrastructure projects. Taking into consideration the social significance of the state of infrastructure, most countries create and modernize infrastructure facilities on the account of budgetary allocations. First of all, the state finances the construction of roads, the creation of public utilities, the modernization of social infrastructure, the development of education and health care, etc. Moreover, as practice shows, government resources alone are not enough to satisfy the demand for infrastructure investments in full [7].

Most countries of the world, regardless of the level of development of their economies, are trying to implement infrastructure projects with the involvement of extra-budgetary funds and private investors, which significantly reduces the burden on the budget and improves the quality and efficiency of infrastructure facilities maintenance. At the same time, the feasibility of implementation of an infrastructure project in the PPP format is evaluated in relation to each specific case. Public-private partnership cannot solve all the problems of infrastructure development in the country but PPP mechanisms in some cases can significantly increase the efficiency of public spending and it is especially important in conditions of state budget deficit. In this regard, the interaction of the state and private business in the implementation of national projects not only in the social but also in the industrial sphere is essential. So far this interaction is limited.

Thus, for the implementation of the industrial policy in the field of the innovative development of Russian industrial enterprises and the implementation of national projects, the format of public-private partnership is a promising area of interaction between the state and private business. For state institutions, PPP is a tool which helps to make industrial policy more perceptive to the changing nature of innovations, as well as social and sanction constraints. For private business PPP will be helpful in developing new markets and making business more valuable.

4 Conclusion

The content of the research carried out in the article does not pretend to be an exhaustive disclosure of all theoretical and practical problems of the Russian industrial policy aimed at the development of public-private partnerships in the implementation of innovative investment projects not only in the social but also in the industrial sphere. The author thinks it is necessary to make more profound study of these problems in both fundamental and applied areas.

Acknowledgements

The study was carried out with the financial support of the Russian Federal Fund of Fundamental Research within the framework of the scientific project No. 18-010-00359 "Theoretical and Practical Problems of the Industrial Policy of the Development of Public-Private and Municipal-Private Partnerships in the Innovation Environment."

References

1. *The Federal Law of 31.10. 2014 № 488-FZ «About Industrial Policy in the Russian Federation»* (2014).URL: http://www.consultant.ru/document/cons_doc_LAW_173119/. Accessed: 09.09.2019. [in Rus.].
2. Rosstat, *Russian statistical yearbook* (2018). URL: https://www.gks.ru/storage/mediabank/year18.pdf. Accessed: 09.09.2019.
3. Yu.V. Yakutin. *Integrated corporate structures in the market economy of Russia* (Free Economic Society of Russia, Moscow, 2009). [in Rus.].
4. Yu. B. Vinslav, *Integrated business management: Theoretical and methodical aspects. Management innovations for large business* (Publishing House «Centre-LitNefteGas», Moscow, 2017). [in Rus.].
5. S.B. Avdasheva, T.G. Dolgopyatova, H. Plyaynes, *Corporate management in JSC with state participation: Russian problems in the context of the world experience* (Publishing House of the State University High School of Economics, Moscow, 2007). [in Rus.].
6. *The Federal Law of 13.07.2015 № 224-FZ «About Public-Private Partnership, Municipal-Private Partnership in the Russian Federation and Making Amendments in some Legislative Acts of the Russian Federation»* (2015). URL: http://www.consultant.ru/document/cons_doc_LAW_182660/. Accessed: 09.09.2019.
7. V.N. Mochalnikov, *Public-private partnership: Domestic experience, world tendencies, the development vector for Russia* (Economics, Moscow, 2011). [in Rus.].

Estimating Prospects of Russian Foreign Trade under Crisis: Sanctions and Countersanctions

D.V. Manushin[1*]
*Corresponding author: predmet22@mail.ru.

[1] Kazan Innovative University named after V.G. Timiryasov, Kazan, Russia

Abstract. Based on econometric modeling, we proved the good relationship between foreign trade turnover and trade balance with the GDP of the Russian Federation. We identified the varying opinions of the Russian and foreign scholars on the impact of sanctions and countersanctions on the Russian economy. Based on the data on the absolute values of the trade turnover, export, import, and trade balance of Russia from 2010 to 2018, we revealed the positive dynamics of these indices, despite their negative values in 2014-2016. This allowed making a conclusion that the overall results of imposing sanctions and countersanctions for the Russian trade turnover and trade balance is rather positive than negative. This refutes the statements of western politicians, representatives of their subordinate international organizations, and some Russian and foreign scholars about the definitely negative role of sanctions and countersanctions. As a result, it is proposed to review the prospects of imposing additional countersanctions on western countries, which would allow developing the non-food sectors of the Russian economy.

Keywords: trade turnover, sanctions, export, import, balance.

1 Introduction

Today, the overall results of foreign trade are relatively poorly studied. The main attention in this sphere is paid to the export of energy carriers, which is studied in many macroeconomic works (for example, A.S. Ivanov and I.E. Matveev, 2018) [1]. At the same time, the importance of overall trade turnover (export and import) and its prospects is underestimated. To substantiate the importance of studying the Russian trade turnover, we research its relationship with the GDP of the Russian Federation. For that, we build a regression model based on quarterly absolute sums of the Russian trade turnover and GDP (bln rubles) from 1 January 2010 to 31 December 2018 (the total of 36 observations) using the data of Rosstat, Federal Customs Service and RateStats.com websites (to determine the dollar rate to ruble).
As an explained variable, the model uses the summarized GDP RF, and the explanatory variable is the summarized foreign trade turnover of the Russian Federation (further – FTT RF):

$$\text{GDP RF} = 3922 + 2.03\text{*FTT RF} + e \qquad (1)$$
$$(3.13) \quad (12.64)$$

In brackets below this and further formulas, t-statistics are given. In this model the connection between the studied indices is very good. Its R^2 is 0.82; t-statistics is 12.64; F-value is 0.000 (The acceptable value of probability according to null-hypothesis is 0.05, the good value is 0.01 or less. These values show that one may reject the null-hypothesis saying that there is no relation between these indices and accept the alternative hypothesis that there is such a relationship. The acceptable value of t-statistics is more than 2 (modulo), and the good value is more than 4 (modulo)).. The positive sign in the model indicates the presence of direct connection between these indices. Thus, it follows from the model that the growth of foreign trade turnover of the Russian Federation increases its gross domestic product. All this confirms the importance of researching the Russian foreign trade under modern conditions, given that many scholars consider these conditions to be crisis because of the sanctions imposed on Russia.

In this regard, we set the goal to define the current state and development prospects of foreign trade of the Russian Federation. To that end, it is necessary to estimate the impact of sanctions and countersanctions on it. Estimation will be made both by expert opinions and the data of Russian statistics.

As a result, we plan to answer a number of questions. First, whether the Russian foreign trade is currently under explicit crisis or not. Second, what influence – positive or negative – the sanctions had on the Russian economy in general. To that end, the Russian foreign trade from 2010 to 2018 was studied, as during the period from the end of 2011 till the beginning of 2014 the oil prices were atypically high (mainly due to the civil war in Libya, which temporarily excluded this country from the list of largest oil exporters), and the period from the end of 2014 till the end of 2016 was the period of Russia's adaptation to the new conditions, which does not allow researching the mid-term and long-term prospects of Russia in the sphere of foreign trade.

The main objective of the research is to determine, which of the possible options would be the best: to try to transfer from sanctions to constructive dialogue, to maintain the current status quo, or to head for toughening the sanction policy. An auxiliary objective is to achieve understanding regarding the presence or absence of the necessity to implement anti-crisis measures in the Russian foreign trade. Achieving these objectives would allow the researchers to elaborate an optimal strategy of the Russian foreign trade and policy.

2 Methods

To achieve the set goals, we used abstract-logical, economic-statistical, and economic-mathematical methods. The review of the Russian and foreign sources was carried out via search engines Scholar.google.ru, Yandex and Elibrary.ru by the key words: international trade, trade with Russia, sanctions and countersanctions. Modeling was carried out with the econometric software package QSR Eviews, based on the data presented at the websites of Rosstat, Federal Customs Service, and RateStats.com.

3 Results

6.1 Literature review

Today, the key factor influencing foreign trade is the political factor, first of all sanctions and countersanctions. During the one and a half years of imposing sanctions and countersanctions, many opinions have been expressed. They can be classified into three groups:

1. In January 2015, the US President B. Obama claimed that after the sanctions were introduced "the Russian economy is in tatters". The IMF report on the Russian economy states that at the initial stage sanctions and countersanctions may lead to the GDP decrease by 1–1.5%, while in the middle-term the accumulated losses may reach 9% of the GDP [1]. The panic reaction to these words was the opinion that "sanctions are critical for the Russian economy. Under such conditions, the prospects of crisis management should be associated with as fast as possible normalization of cooperation with the foreign partners" [2, p. 35].

2. "The US and EU sanctions provide a unique opportunity to correct the skews in the Russian financial system with great advantage for it. If a country has a positive trade balance, it needs no foreign credits" [3, p. 21].

3. "The sanctions, worsening the situation in the Russian economy, are not the initial cause of its current state" [4, p. 77]. "For example, a much more sensible shock for the Russian economy is drop of oil prices" [1, p. 33]. The situation was aggravated by a significant increase of the key rate by the Russian Central Bank [3]. As a result, the conclusion is made that the main problem of the Russian economy is not sanctions but high taxes, poor industrial development, low purchasing power of the population, and insufficient financing of the Russian regions [5]. As time has shown, the second and third opinions were more grounded and objective.

Having studied the amounts of the Russian and global trade in the period from 2013 to 2017, S. V. Kazantsev (2019) found that the decline of trade turnover in 2014-2016 was observed not only in the Russian Federation but all over the world. The USA suffered the least from their sanctions. At that, they continue to expand and deepen anti-Russian sanctions, while the losses related to their implementation are born by other countries [6]. At that, many foreign scholars are certain that the negative impact of sanctions and countersanctions on the Russian foreign trade is still rather large, and only a small part of them argue for their lifting. For example, S. Krapohl and A. Vasileva-Dienes (2019) assert that "the economic recession in Russia since 2014, aggravated by the Western sanctions and low oil prices, makes Russia less attractive as a market for exporting from Central Asia" [7].

M. Bělín and J. Hanousek (2019) believe that the Russian sanctions, imposed on importing food from Europe and America, have led to about 8 times reduction of trade flow compared with the sanctions imposed by the EU and the USA in regard to exporting equipment. These results are probably not related to redirection of trade flows through unsanctioned countries. Hence, the difference in the sanctions efficiency can be explained by a broader range of sanctioned goods and a potentially stronger position of imposing sanctions on import compared to export [8].

I. Korhonen, H. Simola, and L. Solanko (2018) assert that sanctions had an obvious negative effect on the Russian economy, although the oil price decrease had a larger impact on the Russian GDP in 2014-2016. The Russian countersanctions hampered food exports from the EU, nut macroeconomic effects in the EU are very weak. At the same time, it is probable that if sanctions continue for a longer period, then the potential of Russia's long-term growth can be diminished [9].

M. Smeets (2018) revealed the reverse connection between the duration and efficiency of sanctions, i. e. the damage caused by them diminishes with time. It is rather unlikely that Russia would refuse its interests in the Ukrainian territory. The situation more and more looks like an accomplished fact. Analysis showed that maintaining the sanctions expecting that they would work long-term is unrealistic and is not proved by evidence. At some moment it will be necessary to make a decision regarding the duration and amount of the sanctions and ultimately their lifting [10].

P. Kurečić (2017) highlights the political reasons for the necessity of sanction policy of the West and marks that some NATO members almost completely depend on the Russian gas imports due to their geographical position, which impedes the possibility for them to follow the NATO and EU official policy regarding Russia. As a result, they attempt to maintain good relations with Russia, pursue their economic interests and ensure the best positions for increasing trade with Russia as soon as the EU lifts the sanctions. At that, oil and natural gas have been relatively cheap in the recent three years, which led to serious economic problems in Russia. That is why Russia cannot currently use gas supplies as a means of "gas wars". Some NATO/EU members from this region are not seriously interested in deterring Russia and protecting their state borders with troops and, most likely, would be willing to lift the sanctions. This does not mean that the NATO unity and resolution to act are threatened, especially as the actual military capabilities of the above-mentioned states are virtually negligible compared to those of Russia, which increases the homogeneity of these states and their dependence on NATO (the USA, first of all) in terms of security [11].

Review of the foreign authors' opinions shows that there is a common opinion on the crisis in the Russian economy as a result of sanctions imposition, reduction of prices for fuel-energy resources, and other difficulties. At the same time, many of these authors do not take into account the new course of the US foreign policy aimed at divergence from the WTO rules and unilateral promotion of US protectionist measures (these aspects are thoroughly commented by A. P. Portanskiy (2017) [12]). This course makes other countries consolidate. Besides, the above-mentioned authors, as a rule, do not estimate the mid- and long-term prospects of Russia's functioning under sanctions and countersanctions.

A. N. Lyakin and M. I. Rogov (2017) note that countersanctions have already provided achievement of the set goals. In a number of aspects, active investment is already going on (for example, vegetables grown under cover) and he market will hardly become open under any changes in geopolitics. For example, even recovery of relations with Turkey did not lead to lifting limitations for products into import substitution of which large investments were made [13].

Estimating the Russian countersanctions, E.L. Zolotareva (2018) emphasizes their negative impact on foreign economic links. At that, the European Union is still the largest trade partner of the Russian Federation. The share of the European Union countries in the Russian foreign trade turnover was 42.2% in 2017, although it decreased a little. Recently, trade cooperation between Russia and APEC countries is activated, mainly countries of South-Eastern Asia: China, South Korea. Vietnam, and others. The unit weight of this group of countries is growing and in 2107 it was over 30%. The share of Chine in the Russia's foreign trade reached 15%. Russia exports to China meat, fruit and various ready foods. Russia imports from China vegetables, fruit, various kinds of meat [14]. When estimating this aspect, one should note that ecological situation is very bad in China; they apply a lot of fertilizers prohibited in Russia, which influences the quality of exported goods. As far as it is possible, Russia should reject importing food from China.

According to S.N. Larin (2019), it is too early to assert that sanction limitations against the Russian economy have reached their goals [15]. The review of works by the Russian researchers revealed that there is no unambiguous opinion about positive or negative influence of sanctions and countersanctions; the topic is rather disputable, which determines the need to study it in detail.

6.2 Experimental research results

To estimate the correlation of the studied opinions with the actual situation in the sphere of the Russian foreign trade, we research its changes in the period of 2010-2018 (see Fig. 1).

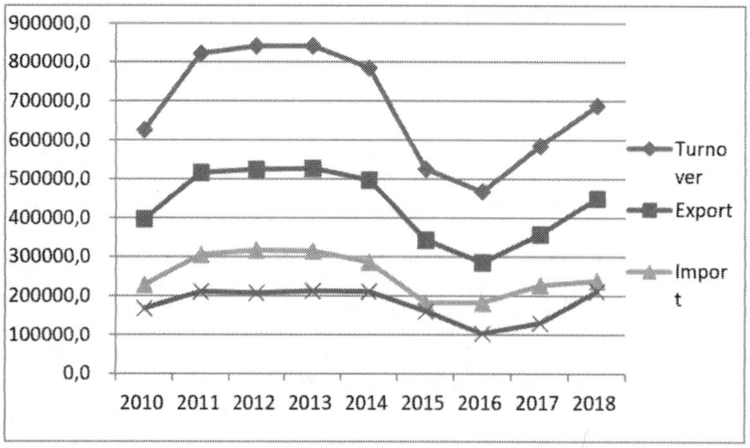

Fig. 1. Trade turnover, export, import, and trade balance of Russia from 2010 to 2018, mln US$ (Source: Authors).

Analyzing Fig. 1, one may conclude that sanctions and countersanctions had significant influence on the Russian foreign trade in 2014-2018. At that, the overall result of imposing sanctions and countersanctions is rather positive than negative. For example, the Russian trade turnover in 2018 surpassed that in 2010 and it is expected that in 2019-2020 it will be comparable to the pre-crisis values of 2012-2013. At that, the trade balance in 2018 is already comparable to that in 2012-2013. In 2018, one may vividly see a significant reduction of import growth rate compared to export growth rate. Apparently, it is the result of successful implementation of the programs of import substitution, import exclusion, and entering new markets of import goods.

As the Russian trade balance demonstrates the best results after imposition of sanctions and countersanctions, it is necessary to estimate its impact of the GDP RF. For that, we build a regression model based on quarterly absolute sums of the Russian trade balance and GDP (bln rubles) from 1 January 2010 to 31 December 2018 (the total of 36 observations), using the data from the websites of Rosstat, Federal Customs Service, and RateStats.com (to determine the dollar rate to ruble). As the explained variable, the model uses the summarized GDP RF, and the explanatory variable is the summarized foreign trade balance of the Russian Federation (further – FTB RF):

$$GDP\ RF = 10013.9 + 4.65*FTB\ RF + e \quad (2)$$
$$(6.05) \quad (5.89)$$

In this model the connection between the studied indices is good. Its R^2 is 0.505; t-statistics is 5.89; F-value is 0.0000. The positive sign in the model indicates the presence of direct connection between these indices. Thus, it follows from the model that the growth of foreign trade balance of the Russian Federation increases its gross domestic product. It is important to note that though the changes in trade balance explain just 50.5% of changes in GDP RF, but this is a good result, showing an apparent and rather significant connection between these indices.

4 Conclusion

The research proves the presence of firm and direct connection between the international trade turnover and balance of the Russian Federation and its GDP. It is shown that the overall mid-term result of sanctions and countersanctions for the Russian trade turnover and balance is rather positive than negative. This refutes the statements of western politicians and representatives of their subordinate international organizations, and some Russian and foreign scholars about the definitely negative role of sanctions and countersanctions. At that, one may expect that the revealed trend will be preserved and there will undoubtedly positive results in the Russian international trade. In this regard, Russia is recommended to make efforts to preserve the key sanctions and countersanctions, as they motivated the Russian government to activate international trade and achieve undoubted success under the crisis caused by Russophobic actions of the western countries. All this demonstrates the lack of evident crisis in the Russian international trade. As for the feasible termination of the agreement on gas transportation with Ukraine, Russia is ready for that. The gas pipelines "The Power of Siberia", "Nord Stream" and "TurkSteam" are already built and ready for operation.

Due to the revealed results and the success achieved in foreign trade, it is proposed that the Russian Federation continues its active policy in the sphere of import substitution, import exclusion, and entering new markets of export and import goods. Provided these activities are implemented, the prospects of the international trade will be advantageous for Russia. At that, it would be reasonable to explore the opportunity of imposing additional countersanctions onto the western countries, which would enable to further develop the non-food industries in Russia.

References

1. A.S. Ivanov, I.E, Matveev, International trade in fuel and energy products: Current state and prospects. *Russian Foreign Trade Bulletin*, **3**, 46-56 (2018). [in Rus.].
2. E.T. Gurvich, I.V. Prilepskiy, The impact of financial sanctions on the Russian economy. *Voprosy Ekonomiki*, **1**, 5-35 (2016). [in Rus.].
3. S.A. Afontsev, Crisis management under economic sanctions: Mission impossible? *Voprosy Ekonomiki*, **4**, 20-36 (2015). [in Rus.].
4. S.Yu. Glazyev, Sanctions of the USA and the policy of Bank of Russia: Double blow to the national economy, *Voprosy Ekonomiki*, **9**, 13-29 (2014). [in Rus.].
5. M.V. Klinova, E.A. Sidorova, Economic sanctions and EU-Russia economic relations. *Voprosy Ekonomiki*, **12**, 67-79 (2014). [in Rus.].
6. S.V. Kazantsev, Evaluation of the sanctions impact on sanctioning countries' trade with Russia. *Finance: Theory and Practice*, **23**(3), 6-15 (2019).
7. S. Krapohl, A. Vasileva-Dienes, The region that isn't: China, Russia and the failure of regional integration in Central Asia. *Asia Europe Journal* (2019). DOI: 10.1007/s10308-019-00548-0.
8. M.Bělín, J.Hanousek, Which sanctions matter? Analysis of the EU/Russian sanctions of 2014. *CEPR Discussion Paper No. DP13549* (2019). URL: https://econpapers.repec.org/paper/cprceprdp/13549.htm. Accessed: 09.09.2019.
9. I. Korhonen, H. Simola, L. Solanko, Sanctions and countersanctions – effects on economy, trade and finance. *Focus on European Economic Integration, Oesterreichische Nationalbank (Austrian Central Bank)*, **Q3-18**, 68-76 (2018).
10. M. Smeets, Can economic sanctions be effective? *Staff Working Paper*, **ERSD-2018-03** (2018). URL: https://www.wto.org/english/res_e/reser_e/ersd201803_e.pdf. Accessed: 09.09.2019.
11. P. Kurečić, The 'New Cold Warriors' and the 'Pragmatics': The differences in foreign policy attitudes towards Russia and the Eastern partnership states among the NATO member states from Central and South-Eastern Europe, *Croatian International Relations Review*, **23**(80), 61-96 (2017).
12. A.P. Portanskiy, Trump presidency changes prospects of international trade. *USA and Canada: Economy, Policy, Culture*, **11**(575), 83-96 (2017).
13. A.N. Lyakin, M.I. Rogov, Sanctions and counter-sanctions: Using political instruments to realize economic goals. *National Interests: Priorities and Security*, **13**(8(353)), 1396-1414 (2017). [in Rus.].
14. E.L. Zolotareva, International food trade: Trends, problems, predictions. *Bulletin of Kursk State Agricultural Academy*, **4**, 158-162 (2018). [in Rus.].
15. S.N. Larin, Sanction limitations of the West against Russia and the Russian contractions: Compensation of damages. *Journal of Economy and Business: Theory and Practice*, **4-2**, 115-121 (2019). [in Rus.].

Application of Foreign Law in Economic Disputes and Implementation of National Interests

N.V. Deltsova[1], *U.A.* Dorofeeva[1], *M.N.* Zubkova[1], and *M.A.* Tokmakov[1,*]
*Corresponding author: maxim.tokmakov@gmail.com.

[1] Samara State University of Economics. Samara, Russia

Abstract. In the context of globalization, the question of protecting one's own national interests is becoming more acute for the state on whose territory foreign law and order is applied. The application of foreign legislation in economic disputes creates various problems and raises questions related to the establishment of the content of foreign law, which must be resolved in the context of Theoretical understanding and law enforcement practice. This study is aimed at identifying the risks of uncertainty in the application of foreign law to relations involving a foreign element in the Russian legal reality and finding ways to overcome them taking into account national interests. Special attention is paid to consideration of the Institute of non-use (limitations) of foreign law is presented in the form of rules on the application of the mandatory rules of the forum (lex fori), the reservation of public order (order public), and eliminating the use of foreign law in case of conflict.

Keywords: national interest, foreign law, lex fori, agreement on applicable law, conflict of qualifications, public order.

1 Introduction

Modern socio-economic integration processes and the blurring of national borders in the information environment of the Internet lead to a close interaction of States in the establishment of regulation of relations arising between entities with different nationalities. At the same time, in the conditions of globalization the questions of protection of own national interests become more acute. At the same time, it should be noted that the main meaning of this term is, first of all, the unity of interests of citizens and society of a certain state, and not the interests of any particular ethnic group-the unity of interests in this case is the result of consensus in the basic, indigenous, interests of all citizens of the state. In other words, sharing V. D. Zorkin's position, national interests consist in inviolability of a constitutional system, sovereignty and territorial integrity, in political, economic and social stability, and also in unconditional ensuring legality and maintenance of a law and order, in development of equal and mutually beneficial international cooperation [1].

At present, economic national security is becoming as important as military security and forms the basis of state sovereignty. Given the breadth and diversity of economic relations, understanding the content of enforceable legal norms is often a difficult task. The variety of regulations, prohibitions, restrictions and requirements creates the need not only to form a common concept of the system of legal institutions defined by the national legal order, but also their relationship and interaction. The solution of the problem of collision, conflict of legal norms is not particularly difficult: the rule of hierarchy of legal norms (the ratio of General and private legal norms), as well as the provisions of judicial practice and legal doctrine, which allow to resolve the issue that arose in the process of application of national legislation, helps to solve this problem. At the same time, practicing lawyers, as well as the actual participants of economic relations, are often faced with the problems of applying the rules of foreign law and order and determining their content.

The application of foreign legislation in relations complicated by a foreign element creates various problems and raises questions related to the establishment of the content of foreign law, which must be resolved in the context of theoretical understanding and law enforcement practice. For a state in whose territory a foreign legal order is applied, it is important to balance the observance of national interests and the interests of foreign States. Such a balance is necessary due to the presence of international relations between persons of different States, the presence of each state's obligation to protect the rights and interests of its own citizens and organizations. The present research addresses issues of uncertainty, application of foreign law to particular relationships, as well as problems of application of conflict rules arising in connection with the operation of peremptory norms, a conflict of qualifications categories of Russian and foreign law and public order, taking into account national interests.

In connection with the above, the following issues are to be resolved within the framework of this study:

- how big are the risks arising from the uncertainty of the application of foreign law to relations involving a foreign element for the subjects of such relations?
- whether adequate and reasonable are such means of protecting national interests as restriction of foreign norms established in the form of reservation of public order, the application of mandatory rules and elimination of the use of foreign law in case of conflict of qualifications?

The purpose of this study is to identify risk uncertainty of application of foreign law to the relations within the form of participation of foreign element in the Russian legal reality and finding ways to overcome them, taking into account national interests, as well as to consider the Institute of non-application of foreign law presented in the form of rules on the application of the mandatory rules of the forum (lex fori), the reservation of public order (order public), and eliminating the use of foreign law in case of conflict of qualifications as a legal means of protecting national interests.

2 Methods

The study analyzed the current Russian legislation, Russian law enforcement practice, as well as the provisions of the Russian and foreign legal doctrine, which allowed to identify the main risks associated with the uncertainty of the application of foreign law in the territory of the Russian Federation, and to establish the main directions of overcoming them, taking into account national interests. The paper also considers and summarizes the main elements that make up the institution of non-application (restriction) of foreign law in the Russian legal order, resulting in the conclusion that it is necessary to find a balance between national interests and the interests of foreign States when using this institution.

3 Results

3.1 Uncertainty of application of foreign law

Uncertainty for each of the parties in the question whether this or that rule to be applied to the dispute in case of its occurrence can significantly complicate the achievement of the economic goal of the party participating in foreign economic turnover. According to J. Basedow, the application of foreign law is not a matter of tolerance towards a foreign state, but rather respect for the expectations of individuals that their relations will be governed by the laws of a particular foreign jurisdiction [2]. Unfortunately, the complete elimination of legal uncertainty in this matter is impossible, since the application of conflict rules depends on the law of the forum (lex fori), i.e. the procedural law of the state where a dispute involving foreign element shall be considered on the merits.

An equally complex situation associated with the uncertainty of the application of foreign law arises when a national conflict of laws rule prescribes the application of foreign law, and this foreign law (its conflict of laws rule) indicates the application of the national law, i.e. refers back (renvoi), or a similar reference is made to the law of a third state [3].

Agreement on applicable law

The possibility of a Russian court applying a foreign legal norm always causes certain risks associated with a variety of circumstances. The economic entities involved in the dispute do not always understand the rules of which state are designed to settle the dispute between them. To some extent, the agreement on the applicable law helps to resolve this issue. A. J. Belohlavek notes that the parties can at any time agree on the applicable law, including during the trial, and such an agreement may differ from the original agreement concluded before the proceedings [4]. Similar rules are provided by the Russian legislation (article 1210 of the Civil Code of the Russian Federation [5]).

At the same time, according to article 249 of the Arbitration Procedure Code of the Russian Federation, if the parties have entered into an agreement in which they have determined that the Russian arbitration court has competence to consider a dispute related to their entrepreneurial and other economic activities, the Russian court will have exclusive competence to consider the dispute, provided that such an agreement does not change the exclusive competence of a foreign court [6].

Nevertheless, the choice of law agreement does not definitively remove doubts as to both the legal system chosen by the court as competent in resolving the dispute and the directly applicable legal rules, including in the case of a plurality of legal systems operating in the territory of the state whose law is applicable to the disputed relations or its individual components.

3.2 Protection of national interests by non-application of foreign law

Application of the peremptory norm

The establishment of the content of a foreign law is preceded by the stage of establishing the presence or absence of grounds for blocking the application of a conflict of laws rule. One of the grounds for refusal to apply foreign law is the effect of a peremptory norm of the law of the country of the court. The validity of recourse to mandatory rules and refusal in this regard from the application of foreign law is illustrated by M. P. Shestakova on the following example. The parties choose as applicable law of the state whose law sets the Statute of limitations for filing claims that is different from the period prescribed by the mandatory rule of a state, which is associated with all of the substantive relations of the parties circumstances. In order to prevent circumvention of the law, in many legal orders there is a provision according to which in such situations the rules chosen by the parties are not applied by virtue of a peremptory norm. That is, in this case, the rule on the limitation period of the country with which the relationship has arisen is subject to application [7]

The conflict of qualifications

The norms of foreign law are also inapplicable if the categorical apparatus of national legislation and the legal system of a foreign state contradict each other, i.e. there is a conflict of qualifications. To illustrate the need to eliminate the conflict of qualifications, one can cite the following example: the institution of trust existing in common law countries is absent in continental law countries, including Russia, its qualification is necessary to establish the content and application of this concept.

The resolution of this conflict in Russian law is carried out by a unilateral conflict of law rule linking the disputed legal relationship to Russian law. Article 1187 of the Civil Code of the Russian Federation establishes the following rule: in determining the law to be applied, the interpretation of legal concepts is carried out in accordance with Russian law, unless otherwise provided by law [5].

The newest interpretation of the concept of a conflict of qualifications is carried out by the Supreme Court of the Russian Federation, which clarified that the rule established in the Civil Code of the Russian Federation represents the primary qualification of a legal norm, since the interpretation of legal concepts is carried out precisely for the purpose of determining the applicable conflict of law norm. After determining the applicable law on the basis of the relevant conflict of law norms, the court interprets the applicable norms of foreign law, guided by the provisions of Article 1191 of the Civil Code of the Russian Federation on the establishment of the content of foreign law [8].

Public Order

In this study, S. Brekoulakis concludes that a state can prohibit any arbitration in its territory, the subject of which is not amenable to arbitration, that is, it contradicts the public order of that particular state [9]. The need for such a rule as a reservation on public order (order public) is undeniable. Of course, there are provisions in the norms of national legislation, the observance of which is objectively necessary, despite the legal system chosen by the court for application in accordance with the conflict of laws norm.

A reservation on public order (order public) is applied in Russian civil law in case of refusal to enforce a foreign judicial or arbitral award, refusal to execute an order of a foreign court or competent authority of a foreign state. It is worth noting that the Supreme Court of the Russian Federation warned the courts against the unreasonable use of ordre public in cases complicated by a foreign element, indicating that the court does not apply the rule of foreign law only in exceptional cases when the consequences of applying such a foreign rule are clearly contrary to the public order of Russia. In addition, the absence of rules or legal institutions in Russian law that are similar to the rules or legal institutions of applicable foreign law is not in itself a basis for the application of ordre public [8]. Thus, in the framework of case No. A60-308/2018, Russian courts, applying the rule of foreign law, noted that the absence in Russian legislation of a rule similar to paragraph b of § 201 of the US Copyright Law [10] does not mean that the ownership of the author's title should be determined based on the norms of Russian law.

Substantive and procedural law

It is also worth considering that even in cases where foreign law is applicable to resolve the dispute on the merits, the court in considering the dispute is guided by national procedural legislation. As A. Surana points out, local substantive legislation may be repealed in specific cases, and foreign law will apply instead. However, local procedural law, as a rule, cannot be violated by foreign law, since the local judicial authority will only apply its own procedural rules (procedural rules of the forum). Accordingly, both internal affairs and affairs with foreign elements are regulated by local procedural legislation [11]. A similar rule is provided for by Russian law - cases involving foreign persons are considered by the arbitration court in accordance with the rules of the Arbitration Procedure Code of the Russian Federation.

4 Conclusion

Each state is interested in the legal certainty provided to the subjects of the civil turnover, in the evidence of the actions taken in the process of carrying out economic activities. Any uncertainty, including that associated with the application of foreign law in the territory of a state, undermines the stability of economic relations and negatively affects their participants. One of the mechanisms to overcome this uncertainty is the agreement of the parties on the applicable law. At the same time, for the state in whose territory the foreign legal order is applied, the observance of one's own national interests, achieved through the application of the law of the country of the court (lex fori) and the non-application of foreign law, is of particular importance. However, it is also obvious that the widespread and unjustified use of restrictions on the use of foreign law and order can lead not to the stability but to a violation of the principle of mutual courtesy, which impedes the implementation of international treaties and agreements aimed at fulfilling obligations by member states to grant foreign citizens regime less favorable than for their own citizens, which is why when applying foreign law, it is so important to find a balance of national interests and interests of foreign States.

References

1. V.D. Zorkin, The role of law in ensuring national interests. *Law and Security*, **1-2** (18-19) (2006). [in Rus.].
2. J. Basedow, The application of foreign law – comparative remarks on the practical side of private international law. In J. Basedow, and K. B. Pißler (Eds.), *Private International Law in Mainland China, Taiwan and* Europe (pp. 85-97). Tübingen: Mohr Siebeck. (2014).

3. A.A. Khan, A. Ghani, F. ur Rehman, M. Ullah, What is renvoi'? Distinguish the theory of 'single renvoi' from 'double renvoi' with reference to relevant case law in private of international law (2018). URL: https://ssrn.com/abstract=3124439. Accessed: 16.09.2019.
4. A.J. Belohlavek, Substantive law applicable to the merits in arbitration. *Romanian Review of Arbitration*, **8**(2 (30)), 1-16 (2014).
5. Civil Code of the Russian Federation (1994). URL: http://www.consultant.ru/document/cons_doc_LAW_5142/. Accessed: 16.09.2019. [in Rus.].
6. Arbitration Procedure Code of the Russian Federation (2002). URL: http://www.consultant.ru/document/cons_doc_LAW_37800/. Accessed: 16.09.2019. [in Rus.].
7. M.P. Shestakova, Obligations arising from unjust enrichment in private international law. In I.O. Khlestova (Ed), *Non-contractual obligations in private international law*. Institute of Legislation and Comparative Law under the Government of the Russian Federation (pp. 80-95). Moscow: Norma, INFRA-M (2017). [in Rus.].
8. Resolution of the Plenum of the Supreme Court of the Russian Federation of 09.07.2019 No. 24 "*On the application of international private law by the courts of the Russian Federation*" (2019). URL: https://rg.ru/2019/07/17/vs-dok.html. Accessed: 16.09.2019. [in Rus.].
9. S. Brekoulakis, Law applicable to arbitrability: Revisiting the revisited lex fori. In L. Mistelis, S. Brekoulakis (Eds.), *Arbitrability: The international and comparative perspectives*. Queen Mary School of Law Legal Studies Research Paper, **21**, (pp. 99- 119). London: Kluwer (2009).
10. US Copyright Law. URL: https://www.copyright.gov/title17/. Accessed: 16.09.2019.
11. A. Surana, International civil procedure and principle of 'lex fori' (2006). URL: https://ssrn.com/abstract=915923. Accessed: 16.09.2019.

Improvement of Formation and Implementation of Deposit Policy in Russian Commercial Banks

V.A. Manyaeva[1], *S.V.* Rykov[2], *M.B.* Tershukova[3], and *L.N.* Milova[4]
*Corresponding author: manyaeva58@mail.ru.

[1] Samara State University of Economics, Samara, Russia

Abstract. Under current market conditions, the problem of attracting available cash assets of natural and legal persons to bank deposits is one of the urgent tasks in the banking sector of the Russian Federation. Under circumstances of high competition and instability in financial markets, the issue of formation, implementation and objective assessment of deposit policy in commercial banks is becoming increasingly important. Reasonable deposit policy should take into account the selected priorities for further growth and improvement of quantitative and qualitative indicators of bank activity, its customers' characteristics, and socio-economic conditions in the country. Deposit policy has a great impact on the liquidity and bank ability to meet payments. The authors studied the modern scientific idea to form and implement deposit policy of commercial banks in the process of flexible management of banking passives under the real state of the Russian financial market, and identified directions to improve formation and implementation of deposit policy in Russian commercial banks.

Keywords: bank deposit, commercial banks, deposit policy, deposit policy evaluation, deposit services.

1 Introduction

One of the main tasks to improve the commercial bank activities in order to attract deposits is to use effective methods and mechanisms to manage borrowings for benefit of all groups of participants: founders, borrowers, depositors and bank staff. The formation of the optimal volume of deposit resources for a long term with the lowest use costs is possible with effective management of active and passive operations by commercial banks. At the same time, the main motivation for the owners who have available cash assets is the amount of deposit payment and return guarantee.

Reasonable deposit policy should take into account the selected priorities for further growth and improvement of quantitative and qualitative indicators of bank activity, its customers' characteristics, and socio-economic conditions in the country. Deposit policy has a great impact on the liquidity and bank ability to meet payments. To strengthen the deposit base, commercial banks need to constantly improve their deposit policy, which is aimed at the deposit attractiveness and provides bank depositors with social and economic protection under the conditions of current loss of funds value. This determines the relevance, purpose and objectives of the paper.

2 Problem Statement

The theoretic and practical analysis of formation and implementation of deposit policy in commercial banks showed that many problems in this area need to be solved taking into account the current conditions of banking system functioning in the Russian Federation: economic sanctions, lack of access to external borrowing markets, decrease in real incomes of the population, limitation of available cash assets in the real sector of economy, etc.

The Russian scientists and others dedicated their scientific works to theoretic researches in banking, and formation and implementation of deposit policy of commercial banks [1-2, 3, 4, 6-9].

To improve the organizational basis for formation and implementation of deposit policy, we need to develop the deposit services quality through its evaluation. National scientists [5, 7, 10] and others, as well as foreign scientists [11, 12] and others devoted their works to this line of research.

Studies on deposit policy assessment are presented in the works of O.D. Zhilan and M.R. Danilova [13], M.S. Maramygin and D.A. Suplakov [14], V.S. Potapov and M.A. Potapova [15], V. A. Savinova and T.E. Borlakova [16], L.I. Yuzvovich and E A. Trofimova [17] and others.

For the successful formation and implementation of deposit policy, commercial banks need new scientific approaches and organizational governance, guidance paper on the deposit portfolio analysis, and evaluation of deposit policy effectiveness that meet the current conditions of the Russian banking system. Precisely, this scientific target involves the development of ways to improve formation and implementation of deposit policy of commercial banks.

3 Research Questions

The following research questions are identified in this study:
- what is the quality content of bank deposit services, what are their quality indicators,
- how is the quality of deposit banking services assessed,
- how is the effectiveness of the bank deposit policy evaluated.

4 Purpose of the Study

The purpose of the study is to develop scientific and methodological foundations and elaborate practical recommendations to improve formation and implementation of deposit policy of commercial banks that meets the requirements of the modern Russian banking system. The developed scientific and methodological regulations contribute to the enrichment of the banking theory. Practical recommendations ensure effective implementation of deposit policy of commercial banks.

5 Research Methods

To develop scientific and methodical bases and elaborate practical recommendations on improvement of formation and implementation of deposit policy of the commercial banks that meets the requirements of modern development trends in the Russian deposit market, we need:

- to interpret the quality content of bank deposit services and define their quality indicators,
- to develop practical recommendations to improve the quality of deposit services in commercial banks on the basis of the SERVQUAL model that provides a survey of commercial banks depositors, using the proposed questionnaires and the order of processing and criteria for the survey results interpretation according to five indicators for the deposit service quality,
- to propose a methodology for performance evaluation of deposit policy of commercial banks that contains a set of evaluation procedures to assess banking soundness and profitability, and also allows to assess the utilization efficiency of deposit resources of commercial banks.

6 Findings

6.1. The quality content of bank deposit services and quality indicators.

When implementing deposit policy, the bank provides clients with deposit services. It sets up bank settlement accounts and current accounts, issues bank certificates and debit cards, and makes fixed-term deposit accounts to persons and corporate entities.

Under the current conditions of increased bank competition in the deposit services market, the question to improve the deposit service quality becomes especially actual. In addition, interest rate cut by the Central Bank of the Russian Federation, in recent years, does not allow banks, as previously, actively apply price instruments to stimulate attraction of financial resources to deposits. That is why the improvement of the deposit services quality plays a crucial role in deposit portfolio formation.

The deposit services quality is a set of characteristics and properties of the deposit, which enables the commercial bank to meet the expected depositor's needs.

It is proposed to define a set of quality indicators for deposit services provided by commercial banks to their depositors. First of all, they are:

- reliability in providing deposit services, that is to say, a guarantee that the client will collect money invested in the bank and accrued interest,
- timeliness, that is to say, deposit service provision strictly at the required time for the client,
- social and psychological indicator, that is to say, polite treatment with depositors by the bank employees, as well as comfort with quick requests and wishes,
- availability, that is to say, enabling the client to use the deposit service offered by the bank without obstruction,
- communication skills, that is to say, providing clients with necessary information, including the deposit services use.

6.2. Quality evaluation of bank deposit services.

To determine the deposit services quality, the algorithm of SERVQUAL model was tested in the study. The branch of public joint-stock company "Sberbank of Russia" was chosen as the object of the empirical research in the survey.

The indicator of deposit service quality is calculated by the formula

$$Q = P - E, \qquad (1)$$

where Q – quality indicator,

P – perception indicator,

E – expectancy indicator.

If the quality indicator Q tends to zero, it means that consumer's expectations coincide with the perceived quality after the fact. If there is a positive expectation value E, the quality indicator Q will be negative. At the highest value of perception, P will be positive.

The calculation of the quality coefficient in reference to the bank customers' expectations, i.e. potential consumers of deposit services, is made by the formula

$$Q_1 = (\ \overline{E} - \overline{P}\ ) \div \overline{E} \times 100, \qquad (2)$$

where Q_1 - quality ratio in reference to expectations,

$\overline{E}$ - average value of expected quality level,

$\overline{P}$ - average value of perceived quality level.

In the study, the basic model questionnaires "Perception" and "Expectation" were developed. They allowed to commensurate the expectations and perceptions of the deposit services quality received by commercial bank clients. The questions are grouped into blocks according to five quality indicators: tangibility, reliability, sympathy, confidence and empathy. According to the model, materiality, reliability, sympathy, conviction and empathy are recognized as evaluation criteria. The above-mentioned questionnaires, developed by the authors, were tested in the course of the survey of various client groups of the branch of PJSC "Sberbank of Russia", which allowed us to identify how the services provided to clients, in their perception, correspond to the expected and planned bank results.

The practical application of the developed questionnaire and methodology, consisting of a number of successive stages, in general, will provide the bank with information about the strengths and weaknesses in formation and implementation of deposit policy, assess the deposit services quality, and identify measures to improve the services quality.

6.3. Assessment of efficiency of bank deposit policy.

The bank deposit policy should be effective, that is to say, it should allow rational and profitable use of the resources involved in deposits, ensuring the necessary level of bank profitability and compliance with mandatory liquidity standards of the Central Bank of the Russian Federation.

The authors defined the stages of performance evaluation of bank deposit policy: assessment of the bank deposit sourcing results, assessment of the deposit resources stability, assessment of the effective use of bank deposit funds, general assessment of the deposit policy effectiveness in order to determine reserves to increase profitability and ensure bank liquidity. During these stages, the bank draws conclusions about the implemented policy strategy to attract deposits.

To assess the effectiveness of deposit policy in the commercial bank, we proposed a set of indicators (table 1).

Table 1. Set of indicators to assess deposit policy of the commercial bank.

Evaluation category	Set of indicators
1. Bank reliability assessment	Rate of deposit risk
	Capital adequacy
2. Bank profitability assessment	Net put and call
	Interest margin
	Average cost rate of attracted deposit funds
3. Assessment of efficiency of deposit resources use	Use factor of attracted deposit funds
	Effectiveness ratio of use of attracted deposit funds
	Effectiveness of deposit operations

Source: Authors.

The deposit policy of the commercial bank will be effective if there is an accomplishment of all these indicators, or their established statutory values, or the values provided for by the business plan. This will allow commercial banks to assess deposit risks, effectively conduct deposit policy in the future, based on the ratio of deposit resources volume with its structure, term structure of deposits and assets, and thereby, ensure profitability of commercial banks and liquidity of their activities.

Conclusion

Scientific approaches, proposals on organizational bases and methodological recommendations on deposit policy assessment are aimed at effective work of commercial banks to attract and use deposit funds, and identify reserves to increase profitability and liquidity of their activities.

The results of the study can be used in the commercial banks activities in the process of formation and implementation of deposit policy. They can also be used in research and higher education institutions for the further development of the banking theory.

References

1. G.N. Beloglazova, Development strategy of the regional segment of the banking system. *Banking*, **2**, 28-31 (2011). [in Rus.].
2. V.E. Kroliveckaya, L.P. Kroliveckaya, I.V. Soldatenkova, Credit and banking system of Russia at present stage. *Journal of Legal and Economic Studies*, **1**, 128-132 (2018). [in Rus.].
3. O.I. Lavrushin, On modernization of regulation and new models of banking business development. *Economy. Taxes. Law*, **11**(3), 14-19 (2018). DOI: 10.26794/1999-849X-2018-11-3-14-19. [in Rus.].
4. S.B. Kovalenko, Topical issues of banking sector development. In Sukiasyan A.A. (Ed.), *Improving the methodology of knowledge for the science development: Collection of articles of International Research-to-Practice Conference* Part 1 (pp. 108-110). Ufa: Aeterna (2017). [in Rus.].
5. C.B. Kovalenko, P.G. Shernin, Deposit insurance systems in the Russian Federation, Baltic countries and CIS countries. *Bulletin of Saratov State Socio-Economic University*, **1**(65), 34-41 (2017). [in Rus.].
6. A.A. Martens, N.O. Derkach, Deposit policy as a basic element of commercial bank funding. In O.N. Shirokov (Ed.), *Education and science: Modern trends* (pp.140-150). Cheboksary: CNS «Interactive plus» (2018). [in Rus.].
7. A.A. Maslova, N.N. Martynenko, Deposit policy of a commercial bank and stereotypes of its development in an unstable economy. *Modern Science: Topic Issues of Theories and Applications*, **7**, 55-57 (2017). [in Rus.].
8. A.N. Nikulina, A.A. Abalakin, Deposit policy of commercial banks: Tools, problems, prospects. *Economy and Management Issues*, **4**(44), 62-65 (2015). [in Rus.].
9. G.S. Panova, Banks and banking business in Russia in the face of international sanctions. *Obozrevatel – Observer*, **12**(311), 94-105 (2015). [in Rus.].
10. E.V. Novatorov, V.D. Novitskaya, Modification and empirical testing of SERVQUAL service quality measurement methodology with reference to banking services. *Marketing and Finances*, **1**, 22-32 (2012). [in Rus.].
11. A. Parasuraman, V.A. Zeithaml, L.L. Berry, A conceptual model of service quality and its implications for future research. *Journal of Marketing*, **49**(4), 41-50 (1985).
12. A. Parasuraman, V.A. Zeithaml, L.L. Berry, SERVQUAL: A multiple-item scale for measuring consumer perceptions of service quality. *Journal of Retailing*, **64**(1), 12-40 (1988).
13. O.D. Zhilan, M.R. Danilova, Deposit policy influence on financial sustainability of commercial banks. *Baikal Research Journal*, **7**(4), 4 (2016). DOI: 10.17150/2411-6262.2016.7(4).4. [in Rus.].
14. M.S. Maramygin, D.A. Suplakov, Managing the liquidity risk of a bank. *Upravlenets*, **9-10**(37-38), 46-50 (2012). [in Rus.].
15. V.S. Potapov, M.A. Potapova, Management of attracted resources as a basis of the deposit policy of commercial bank. *Economy Vector*, **6**(12), 38 (2017). [in Rus.].
16. V.A. Savinova, T.E. Borlakova, Evaluation of the development of the bank's operations to attract financial resources. *Economic Sciences*, **165**, 96-102 (2018). [in Rus.].

L.I. Yuzvovich, E.A. Trofimova, M.S. Maramygin, A.V. Lebedev, *Improving the assessment of commercial bank performance in the context of interbank competition* (Ural University Publishing House, Ekaterinburg, 2018). [in Rus.].

Digital Transformation of Enforcement Proceedings: Russian Experience

G.E. Ageeva[1,*]
*Corresponding author: galinaageevva@mail.ru.

[1]Samara State University of Economics, Samara, Russia

Abstract. In this study, the author considers the main problems of digital modernization concerning enforcement proceedings in the Russian Federation. Generation and further enhancement of the digital space have determined the vector of legal regulation in the implementation of any social processes several years ago. Various aspects of the economic, legal and other spheres of public life were reformed. Changes in various areas of enforcement proceedings occurred in stages, sequentially and in plain consideration with the specific features of legal relationships. However the digital transformation of enforcement proceedings has not yet come to its final stages. Many researchers note the "unavailability" of both legislation and law enforcement procedure for the changes that are taking place as well as to the final establishment of such changes as the basic and fundamental rules of law enforcement. In addition, the existing fragmentary regulation in the process of law enforcement gave rise to many problems. In this article the author attempts to identify the positive and negative experience of digital modernization of Russian law enforcement practice and define suggestions that could improve the process.

Keywords: law enforcement, digital transformation, digital economy, legal process.

1 Introduction

Digital transformation as a process of changing the technological and cyberspace, its ubiquitous implementation into various social processes is based not only on digital technologies, but rather on new principles of human consciousness and communication [1]. Of course, neither law enforcement practice nor legal process as a whole shall remain aloof from the changes.

The digital transformation of the legal regulation of the economic relations of Russian society entrenched not so long ago. In 2017 the Government of the Russian Federation approved the program "Digital Economy of the Russian Federation [2], which soon expired [3] due to the emergence of a national passport in the domestic legal system dedicated to the digital economy [4]. Attempts to consolidate the new informational and technological implementation for law enforcement processes in Russian society have taken place more than a decade ago.

Judicial proceedings in the Russian Federation, as one of the most fundamental blocks of law enforcement, underwent legislative reform in the light of the formation of uniform technological processes in Soviet times [5]. It is, first, about the need to create a single digital base that meets the requirements of technological advancements, information capability and uninterruptedness. The process has been continuously improved, the course on the use of information and communication technologies in the administration of justice has been clearly set now [6]. Scientific discussions on this subject also do not lose their relevance [7].

The digital transformation has affected not only legal proceedings, but also other law enforcement practices. However, despite close attention to the digital content of various legal procedures, they do not occur in any area of law enforcement without any problems.

2 Problem Statement

The digital transformation of domestic law enforcement practice is consistently considered to be more than just the goal of legal regulation. These are the realities of our society. The goal is rather its further dissemination, improvement and the creation of the rules for the unhindered implementation of new information and technological methods for the interaction of subjects of law.

Unfortunately there is now a significant number of problems of legal regulation uncertainty, unclear enforcement process using new digital technologies [8]. In order to resolve them it is necessary, first of all, to understand the legal nature of modern technologies, the essence of digital legal relations generated by them and to develop common models of legal regulation and principles of law enforcement in the context of the digital transformation of Russian society [9].

3 Research Questions

In the process of researching the features of the digital transformation of the law enforcement proceedings it seems necessary to analyze the peculiarities of the info communicative interaction of legal entities, one of which is an individual or collective person having state-power functions. Modern scholars note a number of significant problems in such legal procedures. This, firstly, is due to the fact that legal regulation does not keep pace with the actual dynamics of social legal relations. In addition there are no unified and clear criteria at the state level for determining either the place of modern technologies in the legal array or the legitimate ways to use them. For example, for a long time in our country law enforcement agencies refused to recognize the legal status of virtual currency (cryptocurrencies) as instrument of payment [10]. Litigation here took the lead prompting cryptocurrency to become a financial instrument [11]. However, it is currently impossible to use the specified means of payment in the practice of enforcement, since the procedure for the enforcement of foreclosure on it has not yet been provided. This is explained by the lack of technical ability to access the owner's crypto wallet due to the unforeseen procedure for technological circulation in the financial markets of previously unrecognized cryptocurrencies.

4 Purpose of the Study

The purpose of the study of the features of the Russian experience of digital transformation of law enforcement practice is the need to analyze and generalize the available data baggage. The above is a relevant and necessary element for the development of unified rules for the legal regulation of information and communication interaction of entities. In addition it is important to the research process that the state-imperious participant is a mandatory subject of law enforcement considered as a type of legal process. That is why it seems significant to generalize and analyze the experience of legal activity with the participation of a public administration entity, the regulation of which is currently fragmented [12].

5 Research Methods

In this study the author used the method of materialistic dialectics as a fundamental way of cognition and analysis. In the process of writing the article an extensive layer of law enforcement practice in various fields was studied. For the research such methods of scientific study were used: analysis and synthesis; induction and deduction. In addition, the method of comparative jurisprudence had a significant impact on the formulation of the provisions of this work, which allowed us to identify the general and distinctive features of the digital transformation of law enforcement practice, and due to it, it became possible to draw the main conclusions.

6 Results

Legislation and law enforcement aren't keeping up with the development of digital technologies in modern society. This conclusion was reached by experts of the BRICS Antimonopoly Center and leading world scientists [13]. The above confirms the problem of digital transformation of a particular case of law enforcement practice. However, this expression can reflect any legal relationship.

The digital transformation of certain spheres of public life which began several years ago and is currently incomplete inevitably affects a person, since it is, above all, about new principles of human consciousness and communication. Legal regulation in this case should not be fragmentary, but consistent: first, it is necessary to develop common approaches to understanding the status and place of the latest technologies in the legal sphere.

The first thing to focus on is the problem of digital literacy of legal entities [14]. For successful and effective use of modern technological methods of interaction subjects should be at least familiar with them. In our opinion, it is necessary for legal professionals to introduce mandatory digital literacy courses for further implementation in law enforcement practice.

However, it is important to emphasize that while some entities, lacking sufficient digital literacy, are trying to participate in law enforcement using modern information and communication technologies, others are already able to abuse this literacy. In view of this, at the present stage of development of Russian society, an increase in the level of the fight against cybercrime and ensuring information security is also required.

In addition, another problem of the digital transformation of domestic law enforcement is the inaccessibility of certain technologies. So, for example, e-justice is gaining momentum, in the process of judicial proceedings, subjects are increasingly interacting remotely. However, in some cases, not throughout our country, uninterrupted provision of electronic digital broadcasting is possible. And in some cases, this requires additional resources: both technological and financial. For example, to carry out a number of procedural steps, it is necessary to use an electronic digital signature, which, due to its high cost, is not yet widely used [15]. In our opinion, it is necessary to comprehensively develop and disseminate modern digital technologies in our country, as well as reduce their cost.

To reach the unhindered implementation of law enforcement practice, in consideration with its digital content in our country at the present stage, a comprehensive, science-based discussion of the essence of digital technologies in law and the procedure for their clear and consistent implementation is necessary. The regulation should differ in certainty and compliance with the actual legal relations, the interests of society and the state.

7 Conclusion

The digital transformation of law enforcement practice in the Russian Federation has in fact happened but has not yet been completed and this is the reality we are facing now. However, there is currently many problems with the implementation of many procedures using the latest technologies. Law enforcement practice is ahead of legal regulation that is not complete in reflecting actual state of things. Currently, fundamental work is required in this area to develop unified positions, firstly, in understanding of information and communication tools in the legal array, secondly, in the procedure for using these technologies, and thirdly, in ways to ensure safety of digital interaction of legal entities. However, for starters, in our opinion, it is necessary to increase the digital literacy of legal entities and, if possible, provide them with unhindered access to modern technologies.

References

1. Institute of State and Law RAS, *The transformation of law in the digital age* (2019). URL: http://igpran.ru/download/%D0%A4%D0%BE%D1%80%D1%81%D0%B0%D0%B9%D1%82%2021-03-2019.pdf. Accessed: 05.09.2019. [in Rus.].
2. The order of the Government of the Russian Federation of July 28, 2017 N 1632-r "On approval of the program "Digital Economy of the Russian Federation" (2017). URL: http://www.consultant.ru/document/cons_doc_LAW_221756/. Accessed: 05.09.2019.
3. The order of the Government of the Russian Federation of February 12, 2019 No. 195-r "On recognition of the expiration of the order of the Government of the Russian Federation of July 28, 2017 N 1632-r" (2019). URL: http://www.consultant.ru/document/cons_doc_LAW_318450/. Accessed: 05.09.2019.
4. The passport of the national program "Digital Economy of the Russian Federation", approved by the Presidium of the Presidential Council for Strategic Development and National Projects, protocol No. 16 of December 24, 2018 (2018). URL: http://www.consultant.ru/document/cons_doc_LAW_319432/. Accessed: 05.09.2019.
5. M.V. Chizhov, *The use of information and communication technologies in judicial activities in the context of the formation of the information society*. Ph.D. thesis. Moscow: Russian State University of Justice (2013). [in Rus.].
6. Decree of the Government of the Russian Federation of December 27, 2012 N 1406 (as amended on December 24, 2018) "On the federal target program" Development of the Russian judicial system for 2013 - 2020" (2012). URL: http://www.consultant.ru/document/cons_doc_LAW_140638/. Accessed: 05.09.2019.
7. D.V. Belousov, *E-justice and its role in the organization of judicial protection in Russia, Administrator of the Court*. URL: http://xn----7sbbaj7auwnffhk.xn--p1ai/article/13858. Accessed: 05.09.2019. (2015). [in Rus.].
8. V.B. Naumov, Law in the era of digital transformation: in search of solutions. *Russian Law: Education, Practice, Science*, **6**, 4-11 (2018). [in Rus.].
9. G.E. Ageeva, Legal process and "digital right". In G.R. Khasaev. S.I. Ashmarina (Eds.), *Proceedings of the VII All-Russian Scientific and Practical Conference «Russian Science: Actual Research and Development»* (pp. 324-328). Samara: SSUE (2019). [in Rus.].
10. G.E. Ageeva, P.P. Lang, A.V. Loshkarev, T.V. Chugurova, E.N. Churakova, Peculiarities of protecting the rights of participants of financial markets in court. In E.G. Popkova (Ed), *The Future of the Global Financial System: Down-Fall or Harmony. Lecture Notes in Networks and Systems*, **57** (pp.545-552). Cham: Springer International Publishing (2018).
11. E. Litova, A. Mikheeva, A. Krivorotova, The court first recognized cryptocurrency property. *RBC*. URL: https://www.rbc.ru/finances/07/05/2018/5af0280d9a7947165a6e8c22. Accessed: 05.09.2019. [in Rus.].
12. N.E. Dmitrieva (Ed.), The digital transformation of public administration: Myths and reality (HSE, Moscow, 2019). [in Rus.].
13. Federal Antimonopoly Service (FAS), *The development of the digital economy is ahead of legislation and law enforcement* (2019).URL: https://fas.gov.ru/news/28360. Accessed: 05.09.2019. [in Rus.].
14. N.D. Berman, On the issue of digital literacy. *Modern Studies of Social Problems*, **8**(6-2), 35-38 (2017). [in Rus.].
15. O.V. Shemetova, S.V. Chugunova, Some problems of application of the digital signature in administrative activity of the enterprises and way of their decision. *Actual Problems of Aviation and Astronautics*, **3**, 425-427 (2017). [in Rus.].

Financial Technologies of International Banking Regulations

G.S. Panova[1]
*Corresponding author: gpanova@mail.ru.

[1] MGIMO-University, Moscow, Russia

Abstract. The author considers the problems of banking regulation in the context of globalization. An analysis of relevant issues indicates the need to improve financial technologies for banking regulation. Basel innovations, designed to ensure the stability and uninterrupted operation of the global banking system, have led to the creation of counter-innovations by the banking sector. Basel Accords led to the development of the so-called "regulatory rally", when increasingly sophisticated methods of regulation gave rise to increasingly inventive ways to protect the gains of the banking business. These ways sometimes became an indirect source of rising risks, and were initially taken as effective protection against these risks. The author analyzes the main advantages and disadvantages of the latest Basel Accords on Banking Supervision (Basel III) and identifies specific directions for its improvement, taking into account current practices of national and international approaches to regulating the activities of credit organizations 10 years after the global financial and banking crisis. The importance of the study is determined by the need to develop financial technologies for international banking regulations, as well as theoretical and methodological approaches that determine the interconnectedness and interdependence of financial markets. It is also important to evaluate the effectiveness of measures to regulate the activities of financial and credit institutions at the national and international levels to develop strategies and tactics for the optimal progressive development of financial markets. The purpose of the study is to develop theoretical and methodological approaches to assessing the impact of international standards on activities of Russian credit organizations.

Keywords: Basel III, financial technology, banking regulation.

1. Introduction

The global financial and economic crisis of 2007-2009 drew particular attention to exposure of the banking sector to systemic risk, which undermines the stability of financial markets. The changes in world financial markets in recent years, such as increased competition, globalization, new information technologies, new high-tech and structured banking products have lead to increased risks. As a result, there is a need for standardization and tightening of requirements to financial stability of financial and credit institutions. In the course of transition of the domestic banking system to international accounting standards, implementation of the recommendations of Basel Accords (Basel II and Basel III), Russia's accession to the WTO, the most urgent task was to provide banking management with analytical information reflecting the influence of various factors on the stability of financial markets. For Russian commercial banks, most of which are still inferior to leading foreign financial and credit institutions in terms of the size of the banking business, the solution of problems ensuring their financial stability is especially relevant. The constantly changing market environment, the volatility of market conditions, the place and role of banks in the economy, as well as the insufficient knowledge of the issues of financial stability of banks in the globalized economy, indicate the relevance of the issues raised. The development of financial markets has a direct impact on the state of the world economy and national economies of individual countries. In the current stage of development of the global economy, financial and credit institutions are urgently faced with the question of minimizing risks associated with the foreign economic activity of their clients (foreign exchange, interest, credit, operating, etc.). An analysis of the opportunities for banks to reduce risks and ensure stable returns in the post-crisis period determines the relevance of the issues raised.

2. Problem Statement

Regulation of global financial markets is interconnected with a complex of problems and challenges, including contradictions in state regulation of banking at the national and global levels, as well as contradictions between the global functioning of financial markets and the interests of individual states, contradictions between global financial and credit institutions in the sphere of banking regulation and contradictions between supporters of globalization and / or regionalization of financial markets. Financial regulators strive to solve a three-pronged task, ensuring financial stability in the environment of economic growth and free competition. The consistent implementation of international banking standards in national jurisdictions implies the simultaneous development of free competition and financial innovation.

An urgent problem consists of the fact that international financial institutions, developing international standards and regulating banks' activities, do not have sufficient authority to ensure the implementation of these innovations by national regulators. Similar problems exist at the regional level (within the EU and other associations). Therefore, the ability of the Financial Stability Board (FSB) to monitor the state of global financial markets can be seen as a step forward in solving this problem. However, the Board's tools are very limited. These are: creating a model of behavior, peer review and a special set of measures of influence, mainly of a moral nature, as a model for national jurisdictions. At the same time, international institutions seek to expand their influence on national regulators of financial markets. For example, the International Organization of Securities Commissions (IOSCO), as a prerequisite for a member to join the 2001 Memorandum on International Cooperation in Combating Violations in Securities Markets, demands the adoption of national legislation that counteracts market manipulation, which is one of its standard recommendations [1]. Indeed, the lack of coordination and standardization of banking policy at the national level creates the prerequisites for regulatory arbitration. It is no accident that when the Dodd-Frank Act on financial reform and protecting the interests of consumers of financial services was passed, ex-US President B. Obama insisted that other countries adopt similar norms in 2010.

Another problem is the lack of clear rules of subordination between global institutions (for example, the G-20 and the Financial Stability Board), on the one hand, and organizations that develop and establish international standards for the activities of financial and credit institutions (Basel Committee, IOSCO, etc.), on the other hand. For the latter, their own, "specialized" vision of the main tasks and their solutions is often a priority, without taking into accounting the consequences for macroeconomic stability. Overcoming this contradiction, highlighted during the financial and banking crisis of 2008-2009, has become one of important tasks of the modern stage of economic development in the world.

Centrifugal forces also act along with the objective tendencies for coordination and harmonization of financial regulation, whose representatives note that since national financial markets can vary significantly, international standards must be adapted to national specifics. In addition, international standards adopted on a compromise basis, due to the peculiarities of their development, may not always offer the best solution to the problem. Given the differences in the level of maturity of financial markets, as well as cultural, historical, geographical and other features of countries, differences in the operating activities of large and small banks, the very existence of a single international standard for all countries and institutions is being questioned.

To solve pressing problems, the economic press suggested the creation of the World Financial Organization (WFO), by analogy with the World Trade Organization (WTO), to regulate global financial markets, establish international standards for regulation and prudential supervision, monitor their implementation, and create a mechanism to resolve disputes [2].

The idea of adopting an international convention or international cooperation agreement to harmonize regulatory rules in leading financial markets has been repeatedly proposed by the Institute of International Finance (The Institute of International Finance is a research, consulting and information interbank center, created by the largest Western banks in 1983. Its members are more than 180 private banks) [3]. As an argument in favor of creating such a structure, it is possible to use it as an important tool for effective cooperation, able to resolve potential conflicts of interests of individual countries and remove barriers to cross-border business, which is especially important for the largest banks in the world. However, so far this idea has not been discussed at official multilateral meetings. Even for its supporters, it seems to be a very distant prospect since the existence of the IMF and the World Bank partially solves the problem of monitoring financial markets. At the same time, the main reason why it is not possible to create the World Financial Organization as a world regulator with legal and enforcement functions is the unwillingness of national governments to sacrifice part of sovereignty and transfer the right to control their financial markets to an international organization, the decisions of which they cannot fully control. Events 2008-2009 showed that in the event of a crisis, bank rescue was carried out mainly at the expense of national budgets. According to Mervyn King, who runs the Bank of England at that time, financial institutions are global while they are alive, but become national when they are dying [4].

The crisis of 2008-2009, accompanied by the bankruptcy of many banks and the exposure of dishonest behavior of bank employees and top management, led to a drop in confidence in banks. In these conditions, the importance of ethical standards and public interests in the banking business, ensuring accessibility, security, transparency and comprehensive efficiency of bank operations, is increasing. The answer to these requirements and the step to restore confidence in banks and financial markets was developed by the joint efforts of the OECD, FSB and other international organizations, the High Principles for the Protection of Consumers of Financial Services, which were approved at the G-20 summit in Los Cabos in June 2012, and then adopted by the OECD as recommendations for member countries in July 2012. These principles address issues of openness and publicity in the provision of financial services, responsible behavior of banks, complaints and conflict resolution.

Currently, international practice officially provides for the participation of representatives of financial community, as well as trade unions, consumers of financial services in the discussion of international standards being developed. So, within the framework of the G-20, there are: B-20 (Business Twenty, representing the interests of business, including financial, of the participating countries), L-20 – (Labor Twenty) and C-20 (Civil Twenty), representing the interests of trade unions and civil society. As for international financial organizations and other international structures, there are special advisory committees and commissions whose activities are aimed at developing recommendations, including in the financial sector by business and consumers of financial services. Active participants in discussions on financial security are representatives of associations of financial institutions and their self-regulatory organizations (for example, the Global Banking Association (GBA), the International Swap and Derivatives Association (ISDA), the International Investment Funds Association (IIFA), etc.).

To build confidence in banks and financial markets, it is necessary to strengthen regulatory measures and especially supervision. The development of international cooperation and harmonization of national regulatory practices are significant for global financial institutions as a prerequisite for their sound functioning and cross-border business.

3. Research Questions

Consideration of methodological tools for analyzing current trends in global financial regulation of financial and credit institutions and assessing the main purpose – strengthen the stability of financial markets to ensure sustainable economic growth – allows you to focus on the need for further scientific and practical development of identified issues. The main areas of research are:

- Analysis of the main problems and contradictions of the modern international banking regulation system;
- Assessment of the consequences and prospects for the implementation of recommendations of Basel Accords (Basel III) in the short and long term.

4. Purpose of the Study

The purpose of the study is to develop the main directions for improving Basel Accords for the regulation and supervision of credit organizations. The issues of improving banking regulation are debatable, since excessive strengthening of supervision can undermine competition and vice versa: weakening of the regulation can negatively affect the stability of financial and credit institutions. In these circumstances, it is necessary to monitor prudential reforms in order to minimize negative consequences for the economy and at the same time maintain healthy competition in the financial markets.

5. Research Methods

The study suggests practically oriented approach to the development of Basel Accords to ensure financial stability at the macro- and microeconomic level. In the course of research, relevant scientific tools were used, in particular, such methods as scientific abstraction, systematization and grouping, data extrapolation, comparative and expert analysis.

In accordance with the set goal, the author proposes a practically oriented approach to the development of the Basel agreements to ensure the financial stability of banks at the macro – and microeconomic level. In the course of the research, the author used relevant scientific tools for the analysis of banks activity, in particular, such methods as scientific abstraction, systematization, analysis and synthesis, classification and grouping, extrapolation of data. Retrospective, trend and factor analysis were also used in the research process. Comparative and expert analysis of the Basel agreements (I, II and III) allowed assessing the adequacy of measures applied by financial regulators to ensure the stability of banking systems and individual financial and credit institutions in the national and global dimensions.

6. Findings

The need to tighten existing regulatory requirements became apparent during the implementation of Basel II against the backdrop of the global financial and economic crisis that began in 2007 as a mortgage crisis in the United States and then spread to other countries. It gave rise to many new risks, including the risks of "infection" and insolvency, which revealed the failure of existing risk management systems and the lack of regulatory and supervisory coordination between financial, currency and macroeconomic regulators. Despite the fact that recommendations of Basel III were aimed at solving the above problems, a number of issues remain relevant for finding optimal solutions. In this regard, it is advisable to highlight the following main areas for improving Basel III [5].

1) Increasing the normative indicator of financial leverage. In countries that have introduced increased requirements for the debt ratio (EU, USA and some Asia-Pacific countries), as a rule, it is 5% instead of 3% proposed by the Basel Committee on Banking Supervision. Proponents of toughening this standard argue that uncertainty in financial markets is more appropriate to measure with simple and commensurate indicators than using complex probabilistic statistical models [6].

2) Simplification and transparency of advanced IRB. To use the advanced IRB system, banks need external, statistically sound information about historical default levels for different groups of borrowers. But at present, banks do not have sufficient empirical data on this issue, which has led to the rejection at the present stage of the widespread use of such risk assessment methods.

3) Increasing the role of liquidity ratios in assessing the results of stress tests. The devastating consequences of the global financial and economic crisis of 2007-2009 and, in particular, the collapse of the investment bank Lehman Brothers demonstrated that risk, which begins as market or credit, develops in a crisis of inadequate liquidity and becomes a problem for financial markets. As a result, issues of ensuring the liquidity of banks are currently receiving increased attention in national and international financial markets.

4) Further tightening regulation of global systemically important banks, including an increase in the premium for systematization. In this regard, the experience of developing an orderly bankruptcy procedure (living wills) for systemically important financial corporations in the USA, implemented by the Federal Deposit Insurance Corporation if necessary and not involving additional expenses of taxpayers is interesting. The law requires bank holdings with assets

in excess of $ 50 billion, as well as banks and non-banking financial companies whose bankruptcy can seriously damage the US economy, to provide regular reports on how quickly and correctly conduct liquidation procedures. It is believed that this requirement should create incentives for companies to simplify their organizational structures and solve the problem of the emergence of too large financial institutions, the bankruptcy of which could negatively affect the state of the economy as a whole. In the Russian Federation, this approach was implemented by the Bank of Russia in the process of creating the Banking Sector Consolidation Fund and adopting rules for the development and implementation of financial recovery plans for commercial banks.

5) Decrease in the level of procyclicality and provision of countercyclical reserves in the form of capital buffers. In particular, the Bank of Russia uses a credit gap - the difference between the ratio of the volume of credit supply to GDP and the long-term trend of this ratio, determined in accordance with the methodology proposed by the Basel Committee on Banking Supervision. In addition, the Bank uses a long-term equilibrium model of lending, a model for assessing the length of the credit cycle, a model of supply and demand for credit.

In general, the activities of the Basel Committee on Banking Supervision are aimed at the development and continuous improvement of supervisory tools and standards of macroprudential regulation, including the expansion of cross-country information exchange and mutual supervision. Thus, Basel III aims to strengthen international standards for managing capital and liquidity of banks to increase the stability of the banking sector, as well as to improve the ability of banks to overcome the effects of financial and economic crises and thus reduce the risk of these problems spreading from the financial to the real sector of the economy.

Basel III is innovative in nature, as it provides for changes in approaches to financial management and risk management by improving risk management standards, the degree of integration of financial and risk management of banks. It aims to provide a stable base and the necessary reserves to counter potential risks and crises, as well as to form a more complete, consolidated business model for shareholders and management. These changes can be divided into the following areas:

Improving the quality and amount of capital - the document substantially toughens the requirements for the structure and quality of the bank's capital base to increase the ability of banks to absorb losses in both normal and liquidation scenarios, proposes new minimum requirements to equity and first-level capital, new requirements to capital of the first and second levels, a gradual departure from various hybrid instruments, a list of possible deductions from the capital base. New standards for liquidity management, increased coverage of short-term liquidity - Basel III defines liquidity requirements, liquidity standards, minimum liquidity requirements, and a system of monitoring tools [7]. Increasing stable long-term financing – Basel III introduces the Net Stable Funding Ratio (NSFR) to stimulate and motivate banks to attract stable sources of financing for their own operations, as well as to reduce their dependence on short-term funding.

Full-scale risk coverage and risk coverage. Basel III provides for full coverage of risks, increasing capital requirements for covering risks that are not adequately reflected in Basel II, and also provides for an increase in requirements for bank trading portfolios, securitization operations and transactions with derivative securities [8]. To regulate systemically important banks, additional requirements are introduced to the level of adequacy of their capital.

7. Conclusion

Basel III acts as the result of the interaction of various forces and interests. The analysis shows that international financial institutions have been able to achieve some progress in developing international banking regulation standards. The growth of global banks' capitalization has been mainly due to retained earnings. Large international banks increased the average capital adequacy ratio of CET1 in accordance with Basel Accords III from 7 to 11%, which exceeds the normative level of 2019. This process is accompanied by a reduction in risk-weighted banking assets, which indicates an increase in the stability of banks. At the same time, the share of risky assets in the currency structure of the bank's balance is decreasing, which confirms the tendency to transfer part of operations to the non-banking sector. In order to ensure guaranteed profits for capitalization growth, many banks have preferred the traditional banking business model in recent years. Disputes about the possible consequences of new regulatory measures for the development of the global economy do not subside. The Basel Committee predicts the absence of a significant long-term negative effect from tightening banking regulation standards, since positive results will be more significant than costs. The introduction of Basel III standards, emphasized in the BIS Annual Report, will benefit banks and the financial system as a whole, and, therefore, serve as the basis for economic growth.

The analysis of the main problems and contradictions of the modern international system of financial regulation allowed us to draw the following conclusions. First, in order to strengthen the stability of the financial sector, it is necessary to cooperate and be able to find compromises taking into account the interests of all participants: global and national regulators, banks and their customers, which is of particular importance for ensuring sustainable growth of the global economy. Secondly, despite the differences in quantitative assessments of the effects of Basel III, the banking sector is currently facing serious challenges in the short and long term. In the short term, banks should clear their balance sheets from bad assets, and it requires decommissioning and subsequent recapitalization. This measure will allow banks to restore customer confidence and strengthen their market position. And in the long run, banks must restructure their business models by conducting more cost-sensitive accounting, preferring a strategy of lower, but more stable profits, in order to maintain their financial stability while working in the new regulatory environment.

Basel III represents a fundamental strengthening of global banking standards and is the foundation of long-term financial stability and economic growth. However, tight regulation threatens economic growth. According to the estimates of the Basel Committee, the average global GDP growth rate will be lower by 0.04% over 4 - 5 years due to the introduction of new rules. In addition, there is a threat of financial sector contraction, as not all banks will be able to withstand new capital adequacy standards. The adoption of Basel III, assuming the attraction of additional funds by banks, decreases their profitability and increases the cost of borrowing. In this regard, an increase in interest rates on loans is possible, which may have a negative impact on economic recovery. As a result, it is possible to intensify the process of bank mergers and acquisitions, which will inevitably weaken competition in the banking sector. On the other hand, the standards of Basel III in the calculation of liquidity ratios imply an increase not only in capital, but also in assets. The introduction of new capital requirements in the United States and Western Europe may provoke a capital shortage and a decrease in bank leverage. Under these conditions, banks will be forced to slow down lending, and existing assets should be taken into account on the balance sheet (especially within financial holdings). An increase in the requirements for equity will lead to a decrease in its profitability, which means a simultaneous decrease in the attractiveness of investments in bank capital and an increase in the need for capital. Thus, the introduction of these measures is pro-cyclical in nature and may lead to a deepening recession in the economy. Moreover, in the new international standards of banking capital, the supervisory component is not sufficiently developed. And to ensure the financial stability of banks, it is not enough to tighten the requirements for equity. Therefore, in addition to raising capital standards, it is necessary to take measures aimed at improving the quality of assets, creating a liquidity reserve and improving the management of banks. However, the economic benefits of Basel III prevail. The main goal of Basel standards is to increase the stability of banks and protect the rights of consumers of financial services. It means that international standards are aimed at ensuring the stability of financial markets, reducing risks and reducing the likelihood of future crises, which will allow banks to independently (without government support) withstand financial shocks of such magnitude as crisis of 2008-2009. These advantages are not only a factor in strengthening the banking system in the long term, but also increase confidence in the financial stability of banks.

References

1. L.S. Khudyakova, The international experience in the struggle with manipulated transactions in financial markets. *Money and Credit,* **7**, 65-70 (2011). [in Rus.].
2. B. Eichengreen, International financial regulation after the crisis. *Daedalus Fall,* **139**(4), 107-114 (2010).
3. C.H. Dallara, Containing extraterritoriality to promote financial stability. *Financial Stability Review*, **17**, 47-59 (2013).
4. H. Davies, Global financial regulation after the credit crisis. *Global Policy*, **1**(2), 185-190 (2010).
5. Basel Committee on Banking Supervision, *Basel III: A global regulatory framework for more resilient banks and banking systems* (Bank for International Settlements Communications, Basel, 2011).
6. KPMG, *Basel 4 revisited: The fog begins to clear* (2015). URL: https://assets.kpmg/content/dam/kpmg/pdf/2015/10/basel-4-report-concepts.pdf. Accessed: 21.10.2019.
7. Basel Committee on Banking Supervision, *Proposal to ensure the loss absorbency of regulatory capital at the point of non-viability* (Bank for International Settlements Communications, Basel, 2010).
8. Basel Committee on Banking Supervision, *Proposal to ensure the loss absorbency of regulatory capital at the point of non-viability - consultative document* (Bank for International Settlements Communications, Basel, 2010).

Euro-Asian Cooperation in the Field of Training on Economic Security

V.A. Piskunov[1], *E.P.* Pecherskaya[1], *T.M.* Tarasova[1*], and *S.A.* Kozhevnikova[1]
*Corresponding author: tarasova2004@inbox.ru.

[1] Samara State University of Economics, Samara, Russia

Abstract. The modern models of training and retraining of personnel, which are created for the all-Russian system have certain lacks, regarding the creation of models and the determination of connection between them. The article reviews the system of personnel training to combat the money laundering and financing the terrorism (AML/CFT) in the Russian Federation. The article shows the necessity of the international integration for the training of highly qualified personnel and the advantages of creation of the international consortiums uniting universities, scientific centers and non-profit organizations. This will allow to develop the practice, science and education in the field of economic security.

Keywords: economical security, AML/CFT, anti-laundry campaign, consortiums, specialist, training of personnel.

1 Introduction

The necessity of economic security of the country, as a part of the national security, significantly increases in terms of globalization of the international economic system and the intensification of the economic relations between the countries. The sum of annually laundered money, which is received in the criminal way, is about $1.5 trillion, which makes about 1.9% of world GDP.

Money laundering and financing of terrorism are the threats for the international financial system and the economy of any state. These phenomena are inseparably linked with other crimes: corruption, drug trafficking, human trafficking, fraud, thefts. The list of crimes can be continued, because the criminal legalized income is a consequence of any illegal activity. In terms of real terrorist threats and the new schemes of money laundering arises the question of the competence of the staff of organizations working with money and other property which can be used for both laundering of income and financing of terrorism, their knowledge and readiness for identification and prevention of illegal financial flows which are used for terrorist activity.

Experts in economic security are very demanded today in the labour market. However the level of their preparation in this field doesn't meet sometimes the international standards. The constantly changing economic situation and peculiarities of inter-country interaction demand profound knowledge of specialized disciplines and analytical skills.

2 Problem Statement

In modern economic literature there are many works devoted to the analysis of topical issues of improving the system of economic security of the state, organization and personality - this is primarily the work of T. Bobrovskaya et al. [1], S. Mojsoska & N. Dujovski [2], J. Tymchyshyn-Chemerys [3], J. Heinz [4], C. Chen et al. [5], V. Klimuk et al. [6], E. Pecherskaya et al. [7]. However, today there are still a lot of unresolved issues.

Informatization of society has led to such a concept as "information wars", when the informational space becomes the arena for spheres of influence. Information has become one of the most valuable resources and especially subjected to threats and risks. Computer swindlers commit even more economic crimes, that is facilitated by the development of electronic commerce and payment systems in the Internet. Cracking of databases of the state and business firms by means of computer technologies brings multimillion losses. Today we can see the urgent need of specialists in the field of information protection, counteraction to money laundering and financing of terrorism.

Now over 6 million employees are involved in the world anti-money laundering system. In Russia, in spite of the fact that this sphere is new, already 120 thousand people are involved in AML/CFT. However there are still a lot of unresolved questions and problems which demand new knowledge and abilities from the experts.

The models of training and retraining of personnel in the field of anti-money laundry have certain lacks with regard the creation of these models and their interaction. These lacks are:

1. Definition of conceptual bases of the creation of models.
2. Choice of structure of models.
3. Differences in training of specialists with the higher education and the training of specialists within the system of additional professional education.
4. Definition of the order of the development of models.
5. Identification of interaction between the developing models and the educational standards.

3 Research Questions

Let's analyses the specified lacks and the means of their elimination in more detail.

Definition of the conceptual bases of the creation of models. Based on the experience of development and the realization of the state educational standards of various generations it is possible to formulate basic requirements to the created models:

- The models have to reflect the level structure of the national system of the training of specialists.
- The models have to describe various trajectories of training of professionals in the field of AML/CFT.
- The structure and content of models have to provide continuity of educational standards with models.
-The content of models has to reflect a competent approach to the qualification of experts.

Choice of the model structure.

The continuity of educational standards and models will be reached, if the following features are defined in the structure of each model (bachelor, master, expert):

- Tasks of professional activity (by types of professional activity).
- Objects of professional activity.
- Types of professional activity.
- Professional competences (by types of professional activity).

Differences in models of training of specialists with the higher education and the specialists within the system of additional professional education. The main difference is the lack of regulatory requirements for the necessity of the development of the state educational standards of additional professional education. Thus, the continuity between the educational standard and the model is not required.

Therefore the structure and the content of the model of professional development and retraining of experts of AML/CFT can be different in comparison with the structure and the content of models of training of bachelor, master or expert. However, the necessity of the ensuring the continuity of education in the field of AML/CFT requires the development of the educational standard of additional professional education of the branch (not state) level.

4 Purpose of the Study

Proceeding from the existing problems, it is possible to formulate the main scope, without the achievement of which, it will be impossible the development of the effective model of professional development and retraining of specialists of AML/CFT. It is the development of the industry educational standard of additional professional education. It should be noted, that we have such an experience: the educational standard of additional professional education of civil servants is already developed and used.

5 Research Methods

If during the development of the industry educational standard of additional professional education the continuity is provided with Federal state educational standards of the third and the subsequent generations, then the structure of models of training of specialists with the higher education and structure of model of professional development and retraining of experts of AML/CFT will be identical.

The uniform conceptual bases of models of preparation and retraining of personnel for AML/CFT, which are formulated above, allow us to create the generalized algorithm of performance of the main development stages of such models:

- The trajectory and educational level of training of experts (bachelor, master, expert) or level of training and retraining of personnel is chosen.
- The field of professional activity is defined for the chosen level.
- Objects of professional activity are defined for the chosen level.
- Professional competences (by types of professional activity).
- Types of professional activity are specified.
- The tasks of professional activity are formulated on the basis of generalization of the specified tasks described above.
- One or several professional competences are formulated for each task of professional activity.

The presented algorithm has universality and does not depend on educational level of training of specialists.

6 Findings

The work on the development of the system of training for the organizations which are carrying out operations with money or other property, that, according to the current legislation, are obliged to undertake measures for AML/CFT, is carried out since 2009. Personnel training of financial institutions and organizations are an important task. Not less significant is the training of public authorities in the sphere of AML/CFT: law-enforcement, control, supervising and others, which fulfil measures for AML/CFT. For this reason the whole blocks, sections and separate subjects on AML/CFT are included in the programs of professional development of government employees. Such integrated professional training promotes the elaboration of general, integrated approaches in combatting the money laundry and financial terrorism. It also strengthens interdepartmental interaction and collaboration of structures and departments, being a basis of system approach in work of the state and its bodies.

One of the significant subjects in the system of professional retraining and professional development of personnel for the national AML/CFT system is the autonomous non-profit organization International Educational and Methodical Center of Financial Monitoring (IEMCFM).

The research works of IEMCFM are performed for ensuring the participation of the Russian Federation in activity of The Group of development of financial measures of combatting the money laundering (FATF), as well as the regional groups created on its type (RGTF), also the group of divisions of financial investigation organization "Egmont" and other international organizations dealing with AML/CFT.

IEMCFM was also an initiator of creation of the International Network Institute (INI) in the sphere of AML/CFT. The main advantages of its creation are presented in Brochure "Network Institute in the field of combating the legalization of criminal proceeds and terrorism financing" [8].

The uniform system of use of intellectual, material, information resources and innovative technologies of the International Network Institute allows the students to get the most relevant, modern knowledge in the sphere of AML/CFT and to participate actively in various international actions and research projects which are carried out by participants and organizations of the Network Institute irrespective of the location of the universities.

Among the participating countries of the International Networking Institute (INI) in the sphere of AML/CFT there are Abkhazia, Azerbaijan, Argentina, Armenia, Belarus, Bolivia, India, Kazakhstan, China, Kyrgyzstan, Mexico, Moldova, Mongolia, Peru, Transnistria, Russia, Serbia, Tajikistan, Uzbekistan, the Republic of South Africa, South Ossetia. INI represents the modern scientific and educational consortium uniting the universities, scientific and education centers of the states of the Euro-Asian region.

Among the foreign institutions of higher education which constitute the International Networking Institute, it is possible to distinguish the Academy of public administration at the President of the Kyrgyz republic, the Kyrgyz-Russian Slavic university named after the first President of the Russian Federation B.N. Yeltsin, the Kyrgyz national university of Zhusup Balasagyn, the Educational and methodical center of Public service of financial investigation at the Government of the Kyrgyz republic, the Belarusian state university, the Belarusian state economic university, JSC Financial Academy of the Ministry of Finance of the Republic of Kazakhstan, Research center in the sphere of AML/CFT of the Fudan University (China), Association of business schools of the countries of BRICS, the Karaganda state university named after the academician E.A. Buketov, the Russian-Tajik (Slavic) university, Financial and economic institute of Tajikistan, Tax academy of the state tax committee of the Republic of Uzbekistan, the Tashkent state economic university, the Tashkent state legal university, the Tashkent university of information technologies, the Tashkent financial institute.

The new format of successful interuniversity and scientific interaction will provide the development of practice, science and education in the sphere of AML/CFT and it will also provide Russia and the partner states with personnel and scientific potential.

7 Conclusion

Modern reality requires the effective concept of economic security as the basis for the effective management. Within this approach it is possible to attract investors and to provide sustainable development and the high-quality growth.

The condition of economic security of the state is essential for its functioning. Active and stable economic growth, responsible use of national finances, national resources and, first of all, protection of economic interests within the international cooperation are necessary for the state normal functioning. The creation of international consortiums will allow Russia and the participating countries to provide highly skilled personnel and scientific potential.

References

1. T. Bobrovskaya, I. Sannikova, A. Litvinenko, The problems of specialists training in the field of economic security. *Society. Integration. Education. Proceedings of the International Scientific Conference*, **1**, 67-77 (2019). DOI: 10.17770/sie2019vol1.3842.
2. S. Mojsoska, N. Dujovski, Economic security and economic security index as a measure of economic security. In: G. Meško, B. Lobnikar, K. Prislan, R. Hacin (Eds.), *Criminal Justice and Security in Central and Eastern Europe: From Common Sense to Evidence-based Policy-making* (pp. 274-282). Maribor: University of Maribor Press (2018).
3. J. Tymchyshyn-Chemerys, Features of economic security of the region as a composition of national economic security. *International Scientific Journal "Internauka"*, **4** (2019). URL: https://www.inter-nauka.com/issues/2019/4/4826. Accessed: 04.09.2019.

4. J. Heinz, *United States strategic trade: An export control system* (Westview Press, Colorado, United States, 1991).
5. C. Chen, E. Kempson, D. Park, J. Yoong, Gender and old-age economic security. In H.P. Kai, G.G. Lee, T.Y. Mui (Eds.), *Ageing in Asia: Contemporary trends and policy issues* (pp. 95-123). Singapore: World Scientific (2019).
6. V. Klimuk, V. Piskunov, E. Pecherskaya, T. Tarasova, Methodical approaches to assessing the economic security. In: V. Mantulenko (Ed.). *Proceedings of GCPMED 2018 - International Scientific Conference "Global Challenges and Prospects of the Modern Economic Development"* (pp. 283-290). *The European Proceedings of Social & Behavioural Sciences EpSBS*, **LVII** (2019). DOI: 10.15405/epsbs.2019.03.29.
7. E. Pecherskaya, V. Klimuk, T. Tarasova, Analysis of European approaches to improving the life of the population through the implementation of a mechanism of independent assessment of qualifications. In Ashmarina S., Vochozka M. (Eds.), *Sustainable Growth and Development of Economic Systems. Contributions to Economics* (pp. 249-257). Cham: Springer (2019).
8. International network institute for combating the legalization of crime proceeds and terrorism financing, *Brochure "Network Institute in the field of combating the legalization of criminal proceeds and terrorism financing"* (2018). URL: https://www.sevsu.ru/univers/pod-ft. Accessed: 04.09.2019. [in Rus.].

Stock Market Crisis Identification as a Factor Ensuring Country's Economic Security

M.E. Konovalova[1*], *O.Yu.* Kuzmina[1], *V.Ya.* Vishnever[1], and *S.Yu.* Salomatina[1]
*Corresponding author: mkonoval@mail.ru.

[1] Samara State University of Economics, Samara, Russia

Abstract. The article reviews the issues of the economic security of the country from the standpoint of identification of financial crises. The essence and content of a financial crisis are given, characteristic traits of a stock market bubble as an integral part of a stock market crisis are singled out. Special emphasis is laid on the study of the existing methods of identification and measurement of bubbles. The authors analyze the modern stock market crises, describe the Russian stock market vulnerability factors. The article evaluates the possibility for the origination of a stock market crisis in Russia on the modern stage resulting in the creation of a multi-factor stock market bubble identification model. The authors justify proposals and recommendations for the elimination of the existing Russian stock market problems aimed at removal of the economic security threats of the country.

Keywords: financial crisis, stock market, stock market bubble, financial crisis ratios, economic security.

1 Introduction

A stock market is meant to perform one of the most important functions, which lies in the attraction of investments in the real sector of the economy. However, when the market functioning mechanisms get broken and the market is no longer assessing the value of financial instruments adequately, the probability of the origination of such phenomenon as a stock market crisis is high. Stock market crises take place fairly often, bear a systemic character and may distribute globally in case of turmoil at one of the markets.

2 Problem Statement

The article is dedicated to research of the issue of the identification of financial crises for minimization of their negative influence on the economic security of the country. It is the stock market sector, that many financial crises start from, thus, an analysis of the regularities of stock market crises seems the most interesting [1]. To overcome a crisis, one needs to be ready for it; timely identification of a stock market crisis will not only enable the investors to make correct decisions, but will help state authorities mitigate its negative effects or prevent it altogether.

3 Research Questions

The following tasks need to be fulfilled to solve the problem set in the course of the research. Firstly, to single out characteristic traits of a stock market bubble and review the existing methods of its identification and measurement on the securities market. Secondly, to carry out a lookback analysis of the reasons and factors affecting the origination of crises on the Russian stock market. Thirdly, to determine the Russian securities market crisis vulnerability factors and propose instruments of raising its financial stability.

4 Purpose of the Study

The main purpose of the study is the development of the theoretical and methodological provisions of the financial crisis establishment and development process. Neither foreign nor Russian scientific publications have the unified and generally accepted approach towards forecasting of securities market crises and bubbles forming the request of the society for fundamental and exploratory research of the issues related to early identification of financial bubbles. One also needs to justify the main areas of the improvement of financial relationships causing an increase in the stock market functioning efficiency.

5 Research Methods

The following general scientific methods of cognition of phenomena and processes serve as a methodological basis for the research: methods of logical, dialectic and historical analysis, the systemic approach towards the analysis of crisis phenomena on financial markets and the development of methods of their early identification. Methods of statistical, econometric analysis, economic and mathematical modeling are also widely used. These methods jointly with theoretical provisions and economic information allow securing the reliability and justification of the results received and conclusions made in the course of the research.

6 Findings

6.1 Theoretical Bases of the Study of the Content of Financial Crises

Having reviewed various theoretical perceptions of the nature of financial crises and their quantitative characteristics, we have come to the conclusion that no unified approach towards the understanding of the problem has been formed yet. Indeed, quantitative characteristics of bubbles are quite subjective and significantly vary depending on the market character, national identity, scale and many other factors [2]. Not too much studies of this sector have been carried out in Russia so that some bubble parameters could be applied with confidence, and the use of foreign methods does not seem correct in view of serious differences between the markets. Timely identification of a stock market crisis will not only enable the investors to make correct decisions, but will help state authorities mitigate its negative effects or prevent it altogether [3]. The development of the methodology required for crisis diagnosing will solve this problem.

6.2 The Stock Market Bubble Concept and Identification

Judging by the history of stock market crises, it becomes obvious that they are most often triggered by a stock market bubble. The term 'bubble' is used to describe stable excess of the asset price over its fundamental value. That means that in order to identify it, one needs to know the fundamental asset value, which is not simple and obvious to calculate; the period and amount of excess of the asset price over its fundamental value. Multiple attempts of researchers and analysts at specification of these criteria have brought no real results; all received results are rather subjective and cannot be fixed once and for all. In view of this, a generally accepted quantitative definition of a stock market bubble still does not exist [4].

There are many ways to identify bubbles. The main of them are: the macroeconomic ratio method, technical analysis, economic, physical, econometric methods, instruments of the behavioral finance theory [5], experimental modelling of stock market bubble forming. It is insufficient to use just one approach due to the risk of unavoidable omissions intrinsic to one method taken separately. So it is more correct to apply a complex analysis of bubble identification enabling to characterize various stock market aspects: the behavioral factor, macroeconomic environment, systemic mistakes and influence of various factors [6].

6.3 Analysis and Evaluation of the Probability of Financial Bubble Origination on the Russian Stock Market

The financial bubble identification process has been analyzed in three steps: testing of the Russian stock market efficiency, testing of financial asset prices consisting of a macroeconomic, factor and fractal analysis of the prices of financial instruments circulating on the Russian stock market. The received results have been interpreted in the conclusion of the research from the standpoint of the evaluation of the probability of stock market bubble origination.

Since bubble inflation is largely facilitated by the behavior of the participants adopting various decisions based on the information existing on the market, one needs to check the information efficiency of the market with the help of the hypothesis phrased by E. Fama [7]. The Durbin-Watson test detecting the presence or absence of autocorrelation in the asset price dynamics has been used to determine the efficiency degree. The Russian market exhibits autocorrelation between the levels of the range starting at lag 7 (except for separate periods when the dependence has been identified at lower lags) (table 1). Thus, a conclusion can be made that the participants adopting investment decisions have been relying on the information on previous index dynamics, i.e. the players' behavior is characterized with mimical and imitative elements.

Table 1. Values of the Durbin-Watson statistics from 2003 to quarter I 2019.

Period	Lag				
	6	7	8	9	10
2003	**1.046**	**1.065**	**1.098**	**1.115**	**1.1**
2004	**1.151**	**1.109**	**1.178**	**1.156**	**1.042**
2005	1.987	**1.071**	**1.046**	**1.011**	**1.034**
2006	1.989	**1.081**	**1.101**	**1.046**	**1.087**

2007	2	**1.069**	**1.02**	**1.06**	**1.026**
2008	**1.584**	**1.527**	**1.584**	**1.532**	**1.536**
2009	**1.078**	**1.109**	**1.089**	**1.094**	**1.108**
2010	2.009	**1.04**	**1.045**	**1.032**	**1.029**
2011	1.993	**1.079**	**1.036**	**1.118**	**1.088**
2012	1.786	2.045	1.927	1.828	2.059
2013	1.855	1.932	1.963	1.93	1.959
2014	1.729	1.745	**1.604**	2.098	**2.437**
2015	1.828	1.763	**1.597**	**2.411**	**2.444**
2016	1.827	1.935	**1.297**	**2.287**	1.721
2017	2.102	1.739	1.979	2.073	2.067
2018	**1.043**	1.744	2.048	1.974	2.04
Quarter I 2019	**2.286**	**2.291**	**2.311**	2.259	2.102

Source: Authors.

On the second step, we have tested the asset price by three various means. The stock market role in the national economy has been identified using the macroeconomic analysis. The stock market capitalization-to-GDP – MCap/GDP ratio (the Buffett Indicator) has been used as the indicator. The analysis has been carried out in dynamics with identification of the average threshold value for the researched period (62%). The current ratio exhibits the value of 42-43% speaking of the absence of any stock market bubbles and balanced correlation between the stock market and the economy of the country [8].

External and internal factors producing the largest impact on the ratio dynamics have been selected on the factor analysis step, namely: the M2X aggregate, MIACR, Brent oil and UST-10 yield. The correlation and regression analysis has exhibited the absence of co-integration in the MIACR and UST-10 yield ranges with the MOEX ratios showing that the influence of these factors on the Russian stock market dynamics is statistically insignificant. As a result, a two-factor model has appeared. We have further determined the bubble's part in the current index value by calculating the amount of excess of the market ratio value over the generated model values. The indicator value is reaching the top threshold of 30%. This suggests a possibility for the origination of a new financial bubble especially in the conditions when the ratio reaches new historical maximum values for the recent time.

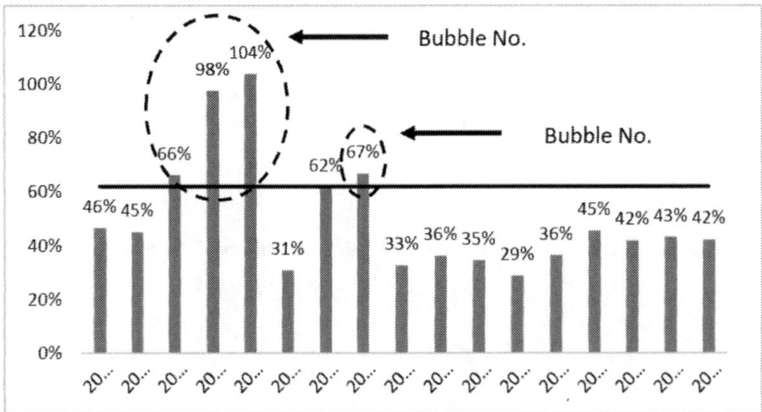

Fig. 1. MCap/GDP dynamics (Source: Authors).

The fractal analysis is used to determine the dependence between the future asset prices and their previous changes. The Hurst exponent serves as the quantitative expression of fractal processes as it measures 'roughness' of a time series and makes it possible to single out ranges of excessive cumulative dependence of observed values [9]. The Russian stock market really has a long memory. As soon as the exponent reaches the threshold value (0.75), the market 'remembers' the previous dynamics and aims at repeating it. The Hurst exponent reached 0.71 in March 2019, which is fairly close to the border, but there is no question of the possibility of a catastrophe yet.

The received results attest to the fact that the stock market is now on the growing stage, without any obvious bubble origination symptoms as only one of the four indicators has exhibited a possibility for the origination of a bubble on the Russian market [10]. The other ratios speak of the absence of such possibility. However, some of them are nearing the border. For example, the ratios received following the factor analysis is just 9% below the threshold. Taking into account the information cascading effect and the market dynamics being quite good lately, these 9% can quickly turn to nothing. The fractal analysis is also demonstrating the proximity of the calculated and threshold values meaning that if the level of 0.75 is reached in the nearest future, there will most probably appear a market bubble.

The results of the macroeconomic environment analysis point to the economically justified correlation between the stock market and the country GDP in March 2019 from the standpoint of bubble identification. It means that the origination of the latter in such conditions is unlikely.

7 Conclusion

Thus, the following conclusions have been made as a result of a complex stock market bubble analysis. Firstly, the built model is adequately reflecting the earlier originated stock market bubbles proving that it is fit for use. Secondly, the model characterizes various stock market sides: identification of any deviations in any sphere will immediately point at the reasons for anomalies on the stock market. For example, the main problem of the share market today is of behavioral character, i.e. the participants are prone to collective actions.

References

1. D.K. Tarullo, Financial regulation: Still unsettled a decade after the crisis. *Journal of Economic Perspectives*, **33**(1), 61-80 (2019).
2. A.J. Makin, Lessons for macroeconomic policy from the global financial crisis. *Economic Analysis and Policy*, **64**, 13-25 (2019).
3. A. Volkova, A review of government programs. *Procedia Economics and Finance*, **27**, 484-493 (2015). DOI: 10.1016/S2212-5671(15)01025-4.
4. G.Zhang, W. Liu, Analysis of the international propagation of contagion between oil and stock markets. *Energy*, **165**, 469-486 (2018). DOI: 10.1016/j.energy.2018.09.024.
5. M.L. Dorofeev, G.V. Samarskiy, Modelling of financial bubble processes on the Russian stock market. *Finance and Credit*, **15**, 45-62 (2016). [in Rus.].
6. A.V. Zinenko, R/S Analysis on a stock market. *Business Informatics*, **3**(21), 24-30 (2012). [in Rus.].
7. E.F. Fama, Efficient capital markets: A review of theory and empirical work. *The Journal of Finance*, **25**(2), 383-417 (1970).
8. Ch. P. Kindleberger, R. Aliber, *Manias, panics and crashes: A history of financial crises* (Peter, Saint Petersburg, 2012). [in Rus.].
9. E .Majewska, J. Olbryś, The evolution of financial integration on selected European stock markets: A dynamic principal component approach. *Comparative Economic Research*, **20**(4), 45-63 (2017). DOI: 10.1515/cer-2017-0027.
10. D. Duffie, Prone to Fail: The pre-crisis financial system. *Journal of Economic Perspectives*, **33**(1), 81-106 (2019). DOI: 10.1257/jep.33.1.81.
11. A. Kireyev, A. Leonidov, Network effects of international shocks and spillovers. *Networks and Spatial Economics*, **18**(4), 805-836 (2018).

New Functionality, Security and Protection of CCTV Systems: Technological Progress and Digital Society Development

R. Vaníčková[1,]*

*Corresponding author: vanickovaradka@gmail.com.

[1] Institute of Technology and Business in České Budějovice, České Budějovice, Czech Republic

Abstract. The objective of the paper is to analyse and evaluate the current state of the project of integrated security CCTV systems including the description of the original plan, the course of multiannual implementation, analysis of partial results. Next discussion of the current state and developmental technological changes in information and communication systems of the 21st century digital society including accepting the trends in developments, changes and forecasts in the field of cyber security. In the analytical part of the contribution, the author analyses and evaluates the planned/unplanned activities, the course of multiannual implementation and achievement of the strategic objectives including the characteristics of the current protection and security of the CCTV systems and the impacts of the installed security and monitoring systems in a banking institution. The final part of the contribution evaluates the project success with respect to the protection and security of the systems.

Keywords: digital society, technological progress, cyber security, project of integrated security, protection of CCTV systems.

1 Introduction

The beginning of the project is connected with the creation of the bank strategy influenced by a new owner entering the banking sector. The concept of the bank management is a more professional approach to the management of the institution, improvement of the products and services portfolio including setting up services parameterization, consolidation of the processes, creation of the control and security mechanisms, effective use of available sources and information. The management of the banking institution decided to map, change, and innovate the situation unclear in terms of processes, organization, and economics, related to the security and protection of the camera and monitoring systems in order to be in line with the current trends in IT. For the purposes of market research and monitoring the state-of-the-art, the methodology of elaborating the feasibility study within the implementation of the integrated security project into the banking sector. A parametrically selected supplier was approached, who acted as an integrator of security services under a framework contract on complex services provision [1]. The subject of the services offered was the modification of the rights and obligations of the contractor and the customer by the security specialists and external suppliers, hereinafter referred to as "complex security services", which includes activities such as physical surveillance of property and persons, monitoring of alarm and technical signals [2, 3], including intervention [9], district car patrols and mobile patrols [5], supply and installation of security systems [6], service for security systems operation [7], technical services for protecting property and persons [8]. Complex activities do not include the activities provided by the customer through the IT department, department of financial analysis of the department of extraordinary investigation. The rights and obligations of the contracting parties, i.e. the contractor and the customer in accordance with the framework contract concluded, are governed by the relevant provisions specified in the contract, including the annexes. The objective of the complex security services in the given extent and partially quoted in the relevant Act and the implementing regulations is to strive for eliminating undesired phenomena, risks, and adverse events that could affect the security and protection of property and persons [9, 10], employee performance, goodwill and the image of the banking institution across the range of the organizational structure. The protection of health and human life of the permanent employees and the clients/customers, external workers, and suppliers is subsidized in integration with the externalities in the customer´s premises and control activities by compliance with legal regulations and regime measures [11, 12]. The customer´s competence includes the protection and surveillance of the property including the prevention of undesirable influences against the customer´s interests and his/her rights related to the protection of the operation in the customer´s workplace [13]. Preventive protection and avoiding the occurrence and consequences of extraordinary events in the customer´s premises is an inseparable part of the contractual arrangements [14]. In the preparation phase of the project, a large number of partial adjustments and specific activities focused on the local and regional market research in the Czech Republic were performed, as well as mapping the technological possibilities and alternatives of negotiations with the producers and suppliers of modern technologies of securing CCTV systems with the emphasis on the feasibility study and situational analysis of protecting the security and monitoring systems security [15]. The project was started in parallel at more levels at the same time and it had an impact on the suppliers and customers departments, e.g. technical department, human resources, economic and financial departments in cooperation with the partner subjects involved [16]. The project team consisted of a project manager supervised by a CFO and directly supervised by the CEO of the banking institution.

2 Materials and Methods

The objective of the paper is to analyse and assess the state of the project of the integrated CCTV systems. For preparation of a feasibility study, the analysis of written documents from primary and secondary sources, and internal documents of the project contracting partners were used, as well as relevant technical literature and sources, literary research, and formulation of theoretical knowledge of CCTV systems. The first activity was a description and analysis of the current state, plan, and the course of the project implementation based on the analysis of the banking institution micro-environment, the analysis of technologies and security services, analysis of specific banking processes and security standards suppliers, the analysis of the main characteristics of the situation. In order to get responses to the survey conducted in April and June 2018 for economically productive men and women of complete secondary and university education in economics (a total of 25 employees of a branch of a bank in the CR, South Bohemia, České Budějovice district), the following research questions were formulated:

What were the outcomes and impacts of the integrated security project?

What was the result of installing CCTV systems in and outside the bank institution?

Can a well-installed security and monitoring device protect or prevent a hacker attack?

Summarized background material for the introduction part of the contribution on the given topic, interpretation and empirical generalization of knowledge and findings in the final part of the contribution.

3 Results

3.1 Project Activity Baseline Characteristics

The start of the project activities was preceded by monitoring and mapping the current situation in bank security systems, and study of security service offers, technological and technical options, including internal survey of the baseline in security and protection of the bank with support alternation of synergy effects. The baseline of security and protection of property and persons was not clear and well arranged with low transparency of the organisational, economic, financial and process structure of systems. The initial findings offered new knowledge of the technologies from at least two dozens of foreign producers. ATMs were connected to the Police of the Czech Republic who gradually desisted the monitoring of private properties. Bank branches were mostly autonomous as regards the management and decision making on the functionality of CCTV systems because the authorised managers and experts contacted businesses providing repair and maintenance services in the local and regional market with different market prices, parametric availability and quality. The organisational structure for cost allocation and controlling of financial and non-financial operations was not defined and not approved at the corporate level by the director of the bank institution in Prague, Czech Republic, but by an authorised team of regional officers. The decision-making autonomy of bank branches and representatives of the property management was not transparent as regards the selection of technology vendors and the associated security of monitoring systems, e.g. security doors and windows. The management of a bank branch made decisions and negotiated with a limited number and scope of rights and powers and with a lacking visible concept because standard procedures, defined price levels, and partial solutions for the selection of suitable contractors, and other necessary business parameters did not exist. Written documents did not have a standardized format, e.g. security guidelines, rules of operation, risk and security studies and other supporting tools. Industrial safety and fire protection was decentralised with no archives and data backups. Insourcing of security specialists, e.g. reducing the number of staff and consequent transfer to the contractor for the sake of integration of security services and autonomy of branches when selecting physical and extraordinary security staff or patrols, was another adverse finding of lacking integrity of security services. For the purpose of entering into and execution of a general agreement with the service integrator, an action plan was developed to map the baseline consisting in priorities of centralized surveillance by security forces of CCTV objects, technological standardisation and integration unification, providing selected buildings with technological backgrounds, building categorisation and passports, cost planning, control and allocation, reduction of the number of contractors and subcontractors, and establishing rules and standards for the protection and security of property and persons. The feared threats included internal risk, most often fluctuation and a high sickness rate of employees as well as external risk of contractors resulting in security incidents of non-updated or informal written documents with high risk of penalties or sanctions caused by ATM disconnection. No less feared risk was corruption in awarding capital projects causing deliveries without consistent control and standardisation of work processes, a plenty of technological contractors specialised in repairs, maintenance, services preferring private interest to public interest, wasting money resulting in uncontrolled efficiency, effectiveness and functionality of spent resources. Decentralised data and information on the price level intended for installation and repairs of technologies and managing possible incidents, e.g. non-standard setting of escalation procedures, was identified as the baseline threats to start activities of the planned project. The bank branch management, the department for protection of property and security of persons urged to change the current situation and implement a new functional security system supported by standardised written documents of partial processes and the implementation of uniform technology aimed at cost savings, in particular payroll, and improvement of work performance of permanent employees.

3.2 Project Implementation Process

It is useful to point out that the comprehensive, articulated and secondary implementation of project goals and objectives took a very long time, more than ten years. Large process changes and time consuming approval procedures were completed over this period of time. The creation of profiles and specialisation of security services was associated with planning of technical implementation and required significant financial subsidies and time distributed within several years. The actual implementation of investment plans and the unification of the related content and formal documents were preceded by challenging agreements between several business partners. The resulting solutions and the format of project plans and project documentation were often changed and modified to reflect innovations in legislation and validation of partial laws and regulations. The initial definition of requirements and service distribution in respect of the client needs, including controlling activities consisted in development of 30 new service reports with pre-defined accounts, cost centres, contact addresses and a set of services. Furthermore, external assistance was available for development of a new billing and information system since it was not manageable to cover 12,000 – 14,000 invoice items during one calendar month. The optimisation of the branch network managed by the business strategy of the bank consisted in reduction and reallocation of branches, upgrading the information and communication technologies with the assistance of the property management department and marketing department and co-operation with public and local authorities. The necessary involvement of designers, design bureaus and technical supervision transformed the summary requirements of the contracting authorities for user practice because sophisticated equipment of computer networks, furniture, including HVAC systems, were required from contractors. Due to the large scope of activities pursued and the number of tasks assigned, the directors and representatives of the controlling departments of the client and the contractor as well as independent consultants and advisors had to discuss and strategically document and evaluate the project outcomes. The yearly balance of services and activities provided in a written document signed by the authorised representatives of the involved parties was a necessary project activity of the authorised members of the project team.

3.3 Evaluation of Project Outcomes

The centralisation of CCTV systems and their unification can be considered as a completed task. The client has gradually changed to so-called on-line remote transmission which allows to carry out the general technical operations, e.g. functionality monitoring, process restart, system setup changes and other activities. Remote technology management has fostered more comfortable control, faster transmission, and the provision of data and events from systems. Technologies are installed to support the technological progress and development and CCTV systems. The stepwise replacement of black and white cameras for colour cameras, later IP (digital) cameras with higher resolution, shooting and zoom options was a major progress. Figure 1 gives a practical example of photos from older installations of security and monitoring systems with lower resolution, black and white images and less identifiable contrejour environment.

Fig. 1. Black and white photos from security and monitoring systems. Source: Author (2018).

Figure 2 shows an enhanced security system with colour resolution and higher sharpness and details.

Fig. 2. Colour photos from security and monitoring systems. Source: Author (2019).

The lowest prices were negotiated for the individual activities of security and monitoring systems as a result of economy measures and a price list of technologies and works was created and is updated every six months to reflect the market price development. In terms of the quantity of products, quantity discounts were negotiated with selected producers and the reduction of material prices (DVR recorder of recording cameras and equipment, IP cameras, etc.) was supported by the trend of the modern technological era in the digital society of the 21st century as well as affordable prices of the electronic security system and equipment producers. According to the approved plan, secured ATMs (automated teller machine) were switched from the reach of the Police of the Czech Republic to the agency branch with a monitoring control centre. Furthermore, the number of contractors and subcontractors was reduced to reflect the parameters of the catchment area and technological capabilities of deliveries.

The contractors could not be checked for services and activities associated with security (physical protection), assistance services and rapid deployment forces because of the permanent lack of qualified IT workers in the labour market. The decision on the scope of partial reconstruction in order to produce a detailed feasibility case study based on a budget and work schedule has fostered a more transparent investment process intended for the involved parties and co-operating departments of the bank who started to co-operate more closely and communicate more regularly, including compliance with task deadlines and reporting of the project outcomes.

The necessary data and information was defined for building passports that can be used for monitoring, classification, billing, reporting and controlling of the cost centre in the administrative building in the format of numerical series transported to the information system; translated, standard or reduced tax codes, building characteristics, e.g. categorisation of the branch, headquarters and archives and ATM (automated teller machine), translation of the address by the current point of contact, including house numbers as well as categorisation by importance and number of permanent employees.

In 2018, the new information systems CAFM and PPS were successfully implemented and installed with the needed functionalities of security, monitoring, protection, and facility management of the bank. The systems also include a module of repair, maintenance, service and inspection requirements, including records (amendments to attachments) in the approval workflow and generating monthly purchase orders for services based on billing of works.

What was the result of installing CCTV systems in and outside the bank? Is a well-installed security and monitoring system able to protect, mislead or prevent hacker attacks? Is it valid to assume that the extension of security camera and monitoring systems helped to reduce crime, which is endorsed by the public statistics of the Police of the Czech Republic demonstrating a clearly decreasing trend of robbery and burglary in the banking institutions. The evidence is based on facts of robbery and burglary in banks, 126 cases and 99 resolved events in 2008 versus 15 cases and 13 resolved events in 2018. In addition, the number of offences has evincible decreased and the identification of offenders and the occurrence of loss or damage and the security and protection of property and buildings have increased. Images from security and monitoring systems helped to identify and catch persons and criminals since the integrated security project has improved the quality and process of security and protection of property and persons, including a timeless security effect.

Fig. 3. Colour images from security and monitoring systems for identification of persons and criminals. Source: Author (2019).

4 Discussion

4.1 Project impacts and current state of security and protection in banking institution

The reasons for the termination of the project were changes in the contractual agreements related to market prices, conditions of part-deliveries, changes in the organizational structure including personnel changes, as well as access functionalities in terms of security, property and persons protection except for customers/clients and suppliers as project partners.

In the integrated security project, there was a stagnation with a gradual decline, to which the supplier (integrator) failed to react, did not maintain the volume of the supplies in the original structure and the stated volume of orders. In the event of any change, the suppler was expected to notify the banking institution and suggest modifications, changes, and innovations during the project implementation phase in order to prevent, protect, eliminate theft, fraud or potential losses or other sophisticated hacker attack. Unfortunately, this did not happen. Client security modified by the new organizational structure allocated the limited powers of the Property Management Department, the organizational structure of the branch was streamlined in favour of the newly created position of the Security Manager, at the expense of the position of the Security Department Director.

https://doi.org/10.1051/shsconf/20197103003

The branch network optimization process consisted in reallocating and reducing the number of branches in order to reduce the number of activities in the field of security and property and persons protection. During several years, the branches from the less frequently visited places were relocated to more attractive, better accessible places, e.g. in shopping centres and squares with a higher number of local inhabitants of the CR, since these multipurpose premises are guarded by security services with a higher number of employees in security and repression services. In cooperation with the Police of the Czech Republic, problem and risky localities were identified, where the branches were not located. The reason was the protection of people, including monitoring, premises protection, and prevention of harmful events. The number of branches was gradually reduced from 650 to 510, with a total number of 1350 ATMs. Within the optimization of the branch network, a new system called Cashless was installed, which reacts to the current trends in the field of cyber security to reduce the volume of cash payments and transactions with a reduced number of clients/customers visiting the branch networks for cash deposit / withdrawal.

Table 2. Advance in security camera systems.

Marking/Format	PAL/NTSC	Pixels
4K / 2160P / UHD / QFHD– aspect ratio 16:9	4096 x 2160	8 847 360 / 9MPx
5M - aspect ratio 4:3	2592 x 1944	5 038 848 / 5MPx
4M - aspect ratio 16:9	2688 x 1520	4 085 760 / 4MPx
3M - aspect ratio 4:3	2048 x 1536	3 145 728 / 3MPx
2M / 1080P / Full HD - aspect ratio 16:9	1920 x 1080	2 073 600 / 2MPx
1,3M / 960P / HD - aspect ratio 4:3	1280 x 960	1 228 800 / 1.3MPx
1M / 720P / HD - aspect ratio 16:9	1280 x 720	921 600 / 1MPx

Source: Author (2019).

FULL HD (1920 x 1080x), widely used in the 21st century digital society, records sharp images in a demanding environment of camera systems, as seen in Table 1 (Advance in security camera systems). Technological advance and development of digital cyber society increased the number of images per second from 100 to 120, with a minimum light exposure / light sensitivity of 0.02 lux. The new product series, especially of panoramic cameras, offer high-quality HD and images with a maximum resolution of 8160 x 3616, 32MP with 30 FPS image resolution for smooth video preview and playback.

5 Conclusion

As it was mentioned in the introduction of the paper, the author focused mainly on the description of the integrated security, property, and person's protection project, recapitulation of the results, and the interpretation of the project impacts. On the basis of the current state and existing possibilities of making adaptations and project changes, the author proposes innovations of electronic, information, and data systems and transfers used progressively in the banking environment without higher financial participation in order to improve and increase efficiency, operability and progressive trends. The individual system modules allow to enter more detailed and structured data, improve the data export into the software system, to create operational, accounting, control, and management statements. The project outputs are linked to an active and responsible approach and the behaviour of the banking institution management and employee behaviour.

After gradual replacement of all analogy cameras with digital ones it was possible to evaluate the benefits and shortcomings of new security camera systems by means of camera functions and additional software for security and protection, business and marketing activities, as well as conceptual management.

Analysing the initial and current situation including the results achieved and partial project activities it can be stated that in terms of the original assignment, the integrated security and property and persons protection project, the project objectives were achieved thanks to interested persons and business partners. The persons involved in the project included ordinary security workers, administrative workers, technology installers, heads of individual departments, external consultants and lawyers, who during the ten years of the bank institution existence contributed to achieve the project goals in accordance with the appeal of higher security and property and persons protection. Although future measurable objectives had not been formulated in advance (e.g. the number of cameras installed, maximum extent of damage, number of attack to buildings or ATMS), there has been a positive change in the last decade of the 21st century digital society consisting in a radical reduction of number of attacks, thefts, and robberies in banking houses, as evidenced by public police statistics [17]. The installation of more modern and high-quality technologies, as well as the overall increase in the number of security and camera systems contributed to reducing the number of attacked banking houses and to solve the attacks, thefts, and robberies. This primary benefit had a slightly negative secondary effect on the security and property and persons protection in the form of reducing the number of jobs in property and object protection, security specialists, and supplier quality specialists.

Another important finding was that with the reduction in the number of attacks, thefts, and robberies, the importance of security department decreased on the side of a client/customer. On the basis of this situation, it can be stated that the field of security and property and persons protection did not mean any added value generating prosperity and economic profit, but the social benefit of responsible behaviour and prevention within the proposed measures preventing risk and milestones at the expense of entry costs of services, technical adjustments, technological innovations and other operational and financial activities. The reduction in the security and protection personnel including the reduction of the budget for project items is the logical result of the existing problem-free situation.

The author of the paper sees that the previously set general goals, such as optimization of the person security and property protection, simplification of the processes, and the improvement of the communication with the subjects appears to be a non-specific strategic vision of the banking institution. The impact and the extent of the competence in terms of person security and property protection has not been analyzed, although the current trend of reducing costs prefers this tendency. In addition, the concept and action plan of the original and the follow-up project in the extent of reorganization changes and other measures was missing. The management of the banking institution was presented with a proposal to optimize person security and property protection, as well as their stabilization in the medium term for the period of 3 – 5 years. Surprisingly unplanned but more beneficial and evolutionary overlap of the project consisted in the passportiszation and categorization of objects and the allocation of the input costs. The plan for implementing operational portals by requirements, related annexes, and written documentation was achieved. Web interface, software applications, and data transfers with the necessary functionalities replace the previous inconsistent forms of ordering services and support activities. The added value of the information systems and operational portals were permanent storage of the input data that have the function of setting the parameters, reporting and statements, export transfer activities and streamlining of services provided and partial supporting activities. The passportisation was complemented by a new costs statement and automatic statistical orders that were executed in accordance with the original plan and in comparison with sophisticated IT platforms. The use of system tools was timeless and beneficial in terms of security and persons and property protection with high added value and explanatory power. The synergy effect of newly installed information and communication technologies was the reduction of services, modification, service and revision of technologies, more user-friendly control of remote management and IP technologies, including individual technical tasks and operational activities.

The volume of related supplies, services provided, and revenues from sales decreased, because the planned extension of the branch network was stopped and conceptually modified. Innovative technologies have replaced the workforce, mainly the physical protection of property, installed web portals, electronic and information systems have replaced email or telephonic orders and time-consuming administration of written documents, e.g. in the case of received and sent invoices and orders, which were replaced by electronic invoices with electronic signature.

Retrofitting was in accordance in the technological advance and development in the area of security and person and property protection. A significant benefit is a gradual replacement of black and white cameras with colour IP cameras with higher image resolution and MPx space size and wide shot with possible zooming. The images from previously installed CCTV systems had lower resolution; the photographic documentation was taken in black and white with inaccurately specified subjects and poor backlight.

New quality CCTV systems operate in colour resolution with the advantage of high sharpness, transmission, and detail. The trend of modern technology in the 21st century digital society is to reduce the costs of the DVR and IP cameras. In the near future (in this or next century), the expert and user community expects new use of CCTV systems with higher resolution. IP camera and software enable to identify persons by their age and sex, recognition of objects and the faces of clients/customers and business partners, identify persons by their number, and other innovative and sophisticated supporting activities. The camera software function enables to programme luggage left in a monitored area, watch the facial expression and movement of persons in the shared areas of user premises and common areas. It is also able to recognize unusual gestures, facial expression and behaviour of individual persons and target groups. New camera functions and additional software is suitable for practical use in the area of security and property and person protection, e.g. in the case of identification of an offender or elimination of risk and damage occurrence.

The interaction between the behaviour of a customer/client, facial expression [18], physical activity, and other factors has a significant impact on the functionality and efficiency of the service system and customer service in terms of obtained input data and information about persons, waiting and check-in time according to the parameters set. A general trend of a circular economy is the reduction of cash transactions in the Czech financial market by reducing the number of payment points and terminals, which influences the number of bank branches and their replacing with cashless banking operations. Through this change it is possible to appeal to the prevention of security risk and milestones, since cash receipt / dispensation is not possible at banking institutions branches. Preventing the transfer of banking operations also prevents the risk of theft, robbery or property damage.

The obtained data and information on camera security and monitoring systems supported by technological advance and the development of the 21st century digital society brings a synergy effect in integrated security and property and person protection, as well as better use of marketing and communication tools for customers/clients in the area of product policy innovation and extended services provided. Multipurpose use and benefits are indisputable, expected future is an operative fact, not only a topic for medium term. There is a big question mark over the further development in the area of security, person and property protection, and a lot of unspoken questions in terms of legal, ethical, moral, social, economic, environmental and business issues, with an accent on maximizing the use of technological opportunities, marketing and communication tools of business policy and strategy in accordance with Industry 4.0. With regard to the current macroeconomic situation in the Czech Republic and the prediction of GDP per capita development, unemployment rate, inflation rate, balance of payments, and other economic indicators of the real economy, Czech citizens do not have to be afraid to invest in banking houses, since CCTV systems provide security and property and persons protection in both active and passive banking operations, since they significantly contribute to the prevention and protection of property and persons. The installation and configuration of CCTV systems is much easier, faster and potentially more economical than it was in the information, not digital society of the 21st century.

References

1. G.J. Hwang, S.M. Yang, A study for the object tracking method between cameras in mass video surveillance environment. *Advanced Science Letters*, **22**(11), 3247-3251 (2016). DOI: 10.1166/asl.2016.7963.
2. L. Lopez-Fuentes, J. van de Weijer, M. González-Hidalgo, H. Skinnemoen, A.D. Bagdanov, Review on computer vision techniques in emergency situations. *Multimedia Tools and Applications*, **77**(13), 17069-17107 (2018).
3. R.D. Putra, A.L. Prasasti, T.W. Purboyo, Analysis of retinex algorithm on digital image from CCTV camera for face recognition. *Journal of Theoretical and Applied Information Technology*, **96**(23), 7942-7962 (2018).
4. S. Cho, E.-M. Yang, C. Seo, Multiple number plates recognition system implementation using HD CCTVs. *Far East Journal of Electronics and Communications*, **17**(3), 643-649 (2017). DOI: 10.17654/EC017030643.
5. A. Vel'As, M. Kutaj, M. Durovec, Influence of changing the parameters of the camera system on video-based motion detection. In *Proceedings of 2017 International Carnahan Conference on Security Technology (ICCST)* (pp. 1-5). IEEE (2017). DOI: 10.1109/CCST.2017.8167829.
6. W.C. Choi, J.Y. Na, Study on real-time crime response service using multi-CCTV collaboration technology. In *ICCBDC'18 Proceedings of the 2018 2nd International Conference on Cloud and Big Data Computing* (pp. 78-81). ACM International Conference Proceeding Series. New York, N.Y.: ACM (2018). DOI: 10.1145/3264560.3264573.
7. V.M.E. Jacintha, S. Jamuna Rani, J. Glory Beula, J. Jeba Johnsly, An extensive resolution of ATM security systems. In: *2017 Third International Conference on Science Technology Engineering & Management (ICONSTEM)*, Chennai, India (pp. 934-938). IEEE (2018). DOI: 10.1109/ICONSTEM.2017.8261340.
8. S. Byon, E. Kwon, E.S. Jung, Y.T. Lee, A study on location information aided re-identification in CCTV environment. In *Proceedings of 2017 International Conference on Information and Communication Technology Convergence (ICTC)* (pp. 817-819). IEEE (2017). DOI: 10.1109/ICTC.2017.8190790.
9. M. Courtney, Stop, thief! *Engineering and Technology*, **10**(4), 70-73 (2015). DOI: 10.1049/et.2015.0440.
10. S. Shammi, S. Islam, H.A. Rahman, H.U. Zaman, An automated way of vehicle theft detection in parking facilities by identifying moving vehicles in CCTV video stream. In *Proceedings of 2018 International Conference on Communication, Computing and Internet of Things (IC3IoT)* (pp. 36-41). IEEE (2018). DOI: 10.1109/IC3IoT.2018.8668135.
11. L.M. Dang, S.I. Hassan, S. Im, I. Mehmood, H. Moon, Utilizing text recognition for the defects extraction in sewers CCTV inspection videos. *Computers in Industry*, **99**, 96-109 (2018). DOI: 10.1016/j.compind.2018.03.020.
12. Y.H. Kim, G.W. Bang, Development of security camera combined beacon signal for transmission of disaster and crime situation as well as tracking location. *International Journal of Engineering and Technology*, **7**(2), 141-144 (2018).
13. H. Gupta, P. Chaudhary, An efficient face parts detection technique for CCTV surveillance. In *2017 International Conference on Innovations in Control, Communication and Information Systems (ICICCI)* (pp. 1-6). IEEE (2019). DOI: 10.1109/ICICCIS.2017.8660814.
14. A.A. Adams, J.M. Ferryman, The future of video analytics for surveillance and its ethical implications. *Security Journal*, **28**(3), 272-289 (2015).
15. A.I. Adrian, P. Ismet, P. Petru, An overview of intelligent surveillance systems development. In *Proceedings of 2018 International Symposium on Electronics and Telecommunications (ISETC)* (pp. 1-6). IEEE (2018). DOI: 10.1109/ISETC.2018.8584003.
16. H. Al-Kawaz, N. Clarke, S.M. Furnell, F. Li, A. Alruban, Advanced facial recognition for digital forensics. In A. Jøsang (Ed.), *Proceedings of the 17th European Conference on Information Warfare and Security. ECCWS 2018* (pp. 11-19). Academic Conferences and Publishing International Limited. (2018).
17. J. Wickes, Facial recognition: Designing 21st century rules. *Biometric Technology Today*, **10**, 7-10 (2018). DOI: 10.1016/S0969-4765(18)30157-7.
18. W. Yimyam, T. Pinthong, N. Chumuang, M. Ketcham, Face Detection Criminals through CCTV Cameras. In *Proceedings of the 14th International Conference on Signal-Image Technology & Internet-Based Systems 2018 (SITIS)* (pp. 351-357). IEEE (2019). DOI: 10.1109/SITIS.2018.00061.

Digital Economy and Anti-Corruption: New Digital Models

E.L. Sidorenko[1,*], and *A.A.* Lykov[1]
* Corresponding author: 12011979@list.ru.

[1] MGIMO University, Moscow, Russia

Abstract. The authors of this paper consider promising areas of the corruption prevention using the latest digital technologies: Blockchain, Internet of Things, Artificial Intelligence and Big Data. The purpose of this research is the analysis of advantages of the digital economy development in terms of solving social problems and crime prevention. The authors also show functional digital models of the anti-corruption compliance are defined. In addition, the research results include the determination of some shortcomings of the proposed models associated with the imperfection of the current legislation.

Keywords: digital public administration; risk assessment; overview of the changes, digital transformation.

1. Introduction

Nowadays, the use of digital technologies for the corruption prevention is a new interdisciplinary point of application of forces of both lawyers and specialists in the field of the sophisticated technology. However, imprudent use of digitalization tools can have significant negative social and political consequences. In this regard, the theoretical development of this area is an important step towards the harmonization of law with the use of new digital tools.

2. Problem Statement

The active development of digital technologies in the framework of e-government and digital society programs poses a number of urgent tasks for researchers related to the search for the most effective directions of digitalization in terms of the optimal ratio of the effectiveness and cost of these reforms. In the last three years, three main priorities have been identified: the development of the financial system through the introduction of transparent and verifiable technologies; the improvement of the public administration system and crime prevention.

And if it is possible to calculate the effectiveness of the first two areas only after 3-5 years, the crime prevention using digital technologies shows its effectiveness during the first two years. Many works of both domestic [1, 2] and foreign authors [3, 4] are devoted to the use of digital technologies while committing crimes, as well as in the framework of their usage for the crime prevention. The researchers pay special attention to the assessment of theft of digital assets and legalization of crimes using cryptocurrency, but ignore another important problem – the prevention of corruption. Meanwhile, the transparency, accessibility and universality of digital technologies make them as convenient as possible for use in the detection and prevention of corruption.

According to the UNO development program, a 1% increase in the e-government implementation index in 2016 leads to a 1.17% reduction in corruption [5, 6]. According to the research of the Center for digital technologies and financial innovations of MGIMO, there is an inverse connection between the dynamics of the public sector digitalization and the corruption crime level: in 2018, in some regions of the country, with an increase in the volume of digitalization in the "state – man" sector by 5%, the level of domestic corruption decreased by 3%. Such dynamics is noted, in particular, in Moscow and the Republic of Tatarstan.

3. Research Questions

According to some experts [7], the introduction of new digital technologies itself leads to a decrease in crime [8] by reducing the information deficit, automating processes and limiting the margin of appreciation by state employees and intermediaries. For example, the introduction of an automated tax administration in Afghanistan has reduced the margin of appreciation by tax officials, with the result that the amount of collected taxes has increased from 250 million to 2 billion since 2004 [9].

In addition, the transparency of public administration provided by digital technologies makes it available for public supervision, which, on the one hand, reduces the level of public distrust in the government and makes the system of public services available, on the other hand, expands the possibility of expert support by establishing feedback with government agencies and allows to clarify official information.

But despite the obvious benefits of the digital technology to ensure the transparency of the public administration system and disclosure corruption offenses, there is no doubt the limited positive effect of this technology is largely determined by the underdeveloped digital infrastructure (lack of broadband Internet, lack of development of online services, digital illiteracy of the population, etc.).

But speaking about the undoubted advantages of digital technologies in preventing corruption, the field of research should include those negative consequences that can accompany the digitalization process. Among them are the following ones:

- weak personal data protection systems and vulnerability of databases for hacking using fundamentally new cyber attacks, facilitated access to the confidential information by database hacking [8];
- absence of an adopted system of responsibility measures for abuses with the new technologies usage;
- widespread dissemination of false information and lack of effective mechanisms to prevent it;
- excessive duplication of information in different databases because of the lack of direct channels between them for the exchange and verification of data, etc. (in this situation, there is a high risk of the information falsification about the corruption facts or overestimation of indicators by their uncontrolled duplication);
- lack of clear criteria for the technology evaluation (due to its novelty, the introduced technology does not have a clear value expression, which significantly increases the corruption risks in public procurement;
- low development of the digital ethics and culture of new technologies users (In Bosnia and Herzegovina, for example, several officials used an electronic registration system to issue fake identification documents [8]; at the same time, the regional public administration school has established numerous cases when civil servants used "e-government" tools for corruption: a Croatian police officer deleted data on traffic violations from the system for bribing wine and fried lamb; Croatian officials registered trucks as cars in the road toll system, the damage from this activity exceeded 2 million euros per year [10].

4. Purpose of the Study

The aim of the study is to analyze the current situation in the context of the considered issues and identify the main functional models of the corruption prevention using digital technologies. The authors also try to assess the effectiveness of these models on the basis of the current legislation.

5. Research Methods

The main research method is analysis. The analysis is based on sociological methods of processing statistical information about the state and dynamics of corruption, as well as a modeling method that allowed to build digital models of the corruption prevention, study and assess the effectiveness, as well as propose ways for their optimization.

6. Findings

Taking into account the previously identified risks, we considered the most effective solutions for the corruption prevention with the use of new digital technologies. Depending on the functionality of technologies, digital models of the corruption prevention can be divided into the following groups:

- ensuring the work transparency of state bodies and the formation of new digital platforms for collecting information about the commission of corruption crimes (transparency model);
- creation of feedback platforms between government agencies and consumers of public services for the rapid receipt of data on abuse of officials (feedback model);
- digital protection for the electronic document flow and public information (open document flow model);
- assessment of possible corruption risks and digital definition of conflict of interest (digital compliance model);
- reduction of the margin of appreciation by individuals potentially exposed to corruption (model for automated decision-making process);
- exclusion of a direct contact between the official and the consumer of public services through the direct access to public services (direct access model);
- monitoring the financial activity of people potentially exposed to corruption, members of their families (financial monitoring model).

The public service transparency model has proven itself well at the launch of crowdsourcing platforms [8]. As an example, we can take the portal "I Paid A Bribe", which allows users to anonymously report facts of bribery in the government agencies (India), as well as the system "Check My School", which allows parents to control the costs of schools (Philippines).

The disadvantage of these platforms is the anonymity of users and the inability to identify them. With such options, it is difficult to verify data on corruption crimes. On the contrary, in the case of the identification of applicants, the effectiveness of the detection of corruption crimes increases by an average of 60%.

The feedback model is implemented through the introduction of public information portals [8]. These portals are created by government agencies or non-governmental organizations in order to make the information about the work of government agencies publicly available (Argentine portal "Dinero y Politica", containing information about the finances of political parties in the country; OpenSpending.org, reflecting the movement of budget funds). The problem of the considered model is that the feedback sites are controlled by law enforcement agencies, which means that the completeness and quality of the posted information depends only on them.

The open document flow model based on Blockchain and Artificial Intelligence technologies has a great preventive potential. The use of new technologies in the traditional document management allows not only to ensure the movement of documents, but also to use the logistics and analytical resources of new networks.

In March 2019, at the OECD conference, a competition of scientific papers was held where it was found that Big Data and Artificial Intelligence technologies can help by detecting corruption [11, 12]. The model for the automated decision-making process is also gaining popularity. It involves the collection of legally significant information [13] and the adoption of legal decisions on this basis.

In the framework of the proposed model, an official as a subject of volitional decisions is excluded from traditional corruption schemes, and therefore there is no need to pay a bribe. In the framework of the proposed model, the decision to issue rare drugs on benefits will not be made by doctors potentially exposed to corruption offenses, but by a system based on data on age, health status, etc. In the optimal case, the entire procedure of decision-making and execution should be automated through the use of smart contracts. For example, entering data about the dismissal or death of a person into a single information system will entail the automatic termination of social and other payments in relation to him. But despite the obvious advantages of this model, there are a number of risks associated with the assessment of legal consequences of the automatically rendered decisions execution, in particular, the responsibility for wrongful decision or errors.

The direct access model assumes the exclusion of direct contact between the consumer of public services and the official making decisions, which significantly reduces corruption risks. The proposed model provides an opportunity for any user through his personal account to carry out online operations with documents in real time and track them 24/7. Similar experience is already being implemented in Russia. With the approval of the digital program, public authorities were the first ones who began to move from paper to digital media.

In addition, specialized websites and applications have been created, for example, the public services portal, which has been operating since 2010 and has been used very successfully. In 2017, the number of registered users of the portal increased by 25 million and amounted to 65 million people, and the share of citizens using electronic public services increased from 25% in 2012 to just over 64% in 2017 [14].

Within the framework of combating corruption, the financial monitoring model has been used quite successfully, this model allows analyzing the financial activity of persons potentially exposed to corruption, members of their families, both during the period of their respective positions and within three years after their dismissal.

This system is based on a constantly updated database, which collects and primarily processes personal information about these people, primarily financial and property status, information about all banking operations, information about payment of fines, housing, loans, taxes and other payments. Each public authority within its jurisdiction defines the types of information that will be received for storage in the database.

The problem, however, is that the proposed model is poorly consistent with the modern system of collection and storage of personal information, and in some cases, it is in the open conflict with the rules on the personal life protection. In addition, the question of what liability will follow in the case of non-reporting or reporting false or incomplete information about the movement of tangible assets is also unresolved.

7. Conclusion

The introduction of digital technologies in the system of public administration and anti-corruption policy of companies allows achieving a significant reduction in the level of corruption by increasing the transparency of the process and ensuring contactless provision of public services. But these advantages of digital technologies are minimized by the lack of a universal electronic document management system, the lack of technical and legal criteria for the verification of revealed facts and the information security system. It is obvious that the important condition for the implementation of the above presented digital models is their consistency with the current criminal, administrative and information legislation, as well as with modern methods of combating cybercrimes.

References

1. E.L. Sidorenko, A.A. Lykov, Prospects for the legal regulation of Central Bank digital currency. In Ashmarina S., Vochozka M., Mantulenko V. (Eds.), *Digital age: Chances, challenges and future. ISCDTE 2019. Lecture Notes in Networks and Systems*, **84** (pp. 613-623). Cham: Springer (2020). DOI: 10.1007/978-3-030-27015-5_73.
2. S.V. Ivantsov, E.L. Sidorenko, B.A. Spasennikov, Yu.M. Berezkin, Ya.A. Sukhodolov, Cryptocurrency-related crimes: Key criminological trends. *Russian Journal of Criminology*, **13**(1), 85-93 (2019). DOI: 10.17150/2500-4255.2019.13(1).85-93. [in Rus].

3. R. van Wegberg, J. Oerlemans, O. van Deventer, Bitcoin money laundering: Mixed results? An explorative study on money laundering of cybercrime proceeds using bitcoin. *Journal of Financial Crime,* **25**(2), 419-435 (2018). DOI: 10.1108/JFC-11-2016-0067.
4. S. Kethineni, Y. Cao, C. Dodge, Use of bitcoin in Darknet markets: Examining facilitative factors on bitcoin-related crimes. *American Journal of Criminal Justice,* **43**(2), 141-157 (2017). DOI: 10.1007/s12103-017-9394-6.
5. Basel Institute on Governance, *New perspectives in e-government and the prevention of corruption*. Working Paper 23 (2017). URL: https://www.baselgovernance.org/publications/working-paper-23-new-perspectives-e-government-and-prevention-corruption. Accessed: 21.10.19.
6. United Nations Department of Economic and Social Affairs, *United Nations e-government survey 2016: E-government in support of sustainable development* (2016). URL: http://workspace.unpan.org/sites/Internet/Documents/UNPAN97453.pdf. Accessed: 01.10.19.
7. L. Ante, Cryptocurrency, blockchain and crime. In: K. McCarthy (Ed.) *The money laundering market: Regulating the criminal economy* (pp.171-198). Agenda Publishing, Newcastle upon Tyne (2018). DOI: 10.2307/j.ctv5cg8z1.10.
8. I. Adam, M. Fazekas, Are emerging technologies helping win the fight against corruption in developing countries? *Pathways for Prosperity Commission. Background Paper Series; no. 21* (2018). URL: http://www.govtransparency.eu/wp-content/uploads/2019/02/ICT-corruption-24Feb19_FINAL.pdf. Accessed: 01.10.19.
9. R. Banning-Lover, Nine ways to use technology to reduce corruption. *The Guardian, Thu 26 May 2016 17.16 BST.* (2016). URL: https://www.theguardian.com/global-development-professionals-network/2016/may/26/nine-ways-to-use-technology-to-reduce-corruption. Accessed: 01.10.19.
10. Regional School of Public Administration, *Abuse of information technology (IT) for corruption* (2013). URL: https://respaweb.eu/download/doc/Abuse+of+Information+Technology+%28IT%29+for+Corruption.pdf/d867df158b864e8c843c15e5eece5016.pdf. Accessed: 01.10.19.
11. I. Fountoukidis, E. Dafli, Analyzing the corruption on a procurement network using Graph theory. *OECD Global Anti-Corruption & Integrity Forum, 2019.* OECD Conference Centre, Paris (2019). URL: https://www.oecd.org/corruption/integrity-forum/academic-papers/Fountoukidis-Dafli-graph-theory.pdf. Accessed: 01.10.19.
12. R. Milani, The impact of corruption on procurement performance. An assessment of Italian public work contracts and suppliers. *OECD Global Anti-Corruption & Integrity Forum, 2019*. OECD Conference Centre, Paris (2019). URL: https://www.oecd.org/corruption/integrity-forum/academic-papers/Milani%202019%20Corruption-Procurement%20Youth%20ResearchEdge%20Papers.pdf. Accessed: 01.10.19.
13. V.M. Koryakin, "Digitalization" of public relations and its impact on the state of corruption in the military organization of the state. *Obosnik (electronic journal)* (2019). URL: http://www.oboznik.ru/?p=63821. Accessed: 01.10.19. [in Rus.].
14. A.K. Moiseevskaya, Implementation of anti-corruption policy in the conditions of digitalization of economy and society. *Scientific Notes of Young Researchers*, **3**, 62-68 (2018). [in Rus.].

Regional Innovation and Investment Development in Conditions of Cluster Industrial Policy

A.V. Zastupov[1,*]
*Corresponding author: oiler79@mail.ru.

[1] Samara State University of Economics, Samara, Russia

Abstract. An important task of the modern industrial policy of the region is to ensure the growth of the industrial component on the basis of innovation and investment development of industries. Issues of post-crisis development and economic growth in conditions of limited resources were considered. The effectiveness of industrial policies in the region needs to be matched to real innovation and investment opportunities. On the basis of the analysis, a model on the formation of directions of innovation and investment development of industrial policy of the Samara region was developed. The relationship between the factors of innovative development of enterprises and the level of investment activity in the industrial sector of the region has been determined. The study identified problems of formation of innovation and investment potential of industrial enterprises of the region and ways to solve them.

Keywords: Industrial policy, innovation, investment, resources, factor, region.

1 Introduction

Today, the problems of post-crisis development and economic growth in the context of limited resources are becoming increasingly pressing for the regions. The main task becomes to ensure the growth of the industrial component on the basis of innovation and investment development of the region's economy. It is necessary to change the structure of the economy, organizational and institutional environment, increase the share of industries with high added value, formation of innovation and investment environment in modern market conditions.

2 Problem Statement

In the current economic conditions, the problem of increasing the innovation and investment development of the region's economy, the development of appropriate scientific-based approaches and mechanisms to increase the innovative potential and investment activity of industrial enterprises becomes an important problem.

Problems devoted to issues of increasing innovative potential of enterprises of the region, growth of innovative activity of industrial enterprises, formation of cluster industrial policy of the region were considered in the works of domestic and foreign authors – B.Y. Tatarskich [3], A.V. Streltsov [3], F. Phillips [1], T. Andersson [2], C. Ketels [4], G. Duranton [5], A. Gawer [8] etc.

According to the analytical review on this issue, the issues of increasing innovation and investment activity of industrial enterprises in the region, as well as organizational and economic aspects of innovation and investment development of industrial industries of the region have not been sufficiently investigated.

3 Research Questions

This study considers theoretical and practical aspects of innovation and investment development of industrial industries of the region, issues of increasing the efficiency of industrial policy of the region through the cluster model of development of the regional economy. The application of theoretical and methodological provisions and practical recommendations of the study will make it possible to make informed decisions in the field of innovation and investment development of industry enterprises.

The practical significance of the study is that the main results can be used by state and municipal authorities and the business community in the development of priority directions for improving the efficiency of the innovative business environment and innovation and investment development of the region.

Further research in this area will allow to expand the understanding of the possibilities of applying theoretical and methodological provisions and practical recommendations in the sphere of increasing investment activity of industrial enterprises and innovation potential of the region as a whole. Research should focus on the results achieved in the development of the region's economy.

4 Purpose of the Study

The following objectives and objectives have been identified in accordance with the issues and questions of the study:

- Explore industrial policy as a factor for economic growth in the region.
- To define directions and approaches to innovation and investment development of industry of the region in modern economic conditions.
- To develop a model of selection of priorities of innovation and investment development of the industrial complex of the region.
- To form proposals of an organizational and economic nature to improve the proposed forms and approaches of innovative and investment development of the industrial economy in the region.

5 Research Methods

5.1 Methods of research

The following methods were used during the study: theoretical (analysis, classification, concretization, synthesis); diagnostic (factor analysis, modeling, prediction, expert assessment method); empirical (observation, comparison, generalization).

5.2 Experimental Study Base

The experimental base of the study was a complex of enterprises of the industrial sector of the economy of the Samara region.

5.3 Study Stages

The study of the problem was carried out in 2 stages:

1. Theoretical:

- Research of industrial policy, investment and innovation potential of the economy of the Samara region.

2. Practical:

- Identification of directions and approaches to innovation and investment development of the region's industry in modern economic conditions;
- Development of a model of selection of priorities for innovation and investment development of the industrial complex of the region;
- Formation of proposals of organizational and economic character to improve the proposed forms and approaches of innovative and investment development of industrial economy in the region.

6 Findings

At present, the development of the Russian economy as a whole is oriented towards measures related to the improvement of market institutions, the increase of openness of the economy and the reduction of state participation in the economy in the long term. These provisions and the rejection of measures of the so-called selective industrial policy are based on the following arguments:

1. Increasing the level of industrial technological development dynamics. This causes a sharp narrowing of the horizons of economic and technological forecasts.

2. Growth of resource and technological needs of modern society. This creates greater complexity in technological solutions and increases the overall level of uncertainty.

The use of industry priorities is characteristic of traditional industrial policy options. It is rejected because of inefficiency and because of the high uncertainty in the innovative digital economy [1]. According to this view, the traditional model of concentration of material, labour and other resources by the State for its own purposes is now losing its relevance. This is due to the acceleration of economic processes and the short planning horizon. Today, in conditions of digitalization of the economy, the market and business are able to determine points of potential growth better than the state. This will be helped by factors of borrowing experience and own innovations, based on investment and innovation in the economy.

At the same time, the possibility of the state carrying out a strategy of catching-up growth is not rejected. However, emphasis is placed on the creation of institutional prerequisites and conditions for macroeconomic growth. Development priorities are proposed to be linked to institutions rather than industries. It should be noted that during the 1990s the economies of most developed countries and the well-being of their populations grew quite intensively on an innovative basis. In Russia, gross domestic product, levels of industrial and agricultural production have almost halved, and the level of real incomes of the population has significantly decreased. By comparison, the maximum fall in gross domestic product in the US during the crisis of the 1930s did not exceed 32 percent [2].

Of course, the importance of institutional reforms cannot be underestimated. Moreover, the extent and nature of the impact of the State on the industrial development of a country depends on the characteristics of its institutional structure and the level of development of the most important elements (industrial associations, financial institutions). This is also determined by the level of development of formal and informal coordination mechanisms for these institutions.

State industrial policy is able to detect and eliminate the negative side effects of harsh monetary conditions. It also allows to redirect priorities of financial support from inefficient enterprises to more economically sustainable economic entities, to focus efforts on priority areas of production. The experience of economic development in modern market conditions suggests that the development of production in market conditions without active structural industrial policies can lead to the loss of available technological advantages. Such a structural industrial policy should be based on an individual approach to certain sectors of the economy [3, p. 86-87].

Years of reform have led to the deindustrialization of many of the country's industries. Economic growth in recent years has not ensured the restoration of pre-reform production in many of the most important sectors that determine the trajectory of the country's economic development. There has been a sharp increase in services, especially in trade and financial areas. In addition, the previous structure of services proved to be inadequate to the requirements of modern market economy. The rapid emergence and growth of such activities as banking, insurance, notary, brokerage and others did not correspond to the current level of industrial development.

Therefore, when the problems of ensuring post-crisis development and economic growth in the context of limited resources arise, the main task becomes to ensure the growth of the industrial component on the basis of innovation and investment development.

Let us turn to the regional issue of economic growth. It is obvious that the adoption of industrial policies is more relevant for industrialized regions with a high share of manufacturing industries, compared to the national economy [4]. The need to overcome the consequences of the crisis, deindustrialization in these regions, modernization and transition to a strategy of innovative development of existing enterprises and creation of competitive advantages in domestic and foreign markets can explain this fact.

Regional aspects of industrial policy should be described. Here it is important to note that within its framework the most important patterns of placement of productive forces must be observed, namely:

- Economically viable and efficient placement of industrial production, taking into account the influence of economic, resource, geographical and environmental factors;
- Comprehensive regional economic development of economic entities and the economy of the region as a whole;
- A rational territorial division of labor between regions and within their territories, taking into account the regional aspects of the interest to attract personnel and the creation of high human capacity;
- Achievement of high rates of socio-economic development of the regions, provision of the necessary number of high-productivity jobs and infrastructure development [5].

In addition, restoring the manageability of the economy, changing its structure, changing the organizational and institutional environment, increasing the share of high-value-added industries, forming an innovation and investment environment in modern market conditions is essential.

In this regard, the methods of industrial policy should be consistently applied to a specific stage of development of the real sector of the economy: information, macroeconomic regulation, institutional and other. In order to achieve the effectiveness of industrial policies in the region, it is also necessary to rank them over time, according to the stages of economic development. It is also necessary to specialize methods by activity and to relate them to real innovation and investment opportunities.

We will analyze the economic state and trends of industrial development dynamics in the Samara region over the last ten years.

It can be noted at once that in 10 years significant changes have taken place in the structure of investments. The share of investments in mining, production and distribution of electricity, gas and water has increased significantly. The share of investments in many traditionally leading industrial industries in the Samara region has significantly decreased. These are food production, chemical production, machine-building. The latter aspect appears to be a particularly negative trend, as mechanical engineering held a leading position in the economy of the Samara region (table 1).

Table 1. Structure of investments by type of economic activity industrial economy of Samara region for 2008-2017.

Structure of investments by type of economic activity, percentage	2008	2011	2014	2017
Mining (petrochemical cluster)	9.4	10.7	8.9	12.2
Processing productions	24.4	24.3	44.1	39.7
from them: food products production	3.3	2.1	1.0	0.9
Production of coke and petroleum products (petrochemical cluster)	4.2	6.8	27.2	24.4

Chemical production	3.7	4.7	2.4	3.6
Machinery and equipment production (machine building cluster)	0.9	1.4	1.8	0.5
Production of electric equipment, electrical and optical equipment (space cluster)	1.0	0.5	0.5	0.4
Production of vehicles and equipment (automobile cluster)	4.8	5.4	8.8	10.8
Electricity, gas and water production and distribution	7.4	11.4	8.0	13.6

Source: Author.

For comparison, in 1990 more than 40% of all volume of investment into the industry of the Samara region were the share of mechanical engineering and metal working (machine-building, machine-tool constructing clusters) (table 2).

Table 2. Structure of investments by industry of the Samara region in 1990.

Investment structure by industry	Value (percentage)
Power industry	5.0
Fuel industry	20.1
Chemical and petrochemical industries	9.9
Mechanical engineering and metalworking	41.1
Construction materials industry	3.7
Weaver's industry	0.8
Food industry	3.6

Source: Author.

Dynamics of indicators of table 1 shows that the prevailing share of investments is the share of the processing productions. However, in the processing productions they are distributed extremely unevenly. Practically for all studied period, behind a small exception, the main volume of investment goes to the enterprises of two types of activity – production of coke and oil products, chemical production (a petrochemical cluster), production of vehicles and the equipment (an automobile cluster). Extremely low share of investments into other industrial types of economic activity – machine-building, machine-tool constructing, space clusters of the region attracts attention.

Thus, on the basis of the above, it is possible to form certain methodological recommendations for the formation and implementation of the investment industrial policy of the region (figure 1).

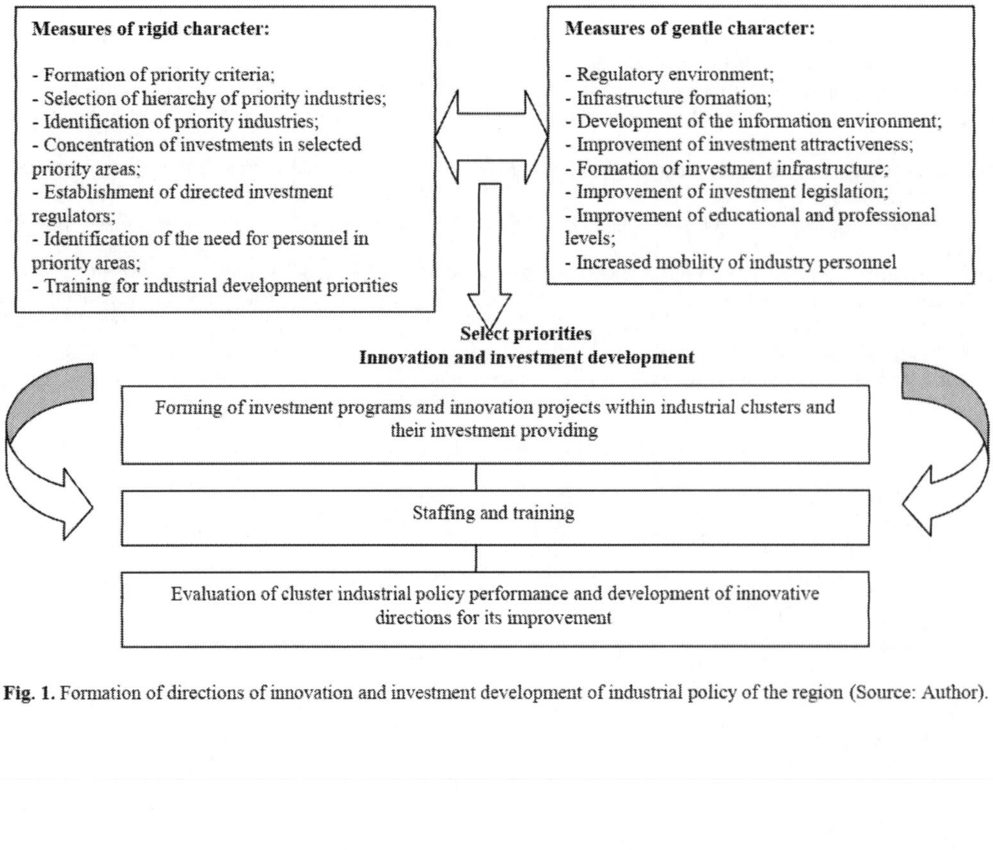

Fig. 1. Formation of directions of innovation and investment development of industrial policy of the region (Source: Author).

Thus, an integrated approach is necessary for the formation and implementation of directions of innovation and investment development of industry of the region. It is capable of creating an innovative product in the industrial sector and attracting investors to finance innovation in industrial industries. Such cooperation and integration of directed actions in ensuring innovation and investment development is most fully and effectively implemented through large industrial forms of organization of production processes at the regional level. Such development is based on the reproductive interaction of economic entities and industrial sectors of the economy of the region as a whole. Such form which is the most effectively answering to the listed parameters is formation of industrial clusters in the region [6, p. 354-355]. Industrial clusters accumulate in themselves traditional approaches in management of region economy. They are based on industry accessory of economic entities and represent modern cross-industry forms of innovation and investment development. In this case the character and depth of the created cross-industry communications provide high-quality innovative development of the territory with attraction of investments into economy of the region and formation of the new markets.

The most important task of innovative development of the industrial sector of the economy is to ensure innovative demand by enterprises of the industrial sector of the economy of the region.

The task of supporting and stimulating innovation becomes ineffective in the face of insufficient demand for innovative resources in the economy. Relevance is lost in the implementation of innovative projects and the creation of innovative products. In this case, investments are aimed at stimulating innovative demand and are becoming ineffective or are becoming very long-term. This can lead to a loss of innovation in the product or technology and inefficient spending of funds. Any innovation requires lengthy research and development. As a rule, the time for its development is much longer than the time for modernization of production facilities. Therefore, the demand for industrial innovation will always be greater than their supply in a normally functioning market economy [7, p. 562-563].

The analysis which is carried out above should be added. It is necessary to consider the state of affairs with innovative development of the enterprises of the industry of the region. Innovations along with investments represent one more major reserve of growth of economy and development of the enterprises of petrochemical, space, machine-building, automobile clusters of the region. At the same time, the most important indicator is the volume of innovative goods, works, services by type of economic activity. Prospects for innovative development of industrial enterprises allow to estimate costs of technological innovation (table 3).

Table 3. Costs of technological innovation by type of innovation activity in Samara region for 2007-2017.

Costs of technological innovation by type of innovation activity, million rubles (in comparable prices)	2007	2010	2013	2017
In total	9 430.9	11 475.4	41 840.2	32 877.9
including: Research and development of new products, services and methods	1 511.6	1 494.6	13 092.6	11 874.6
Acquisition of machinery and equipment	5 506.9	4 529.6	18 042.5	9 752.1
Acquisition of new technologies	8.5	153.3	198.2	109.8
Acquisition of software	121.3	231.4	133.6	241.4
Other types of preparation	1 463.3	549.7	1 827.6	5 657.8
Production Design	383.4	1 816.1	2 374.6	3 710.9
Staff education and training	115.1	68.8	137.7	86.4
Market researches	9.8	5.1	45.3	36.2
Other costs of technological innovation	557.0	5 337.4	9 933.5	9 597.3

Source: Author.

From table 3 it is possible to note that dynamics of volume of innovative products – in general positive, despite the imposed economic sanctions on number of industries of the Russian Federation in 2014. In this regard some reduction of indicators by types of innovative activity in 2017 is observed. Rather high innovative level of the issued goods can be noted on production of vehicles in automobile cluster of economy of the region. If in 2010 the share of innovations in products of automobile cluster (production of vehicles and the equipment) made 72.7 percent, then in 2013 the share of innovations has already made 63 percent, in 2017 – 60.7 percent.

Thus, the presented these loudspeakers of volume of innovative products correspond to indicators of dynamics of the volume of investment by the types of economic activity of economy of the Samara region given in table 1. This shows that the factor of innovative development of enterprises and the region is determined by the level of investment activity in the industrial sector.

Currently, the largest amount of financing for the development of knowledge-intensive technologies is required for the entire cycle of scientific and technical activities, namely:

1) carrying out targeted scientific missions and training specialists on the basis of the leading universities of the country;

2) organization of research work in research institutes, universities, research organizations and scientific and technical centers of leading enterprises;

3) carrying out a scientific and technical examination of the performance of the technology level for monitoring the rating of critical technologies in industrial industries;

4) organization of targeted training of research engineers on innovative management and forecasting of scientific and technological progress, implementation of strategic programs in enterprises of industrial industries of the region.

Forming of strategy of the innovative development of the industrial sector of economy of the region should be based on comprehensive study of the possible directions it implementation. It is necessary to adhere to a certain model of rules on its forming and further implementation in the conditions of educated industrial clusters [8]. First of all, it is necessary to carry out the complex analysis of a status and development of industrial economy in the region. Further, it is necessary to define and structure the created trends and problems, to carry out component analysis, to reveal the key priority dynamic industries which are the driver of growth of economy in the region and are capable to create additional points of growth.

The stage of studying of innovation investment potential of economy of the explored region should become an important stage in carrying out such analysis. For the purpose of implementation of effective industrial policy the choice of priorities of development of the industry of the region around which the innovation industrial clusters should be formed will be result of an analytical research. The choice of the key industries capable to become the engine of development of economy of the region, and the subsequent support of such industries at the level of regional and federal authorities should not contradict industrial policy in general. It is a number of the directions of support of medium and small business, improvement of the normative and legal legislation, infrastructure development, development of cooperation, information environment, providing preferential terms (see fig. 1). Moreover, these measures will allow not only to provide effective development of key industries in the region, but also to create additional opportunities of growth which will be able to provide effective dynamics of sustainable development of allied industries and productions.

Speaking about the peculiarities of development of industrial enterprises in conditions of limited investment resources, it is possible to note the lack of available innovative and investment resources for innovation and investment development of the economy of the region. In this regard, the business community, state and municipal authorities face the most important task of launching innovative industries and producing innovative products, but also of fully involving the scientific community in this process, creating the necessary laboratories and experimental stations, where development and experimental scientific developments on the creation of innovative technologies and products will be constantly carried out.

Thus, the innovation and hi-tech productions created in the region on the basis of the created industrial clusters become a basis in implementation of the innovation and investment strategy of development of industrial policy. They demand providing in the region not only investment resources, but also high scientific potential. Improvement of organizational and administrative mechanisms of innovation and investment development, involvement in the created industrial clusters of the scientific organizations and educational institutions of secondary vocational and higher education are necessary. It can be fully implemented through forming and development of the innovation industrial clusters in the region with participation of the innovation centers and technological parks in the field of high technologies.

7 Conclusion

The article presents the main results, conclusions and recommendations on the problems of increasing the investment activity of industrial enterprises and improving the innovative potential of the region in the context of clustering the economy.

Scientific results of the study fundamentally complement the existing approach to the problem of increasing innovation and investment activity of enterprises of the industrial sector and efficiency of industrial policy of the region as a whole. At the same time, the instruments of cluster approach of innovation and investment development of the region's economy play a key role in this process.

References

1. F. Phillips, H. Linstone, Key ideas from a 25-year collaboration at technological forecasting & social change. *Technological Forecasting & Social Change*, **105**, 158-166 (2016).
2. T. Andersson, S. Schwaag-Serger, J. Sörvik, E. Wise, *Cluster policies whitebook* (IKED - International Organisation for Knowledge Economy and Enterprise Development, Malmö, 2004).

3. A.V. Zastupov, A.V. Streltsov, B.Y. Tatarskikh, V.B. Poluyanov, H.M. Murtazova, Petrochemical cluster as the engine of the region's economic development. *International Journal of Economic Perspectives*, **10**(3), 83-93 (2016).
4. C. Ketels, Recent research on competitiveness and clusters: What are the implications for regional policy? *Cambridge Journal of Regions, Economy and Society*, **6**(2), 269-284 (2013). DOI: 10.1093/cjres/rst008.
5. G. Duranton, California dreamin: The feeble case for cluster policies. *Rev. Econ. Anal.*, **3**(1), 3-45 (2011).
6. A.V. Zastupov, Investment development of enterprises of industrial clusters. In S. Ashmarina, M. Vochozka (Eds.), *Sustainable Growth and Development of Economic Systems. Contributions to Economics* (pp. 349-358). Cham: Springer (2019). DOI: 10.1007/978-3-030-11754-2_25.
7. A.V. Zastupov, Innovation activities of enterprises of the industrial sector in the conditions of economy digitalization. In S. Ashmarina, A. Mesquita, M. Vochozka (Eds.), *Digital Transformation of the Economy: Challenges, Trends and New Opportunities. Advances in Intelligent Systems and Computing,* **908** (pp. 559-569). Cham: Springer (2019). DOI: 10.1007/978-3-030-11367-4_55.
8. A. Gawer, Bridging differing perspectives on technological platforms: Toward an integrative framework. *Research Policy*, **43**(7), 1239-1249 (2014).

Virtual Machines Migration Technologies as a Strategy for Regional Manufacturing Companies

A.E. Bogomolov[1], *A.S.* Kreymer[1], and *L.E.* Popok[1,*]
*Corresponding author: lpopok@gmail.com.

[1] Kuban State Agrarian University, Krasnodar, Russia

Abstract. Modern approaches to the automation of business processes are increasingly confronted with questions of increase of efficiency of use of available computing resources. The transition to cloud computing technology makes the company to revise approaches to the assessment of their own capabilities and look for ways to reduce operating costs, it does not complicate the process of information infrastructure management. In the competitive environment of the organization following the path optimization of capital and operating costs through the introduction of new technologies, as well as through the transition and migration to a more cost-effective technologies.

Keywords: optimization, digital economy, digital technologies, migration, virtualization technologies, IT infrastructure.

1 Introduction

In modern conditions of large companies are increasingly using virtualization platform to reduce operating costs and simplify administration procedures, server systems and equipment as a set of computing resources and their logical combination abstracted from the hardware implementation. Using virtualization platform entails a corresponding additional costs for the purchase of software licenses virtualization platform. In such circumstances, it becomes important criterion TCO platform virtualization software licenses and lower costs for the purchase of these products.

The migration of virtual workloads is a strategically important step in the long-term cost savings associated with the purchase of software licenses for virtualization platform and provide competitive advantages for companies in various sectors in terms of economic instability and rising global competition.

In fact, the countries of the Eurasian Economic Union are in a need for services existing technological monopoly of large software vendors from the United States. In this context, questions to reduce costs for the use of distributed computing infrastructures are of particular relevance.

The relevance of this article due to the fact that most of the major regional companies are faced with the choice of cost-effective and reasonable choice of virtualization software platform manufacturers.

In the article the modern approaches to the implementation of the migration of virtual machines between different virtualization platforms. This article is the comparative characteristic instruments of migration of virtual workloads. The conclusions about the appropriateness of different approaches depending on the facing migration challenges and available resources. Author's technique of migrating virtual machines from VMware vSphere virtualization platform on the Microsoft Hyper-V virtualization platform, allowing to increase the speed and reliability of the migration process and to save on operating costs of the company.

The need for scientific research issues related to the migration of virtual workloads and systems based on Microsoft Hyper-V virtualization platform as the optimal virtualization platform, define the relevance of the topic and the research presented in this article.

Results that can be obtained through the migration of virtual machines, should be considered from the point of view of the transformation of business processes associated with service virtualization platforms and obtain economic benefits from the implementation of these processes. Obviously, the need to use both quantitative and qualitative indicators in determining the impact of migration on the activities of producer organizations.

2 Methods

The authors used such research methods as analysis, synthesis, description and comparison. Their use is determined by the nature of theoretical research, which included the following stages: formulation of the problem, analysis of information on the subject, the calculation of cost-effectiveness, comparison and description of the various migration tools.

As part of the testing tools are evaluated based on several criteria:

- delegation of rights and infrastructure;
- cost of a migration tool;
- support solution by vendor;
- the rate of migration / machine downtime;

- requirements for disk resources for the migration;
- the automation of mass migration;
- rollback and recovery in case of problems.

3 Findings

3.1 The economic rationale for migration of virtual workloads

For the feasibility study of the migration virtual workloads with VMware vSphere virtualization platform on the Microsoft Hyper-V virtualization platform has been developed with inputs to the script, as well as an economic model for the calculation of the following indicators:

- Net Present Value (NPV);
- internal rate of return (IRR) for 5 years;
- payback initiatives;
- the cost of implementation of the initiative;
- savings from the implementation of the initiative.

The scenario describes the migration of virtual machines in 6000 for three years, a platform to host virtual machines are 200 physical servers. This scenario describes the migration of virtual machines with Windows operating system. All prices are in rubles, the model does not consider the situation of fluctuations in the value of licenses, depending on exchange rates.

Input data for economic models are presented in Table 1.

Table 1. Input data for an economic model for calculating expediency migration of virtual workloads.

Parameter	Value	Unit of measurement
The number of servers (4 CPU x 16 Core)	200	psc
The number of virtual machines (2 Core)	6000	psc
Price VMware vSphere licenses 6 Enterprise Plus for 1 processor	0,236	million rubles.
Price Basic Support License / Subscription for VMware vSphere Enterprise Plus for 6 1 processor for 1 year	0,070	million rubles.
The price of Microsoft Windows Server 2019 license Datacenter Single OLP 16License NL CoreLic	0,482	million rubles.
Renewals Software Assurance Single SA OLP 16License NL CoreLic for 1 year	0,160	million rubles.
The price of Microsoft Windows Server 2019 license Single CAL NL Device	0,002	million rubles.
The price of Microsoft Windows Server 2019 license Stnd Single OLP 2License NL CoreLic	0,007	million rubles.
The sales price of the migration project, per year	12	million rubles.
Discount rate	10	%
Inflation rate	5	%

Source: Authors.

Figure 1 provides a summary graph of the results of economic modeling initiatives migration virtual workloads from one virtualization platform to another.

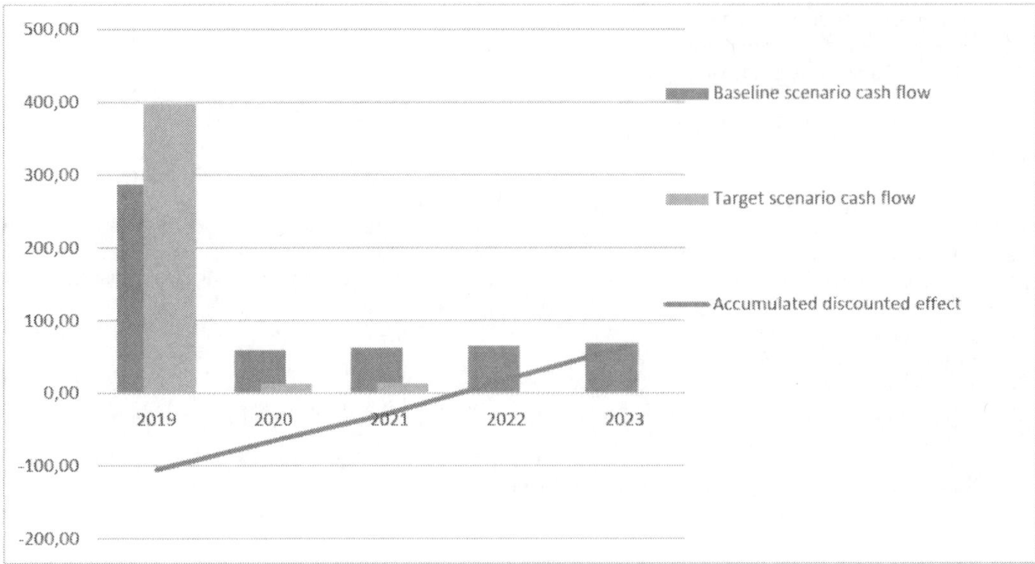

Fig. 1. The graph of the results of economic modeling (Source: Authors).

The graph shows the payback period comes less than three and a half years
Major economic modeling results are presented in Table 2.

Table 2. Main results of economic modeling.

Source: Authors.

Parameter	Value
Net Present Value (NPV), mln. Rub. (5 years)	63
Internal rate of return (IRR),% (over 5 years)	33
The payback period for the discounted effect (DPP), years	3.6
The costs for the implementation of the initiative, mln. Rub. (5 years)	423
The savings from the implementation of the project, mln. Rub. (5 years)	116
Payback period (PBP), years	3.3
Net Present Value (NPV), mln. Rub. (5 years)	63

Based on the results obtained in economic modeling, we can talk about the feasibility and importance of launching initiatives on migration virtual workloads with VMware vSphere virtualization platform to platform Microsoft Hyper-V virtualization.

3.2 Evaluation Criteria migration tools

As part of the testing tools are evaluated based on several criteria:

- delegation of rights and infrastructure;
- cost migration tool;
- support solutions provider;
- the rate of migration / machine downtime;
- requirements for disk resources for the migration;
- the automation of migration;
- rollback and recovery in case of problems.

Below is the basic information about each instrument and score each item on a scale. It is further provided a brief summary table with the scores on the estimated parameters and offered the most effective options.

Microsoft System Center Virtual Machine Manager (SCVMM).SCVMM allows you to connect VMware vCenter management servers from a single console and see all the virtualization nodes and all virtual machines with the possibility of creating, managing and converting machines. In this test only considers aspects of the migration of virtual machines on Hyper-V.

The first step in considering the method of migration is the question of delegation of least privilege migration team for industrial VMware vSphere environment. According to the official Microsoft documentation to add vCenter in

VMM requires an account with full rights to the entire infrastructure of VMware vSphere and local administrative rights on the vCenter. On a scale of SCVMM is estimated at 4 points on the criterion of the delegation and on the infrastructure of the rights due to the requirements of local recording on the nodes [1].

The second important point is to support the decision. SCVMM is part of System Center, the flagship product to Microsoft, periodically updated, supported by the software, and is estimated at 5 points on the criterion of support.

Another criterion is the speed of migration downtime virial machines for the period of migration. SCVMM migrates the virtual machine off, copying it to VMDK disk format to the HTTP protocol node and completing the conversion in VHDX after copying. unavailability of time for a typical stage of a virtual machine test was 1 hour 36 minutes from the start of the process before the end of migration and access to virtual machines over the network. Based on the criterion of the migration rate - 2 points.

From the viewpoint of resource requirements, the destination node requires double the volume allocation of disk space used by the virtual machine: copying disk and its conversion. After conversion, the original disk copy is removed and used a copy of the virtual disk. Based on the criterion of resource requirements - 2 points.

Process automation is one of the key factors for mass migration. SCVMM automatically selects the least loaded node that you are migrating to a virtual machine and a free disk to house its files. The virtual machine is highly available immediately connected to the virtual networks managed by the VMM, receives a specified IP address from a special pool. All operations can be done from the GUI once and save it as a PowerShell script for subsequent modifications. In this case, some of the complex migration scenarios, for example - the virtual machines with non-standard settings or additional requirements, may require rework scripts. Or the development of new scripts, if it would be more appropriate. The object model and scripting SCVMM (PowerShell) language allows such improvements. Based on automation criteria - 5 points.

When a virtual machine migration tool SCVMM source machine on the VMware side is turned off and remains there at the end of the migration, requiring manual removal. This allows simple incorporation of the original virtual machine, as the abolition of migration plan. Based on the criterion of recovery - 5 points.

SCVMM is a paid product, licensed on processor virtualization host, but is included in the package, along with licenses for Windows Server, and, as a rule, is already available in most large organizations. Based on the criterion of cost - 3 points.

Microsoft Virtual Machine Converter (MVMC).MVMC is a separate Microsoft product for machine migration with VMware vSphere infrastructure (either physical servers) on the Hyper-V (or Microsoft Azure) [2].

In terms of delegating enough to MVMC minimum role configured for VMM migration tool. Rights vSphere nodes is required. Based on the criterion of delegation of rights infrastructure - 5 points.

At the current time MVMC is an unsupported product. Based on the criterion of support - 1 point.

During migration, shut down the machine is copied, then converted. It takes 32 minutes for a typical test virtual machine. Based on the criterion of the migration rate - 3 points.

MVMC requires double the amount of disk space on the conversion time. Based on the criterion of resource requirements - 2 points.

Tasks performed by MVMC, partially automated. Connecting to the network, clustering of virtual machines, the installation of component integration, the selection addresses, all this can be done in PowerShell; but the choice of Hyper-V host and the drive to accommodate loads require explicit administrator's decision. Evaluation criteria for automation - 3 points.

When a virtual machine migration tools MVMC original machine on VMware side off and stays there for the end of the migration, requiring manual removal. That enables simple on both the original VM migration plan cancellation. Based on the criterion of recovery - 5 points.

MVMC is a free product and does not require a purchase. Based on the criterion of cost - 5 points.

5nine Easy Converter.5nine Easy Converter (hereinafter 5nine) - this is a third-party solution V2V. It comes in the form of free and paid versions, where the paid version has the ability to automate. The advantage compared to the VMM / MVMC is to support the virtual machines of the second generation, in that it can convert the machine VMware vSphere, use the UEFI boot mode [3].

In terms of delegating side vCenter to 5nine lack minimum role that is configured to perform the migration in VMM. If you are migrating directly from VSphere require superuser (root) on the nodes. Based on the criterion of delegating infrastructure - 4 points.

5nine is a third-party product, the paid version is supported by the developer. Microsoft support incidents do not take to convert a third-party product, but to convert the machine is supported by Hyper-V. Based on the criterion of support - 3 points.

Speed machines and inaccessibility of the migration to 5nine comparable MVMC Migration with vCenter. Turning off the machine is copied, then converted. It takes 37 minutes for a typical test virtual machine vCenter. Based on the criterion of the migration rate - 3 points.

5nine requires double the amount of disk space on the conversion time. Based on the criterion of resource requirements - 2 points.

Tasks performed by the free version 5nine, not automated. For the paid version, there PowerShell module of four cmdlets to automate the migration directly from vSphere. This module does not support the migration from the vCenter, and since, as a rule companies have hundreds of vSphere hosts and virtual machines are automatically moved between them, it imposes restrictions - machines will need to pre-party VMware to migrate to a single node, which is off-balancing technology. Evaluation criteria for automation - 3 points.

When a virtual machine migration tools 5nine original machine on the side of VMware removed. To cancel a migration should first make a clone of the machine (can be automated, but requires a disk resource) or to restore the original virtual machine from a backup. Based on the criterion of recovery - 3 points.

5nine Easy Converter in an embodiment of the automation solution is chargeable and licensed for each VM at prices comparable with Azure; 25-30 $ per virtual machine. Based on the criterion of cost - 2 points.

Scripting solution based on NFS, VMotion and free utilities. As an alternative commercial solution offers an option that allows you to significantly reduce the downtime of virtual machines on the migration period. On a Windows server is one of the drives available in both shared with the SMB protocol (standard Windows shared folders), and the NFS protocol (a protocol for shared disk of UNIX, also supports VMware VSphere). On the side of the VMware vCenter NFS shared volume is added as a storage and virtual machine is transferred to it is turned on using vMotion technology. After the machine files will be on a Windows server, the machine is turned off, and her drive from VMDK format instantly converted first into the VHD, and then VHDX by changing the header files of the virtual disk utilities VHDtool and VHDXtool. Further PowerShell scripts created for the VMM target highly available VM on Hyper-V with the same parameters as the original, it is connected obtained VHDX disk. The virtual disk, even in the off state, the integration components are installed. The virtual machine is turned on, it takes a restart, the virtual machine gets old IP-address and is ready for use after a few minutes of unavailability [4]. In terms of delegating side vCenter requires only two rights - Clone (create a copy for the abolition of migration) and moving disks on NFS. Rights to vSphere is not required. Based on the criterion of delegation of rights infrastructure - 5 points.

The most difficult issue for the described solution is its support. Since conversion is not used for any of Microsoft products, problems with conversion formal support cannot receive. However, if at the end of the conversion of a virtual disk in VHDX format the disk has mounted, it has created a machine Hyper-V has standard support for Microsoft. Based on the criterion of support - 4 points.

The virtual machine is transferred to the Windows server is turned on without stopping services. Converting a virtual machine is performed for a few seconds. Downtime is equal to two times reboots the operating system. Based on the criterion of the migration rate - 5 points. Additional disk space on the process of conversion is not required. Based on the criterion of resource requirements - 5 points. The solution is fully built on scripts using PowerShell module for VMM and PowerCLI from the VMware, as well as third-party freeware utilities nfsutil, vhdtool and vhdxtool. All tasks can be automated [5]. Based on automation - 5 points. When a virtual machine migration tools Storage vMotion VMware source machine files are transferred on the NFS, where on the site will be converted to another format. Disk Copy does not occur at any of the stages, so copies of the original virtual machine remains. To cancel a migration should first make a backup of the virtual machine. Based on the criterion of recovery - 3 points. This approach does not use a fee-based products and does not require a purchase. Based on the cost - 5 points. Test results. The results of the testing migration tools virtual workloads are presented in Table 3.

Table 3. Criteria for evaluating the migration tool virtual workloads

Source: Authors.

Evaluation criteria	SCVMM	MVMC	5nine	scripts
Delegation and the right infrastructure	4	5	4	5
Support Solutions	5	1	3	3
Service downtime	2	3	3	5
Resource Requirements	2	2	2	5
Automation	5	3	3	5
Rollback and recovery	5	5	3	2
Cost	3	5	2	5
TOTAL	**26**	**24**	**20**	**30**

From the resulting table, taking into account the points described by the criteria of evaluation of each approach shows that the best approach - migration via scripting method. You can also recommend the use of the migration method using SCVMM to migrate those virtual workloads that for some technical reasons, cannot be migrated using the scripting approach.

4 Conclusion

Migration of virtual machines with the VMware vSphere virtualization platform undoubtedly brings many economic benefits. However, this migration requires considerable efforts, the launch of a full-fledged design and selection of the correct migration tool to reduce the risks associated with an unplanned "downtime" for mission-critical applications. In addition, the personnel involved in the migration process should be highly qualified and well-known sequence of actions.

It should be remembered that the implementation and debugging of VM migration procedures can be time-consuming, which in the case of incorrect project planning may lead to a further increase in costs - the specific result depends on the choice of migration tools, the nature of the functions of personnel management and related project management.

Thus, there is a large enough reserve in reducing administrative cost of computing infrastructures. This is important in the formation of the integration processes between economic actors from EU countries and the maintenance of a single economic space.

References

1 Microsoft, *Convert a VMware VM to Hyper-V in the VMM fabric* (2019). URL: https://docs.microsoft.com/en-us/system-center/vmm/vm-convert-vmware?view=sc-vmm-2019. Accessed: 30.08.2019.
2 Microsoft, *Microsoft Virtual Machine Converter 3.0* (2017). URL: https://docs.microsoft.com/en-us/previous-versions/windows/it-pro/windows-server-2012-r2-and-2012/dn873998(v%3Dws.11). Accessed: 30.08.2019.
3 5nine Software, Inc., *User Guide .5nine V2V Easy Converter Version 7.0. Description PowerShell API and Microsoft Azure API* (2016). URL: https://www.5nine.ru/Docs/5nine_Easy_V2V_Converter_API_v7.pdf. Accessed: 30.08.2019.
4 VMBlog, *Converting VMware virtual machines to Hyper-V format* (2017). URL: https://vmblog.ru/konvertaciya-virtualnoj-mashiny-vmware-v-format-hyper-v/. Accessed: 30.08.2019. [in Rus.].
5 Microsoft, *Understanding a Windows PowerShell Module* (2016). URL: https://docs.microsoft.com/en-us/powershell/developer/module/understanding-a-windows-powershell-module. Accessed: 30.08.2019.

Some Aspect of Legal Regulation of Special Economic Zones in Russia

Yu.A. Tarasova[1*], *S.A.* Kozhevnikova[1], *T.S.* Bobkova[1], and *O.G.* Porunova[1]

[*] Corresponding author: sangria80@yandex.ru.

[1]Syzran Branch of Samara State University of Economics, Syzran, Russia

Abstract. The authors analyze the Russian and international experience of functioning of territories with special legal status. The multiple advantages of establishing and developing of such territories in the country in the short-term and long term are noted. On the basis of the analysis of the Federal Law "About special economic zones in the Russian Federation" and the other normative legal acts, the advantages and the shortcomings of the legal regulation of the creation, management and closing of the special economic zones are revealed. The factors defining a set of privileges for the foreign investors in various territories with special legal status and also the conditions of the effective work of the special economic zone of industrial and production type are considered. The authors support their analysis of the legislation with the description and comparison of economic indicators of work of two special economic zones of industrial and production type in the Volga Federal District. The authors also consider jurisprudence on the controversial issues arising in the connection with the establishment of special economic zones. The article comes to the end with the consideration of possible prospects of the territories with special legal status transformation.

Keywords: special economic zones, legal regulation, direct investments, benefits and preferences, infrastructure, zone of industrial and production type.

1 Introduction

The Special Economic Zones (SEZ) as the tools of attracting foreign investments are widespread. The economist researchers recognize the benefits of setting up special economic zones on the territory of the country. For example, the report of Th. Farole and G. Akinci indicates the main goals of the SEZ creation - attraction of investment resources, expansion of cooperation with foreign partners, modernization of the economy [1]. Wang notes that special economic zones are the instruments of economic diversification and competitiveness [2]. Ablaev emphasizes the role of special economic zones for the development of certain regions of the Russian Federation [3].

In Europe, where the experience of using special economic zones has proved to be the most successful, they have had a long history. The same Farole and Akinsi note that "free ports" and urban-engulfed free zones that guaranteed free storage and exchange of goods on safe trade routes - such as Gibraltar, Hamburg and Singapore - existed for centuries. The first modern industrial free zone was established in Shannon, Ireland, in 1959. But since the early 1970s, special economic zones have been established in East Asia and Latin America to attract investment in labour-intensive production. These zones have become the cornerstone of trade and investment policies in countries that abandon import substitution policies and seek to integrate into global markets through export-oriented growth policies.

In developed countries, such as Canada and the USA, territories with special legal status serve primarily to increase the competitiveness of products of the national producer.

In Russia, the creation of special economic zones is a large-scale federal project aimed primarily to attract direct Russian and foreign investments in high-tech sectors of the economy, import-substituting industries, shipbuilding and tourism. The advantages of these territories are the special regime of business activity and the possibility of implementing the regime of a free customs zone [4].

2 Problem statement

However, the attractiveness of special economic zones in Russia remains weak, as evidenced by the closure in 2016 of nine economic zones, which during the whole period of their existence did not attract any investors [5]. It can be assumed that the development of legal regulations is one of the main factors in the attractiveness of the territory for foreign investment, as it makes the work of the zone predictable and the possibility of profit guaranteed. But despite the long existence of special economic zones, legislation in this area is still in the process of change.

The purpose of this investigation is to overview the legislation of the Russian Federation on the establishment, management, and legal regime of special economic zones.

3 Methods

Our leading methods come from qualitative approach. A qualitative attitude is based on theoretical analysis of the conceptual questions. We examined the substantial points of SEZ legislation in Russia, compared them with the ones in the Republic of the Philippines legislation on SEZ. We also looked into the realization of SEZ legislation on the example of the two zones of the same type in two different regions of Russia.

4 Results

The implementation and operation of the SEZ in Russia began in 2005 with the adoption of the Federal Law "About special economic zones" 22.07.2005 [6]. According to this law four types of SEZ are identified – industrial production, technical innovation, tourism and recreation and ports. The area of each type is subject to strict requirements. For example, on the territory of SEZ, with the exception of tourist and recreational, accommodation of housing stock is not allowed.

The Resolution of the Government of the Russian Federation "On approval of the criteria for the establishment of a special economic zone", 26.04.2012 [7] eliminated the ambiguity of the criteria for the establishment of a special economic zone, which was present in the core law. Now these criteria include natural economic and geographical competitive advantages; investment projects and potential investors who have confirmed in writing their readiness to implement these projects in the territory of the proposed special economic zone; and at least one economic company intending to perform the functions of a managing company. The aim of returning budgetary investments in the creation of SEZs is stressed. In particular, planned investment projects should correspond in volume to public funds invested in infrastructure creation, and the managing company undertakes to attract at least 50 percent of the planned volume of extra-budgetary investments in the creation and operation of the special economic zone.

The experience of the Philippines in organizing Special Economic Zones is worth to be mentioned. Criteria for establishing a new territory under the Special Economic Zone Act of the Republic of the Philippines (Republic Act No. 7916 25.07.1994), [8] are economic advantages over existing zones, high level of industrial development and strategic position. In the absence of these conditions, the zone can be established, however, financial infusions into infrastructure and other types of material support are placed on local budgets or private investments, without the involvement of the central government. In addition, such a zone should, in five years, demonstrate "substantial development".

In the latest version of the Law on special economic zones, the direct management of the SEZ is carried out by the joint-stock company "JSC" SEZ", the only shareholder of which is the state. Part of the authority to manage the SEZ may be transferred to the territorial subject of the federation. In such a case, the executive authority of the subject of the Russian Federation shall disclose information on its activities and on the special economic zone under its management in the information and telecommunication network "Internet".

Significant legal difficulties arise in the case of the closure of the SEZ. Each zone is established for forty-nine years and agreements with residents can be concluded for a significant period of time [9].

The legislation provides for three reasons for the closure of the SEZ ahead of schedule - the need to protect the life and health of people, the protection of nature and cultural property, the safety of the country and the security of the State; the absence of residents for three years or the failure to implement agreements in accordance with the specialization of the zone. The financing for the creation of facilities for engineering, transport, social, innovation and other infrastructures of the special economic zone is provided from the federal budget, the budgets of the subjects of the Russian Federation and the local budgets [10]. Considering that the period of three years provided for establishing SEZ by the law looks unduly long. It can be considered as the upper limit of the period, the lower limit can be set in a year. Despite the general legislative framework, different sets of investor benefits may be provided in different economic zones. For example, the benefits granted to residents of the industrial production SEZ "Tolyatti" (Samara region) and industrial production SEZ "Alabuga" (Republic of Tatarstan) differ significantly [11]. Compare them in the following table:

Table 1. Benefits and preferences for residents of "Tolyatti" SEZ and "Alabuga" SEZ

Benefits and preferences	"Tolyatti" SEZ	"Alabuga" SEZ
Income tax	2% until 2018 5% from 2019 to 2020. 9% from 2021 to 2022. 12% from 2023 to 2024. 15.5% from 2025 onwards	2% first 5 years 7% next 5 years 15.5% next 10 years
The property tax	0% for 10 years	0% for 10 years
Transport tax	0% for 5 years	0% на 10 years
Possibility of the right to buy a land plot	138,415 RUB per 1 ha.	100 thousand rubles for 1 ha.
Down Factors	No	With the capital investments of at least EUR 10 million - 0.9; With capital investments of at least EUR 20 million - 0.7; With capital investments of at least EUR 30 million - EUR 0.5.

Source: Composed by the authors based on data analysis of SEZ Tolyatti and SEZ Alabuga [12].

The right to purchase land within the special economic zone was introduced by law only in 2015. Despite this innovation, the specifics of land use in SEZ have remained. In particular, the tenant of the land plot and the resident of the special economic zone do not have the right to sub-lease, which is a feature of land use in the territory and does not apply to other categories of land.

In addition, a land plot located within the boundaries of the SEZ may be seized for state or municipal needs in accordance with the procedure established by law, if it is intended to house a linear object constituting the infrastructure of the SEZ [13]. The exception is to place a linear object under easement conditions.

The territories of priority socio-economic development (TPSED) have begun to be a significant competition of SEZ as an institution for attracting foreign investment since 2014 [14]. In the period from 2016 to 2020 the number of TPSED will increase to 45. To ensure their effective functioning in Russia adopted the Federal Law of 29.12.2014 "About territories of advancing socio-economic development in the Russian Federation" [15].

TPSED unlike SEZs, are established on the basis of agreements between specific large investors and an authorized federal public authority, the elaboration of the main direction of economic activity, the determination of the volume of investments and the number of jobs created.

The life of the territory of priority socio-economic development is defined as 70 years, but can be extended. In the case of SEZs, there is no extension of the life of SEZs. A distinctive feature of the TPSED is that the management of any subject of the Russian Federation, unlike special economic zones, has the right to apply for its establishment. In general, the procedure for establishing and managing the TPSED is much easier to create and manage SEZ.

5 Conclusion

There is already a tendency to reorient regional authorities from the establishment of special economic zones in the constituent entities of the Russian Federation to the registration of territories ahead of development. Meanwhile, competition between similar legal institutions can significantly affect the economic sustainability of those real industrial and production complexes that have been established in Russia since 2005. Perhaps this prospect was anticipated by the legislator, limiting the possibility of creating a TPSED in those regions where the Special Economic Zone is already functioning. Nevertheless, the experience of legislation on the establishment, management and development of special economic zones should be taken into account in the implementation of such projects in the future.

References

1. Th. Farole, G. Akinci, Special economic zones. Progress, emerging challenges, and future directions (New York, Washington DC, 2011). URL: https://openknowledge.worldbank.org/bitstream/handle/10986/2341/638440PUB0Exto00Box0361527B0PUBLIC0.pdf. Accessed: 11.08.2019.
2. J. Wang, The economic impact of special economic zones: Evidence from Chinese municipalities. *Journal of Development Economics*, **101**, 133-147 (2013). DOI: https://doi.org/10.1016/j.jdeveco.2012.10.009.
3. I. Ablaev, Innovation clusters in the Russian economy: Economic essence, concepts, approaches. *Procedia Economics and Finance*, **24**, 3-12 (2015). DOI: 10.1016/S2212-5671(15)00605-X.
4. M.E. Suvorova, Creation and management of special economic zones in the regions of Russia. *Strategy for Sustainable Development of Regions of Russia*, **13**, 28-35 (2013). [in Rus.].
5. S. Sosnovskikh, Industrial clusters in Russia: The development of special economic zones and industrial parks. *Russian Journal of Economics*, **3**(2), 174-199 (2017). DOI: https://doi.org/10.1016/j.ruje.2017.06.004.
6. The Federal Law of 22.07.2005 N 116-FZ (an edition of 03.07.2016)"About special economic zones in the Russian Federation" (2005). URL: http://www.consultant.ru/document/cons_doc_LAW_54599/. Accessed: 11.08.2019.
7. Resolution of the Government of the Russian Federation No. 398 of 26 April 2012 "On Approval of Criteria for the Creation of a Special Economic Zone" (as amended and supplemented on 27 March 2013) 92012). URL: https://base.garant.ru/70168878/. Accessed: 19.08.2019.
8. Special Economic Zone Act. REPUBLIC ACT NO. 7916 25.07.1994, Republic of the Philippines (1994). URL: http://www.peza.gov.ph/index.php/about-peza/special-economic-zone-act. Accessed: 11.08.2019.
9. A.V. Pushkin, *Legal regime of foreign investments in the Russian Federation* (Moscow, Alpine Books, 2012). [in Rus.].
10. S.N. Shevtsova, Specifics of management of special economic zones in Russia at the present stage. *Journal of the Belgorod University of Cooperation, Economics and Law*, **4**, 368-373 (2012). [in Rus.].
11. E.N. Zubkova, E.Yu. Dremova, Special economic zone of industrial production type (SEZ PPT) "Tolyatti" as an instrument of regional economic development. *News of the Institute of Management Systems of the SSUE*, **2**(3), 141-144 (2011). [in Rus.]
12. JSC "SEZ PPT" Togliatti, *Data of SEZ Tolyatti* (2019). URL: http://oeztlt.ru/ru/project/. Accessed: 11.08.2019.
13. A.V. Kuchinskaya, S.A. Lipski, Conditions for the creation of special economic zones and the definition of the territory adjacent to special economic zones in the Russian Federation. *Property Relations in the Russian Federation*, **3**(150), 34-38 (2014). [in Rus.].

14. S. Beliakov, A. Kapustkina, Analysis of performance indicators of functioning of territories with special economic status in the Russian Federation. *Procedia Engineering*, **165**, 1424-1429 (2016). DOI: 10.1016/j.proeng.2016.11.874.
15. The Federal Law of 29.12.2014 N 473-FZ "About territories of advancing socio-economic development in the Russian Federation" (2014). URL: http://www.consultant.ru/document/cons_doc_LAW_172962/. Accessed: 17.08.2019.

Method for Monitoring of Managing the Modernization of Regional Economy

T. Krasnova[1], *T.* Plotnikova[2], *A.* Pozdnyakov[1], and *A.* Vilgelm[1,*]

[*] Corresponding author: asvilhelm@gmail.com.

[1]Khakas State University, Abakan, Russia
[2]Siberian Federal University, Khakas Technical Institute, Abakan, Russia

Abstract. This paper proposes a new approach for monitoring of managing the modernisation of regional economic. The model built on proposed methodology will make it possible to smooth out the influence of non-urban areas on the unevenness of economic activity in spatial development. This paper has two goals. The first is to provide a new compilation of data on spatial distribution of economic activity at the sub-regional level. This data set allows us to monitoring of different indicators within macroregions such as Siberia. The second goal is to construct an instrument that helps to overcome the endogeneity problem using new economic geography hypothesis about the mechanisms of distribution of economic activity. Section 2 describes the data and method that we have proposed, discusses the construction of the Theil indexes using these data at the sub-federal and the sub-regional level. Section 3 presents the correlations between spatial distribution of economic activity and local market potential, discusses the robustness of the results; and the last section concludes.

Keywords: economic growth, regional economics, agricultural industry, local markets.

1 Introduction

Spatial inequality is a deterrent of socio-economic development regions of Russia. This is especially true of Siberian regions, most of which have resource dependence in economic development. Investment attractiveness is characterized by such similar economic, geographical and other signs as:

- the spatial location relative to other federal districts and proximity to Asian markets;
- low population density and huge distances between cities (The Siberian Federal District occupies 30 % of the territory of the Russian Federation, while the population does not exceed 13 %);
- low efficiency of agriculture in the field of crop production (risk farming zone), extreme climate, remoteness from economic centres, underdeveloped transport, engineering and social infrastructure;
- wealth of natural resources and untapped mineral reserves.

A number of existing studies on the framework of New economic geography (NEG) consider factors affecting the spatial distribution of resources in emerging local markets. For example, the empirical work of Ezcurra et al. [1], Torres-Gutierrez et al. [2] and Abreu et al. [3] substantiates the influence of trade spending, population density, trucking costs, firm size on spatial lag and firm's productivity in developing country's manufacturing sector.

The modernization strategies of Siberian regions provide for the development of competition, the provision of equal opportunities for the modernization of enterprises, the creation of conditions for attracting and developing human resources, stimulating the development of innovative technologies, the development of cooperation of all types (intra-industry, inter-industry and international). Taking into account the resource opportunities, the development drivers are the mining industry and the agro-industrial complex, since with the share of these sectors in the GDP of Siberia 21 %, the share of investments in the relevant sectors amounted to more than 33 %. Consequently, the study of local markets using output these industries as an example will illustrate not only resource potentials, but also trade and infrastructure potentials.

Enterprises with constant returns to scale and using immobile resources, such as agriculture, may remain concentrated in specific regions [4-9]. To create conditions for the modernization of regional production sectors, the impact of development institutions with business entities on the basis of continuous monitoring indicators is required. Existing methods for estimating and forecasting the potential of local sectoral markets are used in scenario strategic planning, but not in the framework of monitoring.

2 Materials and Methods

The paper proposes a new conceptual basis for monitoring, based on modeling the development of regional processes under the influence of factors studied by the NEG. In this regard, the development of tools for factor analysis of modernization processes in inclusive economics becomes more important. Our hypothesis is based on the assertion about the relationship between national development and spatial inequality, proved by Williamson [10]. Proposed model includes shells to perform monitoring of the social and economic status of spatial and sectoral clusters on the framework of program-targeted management of socio-economic development. The hierarchical structure of the model consists in a gradual transition of the calculation of the Theil index from a municipality to a regional level which can be calculated as follows:

$$T^i = \sum_{i=1}^{N} \frac{x_i}{x} \ln \frac{x_i}{x/N} \quad (1)$$

where i indexes regions, i ranges between 1 and N; x_i is the value of the variable in the region i.

A contribution of the present paper is its empirical approach to estimating local markets distance effects in spatial analyses by carefully mapping the spatial returns to scale function. The main specification is the following:

$$\ln gdp_i^j = a_0 + a_1 \ln pod_i + a_2 \ln pol_i + a_3 \ln trd + a_4 \ln ttd + \\ + a_5 \ln trc + a_6 \ln fins_i + a_7 sbb + a_8 tsb + a_9 \mathbf{x} + \varepsilon \quad (2)$$

where gdp_i is a output of sector; pod_i is a population density; pol_i is a measure of predicted local potential; trd_i is a retail and wholesale trade; ttd_i is a exports and imports; trc_i is a trucking; $fins_i$ is a firm size; sbb_i is a spatial boundary between regions; tsb_i is a spatial transboundary; $\mathbf{x}$ stands for a vector of variables that control for additional predictors that are assumed to have an influence on regional disparities; and ε is a heteroscedastic error. In this study local market potential as relative notions of space put forward by Garretsen et al. [11]. In this case, due to the difference in potentials during the interaction between the settlements due to the exchange of resources, the effect of agglomeration is achieved:

$$pol = \left(\sum pol_q\right)\left(1 + \frac{\sum pol_k}{100}\right) \quad (3)$$

where pol_q is quantitative indicators of state statistical monitoring: labour, output, resource and investment local potential; pol_k is qualitative indicators of state statistical monitoring: infrastructure, innovation and environmental local potential, $pol_k = \sqrt[K]{\prod_{k=1}^{K} pol_k}$ and K is amount of indicators of state statistical monitoring of, e.g., dynamic indices of infrastructure development. The important features of the above model are presented below.

3 Results and Discussion

Preliminary results obtained by applying the conventional monitoring model show that the existence of highly spatial inequality in potential development of the markets considerable distortion of monitoring results is observed that undermines opportunities when choosing management decisions. Our proposed specification on the field Siberian spatial inequality eliminated this lack of monitoring and enhances the effect of choosing a model for modernization of the regional economy, taking into account spatial and sectoral parameters. A modified monitoring model allows you to adapt the monitoring object to the current analysis, which begins by calculating the Theil's measure of inequality for Siberian regions.

As shown in Table 1, the degree of spatial inequality varies considerably across the sample: from the highest values of the index in urban areas to the lowest values in non-urban areas, and over the three-year period, indicators increased in both groups.

Table 1. Theil index for employment.

Indicator	Urban area			Non-Urban Area		
	2016	2017	2018	2016	2017	2018
Mining industry	0.195	0,196	0,196	0,072	0,072	0,071
Agriculture industry	0,083	0,083	0,084	0,074	0,075	0,077

Source: Authors.

The Theil index allows considering two hierarchical levels is Siberian macroregion and municipalities. A study of the Theil index at the regional level allows us to identify areas of convergence of economic activity. For a more detailed study, it is necessary to consider the municipal level.

Using obtained data, it is possible to compute spatial distribution of economic activity of different municipalities within regions. To calculate the municipal-level Theil index, the sample size by the number of municipalities in the Siberian macroregion amounted to 3977 units.

For each municipality j of region i in equation (1) need to calculate the Theil index (T_{ji}) based on the dimensions of diversity identified in this study. The value of the variable in the region x_i from the equation (1) was replaced by x_{ij} where i indexes regions (as above); j indexes municipalities; and x_{ij} is the value of the variable in municipality j of region i.

The obtained regression results are presented in Table 2 for $\ln gdp$ as dependent variable and Theil index as the measure of spatial distribution.

Table 2. The impact of regressors on spatial distribution of economic activity: the second stage of the 2SLS regressions*.

Regressors	(2.1)	(2.2)	(2.3)	(2.4)	(2.5)	(2.6)
Population density	-0.568* [0.411]	0.700* [2.491]	-0.359* [0.756]	0.039* [0.239]	1.503* [2.541]	0.056* [0.204]
Predicted local potential	-0.297* [0.456]	1.221* [2.759]	-0.533* [0.838]	0.094* [0.265]	1.968* [2.814]	0.007* [0.226]
Trade (retail and wholesale trade)	0.414* [0,709]	-4.735* [4.296]	0.661* [1.304]	-0.001* [0.411]	-4.168* [4.381]	0.219* [0.353]
Total trade (exports and imports)	-0.175* [0,437]	2.666* [2.648]	0.077* [0,804]	0.110* [0.254]	2.703* [2.701]	-0.071* [0.217]
Trucking	0.425* [0.372]	1.749* [2.254]	0.531* [0.684]	-0.110* [0.216]	1.485* [2.299]	-0.052* [0.185]
Firm size	0.010* [1.023]	-4.481* [6.201]	0.069* [1.883]	-0.288* [0.595]	-3.869* [6.324]	0.191* [0.509]
Spatial boundary between regions	-0.950* [1.128]	-0.346* [6.831]	0.041* [2.074]	0.194* [0.655]	-0.079* [6.966]	0.191* [0.561]
Spatial transboundary	-0.090* [0.463]	0.504* [2.804]	0.471* [0.851]	-0.121* [0.269]	0.396* [2.861]	-0.097* [0.230]
F-stat (het)	3.731**	1.669*	3.801*	2.523*	8.689***	1.639*
F-stat (hom)	1.935*	1.470*	3.672**	1.686*	3.124**	1.484*
Constant	7.540* [7.770]	51.362* [47.059]	-2.043* [14.287]	0.838* [4.512]	35.526* [47.994]	-2.516* [3.861]
Adjusted R-squared	0.304	0.084	0.667	0.261	0.567	0.225
Control variables	yes	yes	yes	yes	yes	yes
Sample	all regions	all regions	all regions	all regions	all regions	all regions
Regions	12	12	12	12	12	12
Observations	3977	3977	3977	3977	3977	3977

Source: Authors.

Note: * The dependent variable is in (2.4), (2.5), (2.6) Theil index measure, (2.5) is expressed in natural logs. Robust standard errors adjusted for heteroscedasticity in brackets. *, **, and *** indicate significance at the 0.1, 0.05, and 0.01 levels, respectively. „F-stat (het)" reports F-statistics for the excluded instrument from the first stage under the assumption of heteroscedasticity; and „F-stat (hom)" – under the assumption of homoscedasticity.

First, consider regressions in which the left-hand side includes only ln *gdp* (the results are presented in columns (2.1), (2.2) and (2.3), where dependent variable denote, respectively, the natural log of GDP per capita of region, mining sector and agriculture sector). Population density, predicted local potential, trade, trucking and firm size (i.e., number of employees in thousands) are expressed in natural logs. Spatial transboundary and boundary between regions are two dummy variables and their mean is the percentage of observations that take a value of 1. The sample period is spaned from 2005 to 2018 (source: Federal State Statistics Service of Russia Data). The first stage yields the following results. Population density and predicted local potential has a negative significant effect on labour productivity in region (especially for agriculture sector).

Columns (2.4), (2.5) and (2.6) replace the dependent variable for study of influence the same regressors; in order to exclude endogeneity, for second stage of the 2SLS regressions, testing tools were introduced as a dynamic index for the same indicators in retrospect: the international trade showed a positive correlation, bigger addiction, compared with domestic trade, that means influence interaction of firms in local markets than a population distribution, what means confirmation of the NEG hypothesis about the role of local markets in the spatial development within regions. Firm size and retail and wholesale trade has a negative significant effect on spatial distribution of output in mining and positive of agriculture local markets. Most of these control variables have significant coefficients. The only notable difference between the results with and without these control variables is in the F-statistics for the excluded instrument from the first stage, they do drop once each of these control variables is included. Nonetheless, for the measure of spatial distribution, they still remain sufficiently strong not to worry about weakness of the instrument.

The results presented in Table 2 suggest that firm size and domestic trade in mining industry exerts a negative impact on labour productivity but its impact in agriculture firm is positive. This appears to signal the influence of economies of scale in Siberian agriculture sector, but the absent for mining sector. We find that smaller firms are may ousting larger firms from the market. In the same time, the proximity of local markets to the cross-border regions and high trucking indicates the orientation of mining to export local markets (adjusted $R^2 = 0.567$). The results the second stage of the 2SLS regressions suggest that predicted local potential exerts a positive impact on spatial distribution of economic activity, stronger in mining. In addition, we estimated all our instrumental variable regressions with generalized method of moments instead of 2SLS and got very similar results with stronger statistical significance.

Study results indicate that the spatial impact of local market factors is most pronounced in the regional output sectors and less pronounced in the regional and clustered level. In any case, when interpreting the information provided by Table 2, it is important to recall that this analysis allows us to examine the effect of local markets on possible modernization of regional economics. Received results are compatible with a potential positive impact of sectorial modernization processes on regional economic performance in absolute terms.

4 Conclusions

The monitoring system for the modernization of economic sectors is formed from indicators with a feedback function at various levels (regional, interregional, country, intercountry). Based on the monitoring results are factors of influence on the development of socio-economic systems which can be used to predict future development. Moreover, the results of intracluster and intra-agglomeration relations must be taken into account when developing roadmaps.

An example of a project that contributes to modernizing the economy of the Siberian region is Yenisei Siberia. Based on investment capital investments in the real sector of the economy, this project provides the spatial development of regions and agglomerations through modernization and creation of new jobs. It is planned to develop such basic industries for Siberia as mining, construction, energy and agriculture. In this case, these regions are the subject of economic development management through infrastructure projects. Thus, in order to ensure uniform development of the regions, modern economic instruments for managing macro-regions are being modernized.

Taking into account the development trends in agricultural and industrial zones development, it is possible to build a strategic plan for socio-economic regional development based on modern economic monitoring tools of managing macroregions. The above methodology, tested in Russian conditions, is suitable for use in the monitoring system both on a regional scale and on municipal level to identify differences and compare industries growth dynamics and the development of regions, since the level of uneven spatial activity of market entities between regions is greater than within one local market. The use of hypotheses of New economic geography as a theoretical basis in the proposed monitoring system allows to take into account infrastructural and resource potentials for increasing the functionality and predictability of monitoring results which is important for investors and government in management.

Acknowledgements

Authors are grateful to referees and researchers of Samara State University of Economics for helpful comments.

References

1. R. Ezcurra, A. Rodríguez-Pose, Trade openness and spatial inequality in emerging countries. *Spatial Economic Analysis*, **9**(2), 162-182 (2014).
2. T.P. Torres-Gutierrez, J.A. Ordonez, Agglomeration economies and urban productivity. *Region*, **6**(1), 17-24 (2019).
3. M. Abreu, H.L.F. de Groot, R.J.G.M. Florax, Space and growth: A survey of empirical evidence and methods. *Tinbergen Institute Working Paper,* TI 04-129/3 (2004).
4. A. Fracasso, G. Vittucci Marzetti, Estimating dynamic localization economies: The inadvertent success of the specialization index and the location quotient. *Regional Studies*, **52**(1), 119-132 (2018).
5. N. Foo, H. Bloch, R. Salim, The optimisation rule for investment in mining projects. *Resources Policy*, **55**, 123-132 (2018).
6. M. Fujita, P. Krugman, Urban systems and regional development. In M. Chatterji (Ed.) *Regional Science: Perspectives for the Future* (pp. 7-25). London: Palgrave Macmillan (1997).
7. M. Fujita, P. Krugman, A.J. Venables, *The spatial economy* (MIT Press, Cambridge, MA, 1999).
8. P. Krugman, What's new about the new economic geography? *Oxford Review of Economic Policy*, **14**(2), 7-17 (1998).
9. P.P. Combes, T. Mayer, J.F. Thisse, *Economic geography: The integration of regions and nations* (Princeton Univ. Press, Princeton, NJ, 2008).
10. J.G. Williamson, Regional inequality and the process of national development: A description of patterns. *Economic Development and Cultural Change*, **13**, 3-47 (1965).
11. H. Garretsen, R. Martin, Rethinking (new) economic geography models: Taking geography and history more seriously. *Spatial Economic Analysis*, **5**(2), 127-160 (2010).

Ensuring Trading Strategies Profitability with Technical Analysis Tools and Computer Technology

*G.*Abuselidze[1,*], *A.*Slobodianyk[2], and *N.*Reznik[2]

* Corresponding author: abuseri@mail.ru.

[1]Batumi Shota Rustaveli State University, Batumi, Georgia
[2]National University of Life and Environmental Science of Ukraine, Kiev, Ukraine

Abstract. The paper is devoted to the study of the tools of technical analysis using computer technology. It analyzes the current conditions and the essence of the stock market, the basis of its functioning. The advantages of technical analysis are presented including accessibility and relative simplicity of its use. It analyzes the main indicators of the world stock market development and modern constant signs and tendencies of the process. It is revealed that the improvement of the technical analysis and increase of the forecast accuracy is a complex process. The prospects for the effective use of technical analysis in trading strategies for promoting analysis by using historical data of price changes for the asset under study by using computer software are determined. Back testing calculations are presented making clear the fact that the financial result would be given by a trading strategy based on the historical profitability of the asset.

Keywords: technical analysis, fundamental analysis, stock market, trading strategies, average flow method, back testing.

1 Introduction

The transition to the information society sets a number of fundamentally new challenges for the economic science. One of these tasks is to provide the participants of exchange activities with high-quality and timely information, because information in modern conditions is one of the key resources.

A unique source of this information is exchange markets, which, due to the asset prices, signal on the condition and prospects of the global economic system development. In order to ensure the freedom of the activity of any market, it is necessary to analyze it. The precise understanding of the market situation, the ability to forecast market behavior, the analysis of its mood leads to an informed decision-making.

The only question concerning the participants of the exchange activity is how to make a high-quality forecast and achieve a competitive advantage over others. The most common approaches in the environment of exchange practice are considered to be technical and fundamental analysis.

2 Methods

The problems of the world market and the implementation of technical and fundamental analysis were studied in the works of the following scientists: G. Abuselidze, A. Slobodianyk [1, 2], S.V. Arzhenovsky, I.N. Molchanov [3], E.H. Nyman [4] and A.N. Sokhatskaya [5]. The reason for such attention on the part of scientists is the extreme problem actuality as well as the rapid development of monetary relations requiring the detailed study of the market. According to V.V. Acharya and T.C. Johnson [6], asymmetric information and insider trading problems potentially exist in most markets. But the credit derivatives market may be especially vulnerable since, almost by definition, most of the major players are insiders. Insider trading has been the focus of a large body of research in equity markets which has found that insider trading lowers liquidity and increases trading costs (see, e.g., D. Easley, N.M. Kiefer, M. O'Hara, and J.B. Paperman [7], J.C. Bettis, J.L. Coles, and M.L. Lemmon [8], P. Brockman and D.Y. Chung [9], R.P.H. Fishe and M.A. Robe [10], raises the cost of equity capital [11], and increases volatility [12]). Theoretical and methodological bases of the research are scientific works of domestic and foreign scientists in the field of exchange trade, statistics, mathematical modeling. The following methods are used in the paper: descriptive and comparative analysis, synthesis, scientific abstraction, economic-statistical, fundamental, technical analysis.

3 Discussion and Results

The study of the dynamics of the world stock markets associated with the processing of the large amount information based on various data having a rather complex internal structure. The process of improving technical analysis and increasing the accuracy of the forecasts is a complex process and the result can fluctuate only within a limited framework of the technical analysis. Firstly, there is a large number of false signals generated by tools of technical analysis. The number of false signals can be reduced within the framework of the technical analysis methodology, but it is impossible to avoid them completely [1]. Secondly, the methods of technical analysis do not allow to create high-quality long term forecasts, the time horizon of technical analysis is short and medium-term forecasts.

The main task of monitoring the stock market situation is to determine the set of circumstances determining the possibility of a rise or fall in market activity. It is reduced to the formulation and solution of the problem of forecasting the dynamics of asset prices.

Research of the stock market dynamics associated with the processing of large information arrays of various data with a fairly complex internal structure. Moreover, the analysis of the information is based on the implementation of a large number of routine calculations, graphical constructions, providing for the assumption of different levels of the errors and taking a long time in their processing. It is possible to overcome these difficulties by means of computer technology using specialized services and programs.

Each analyst applies technical analysis, forming own decision-making methodology and creating own method of technical analysis, absorbing individual elements of the well-known foundations of technical analysis. It is considered that if the created model allows you to get the correct forecast in 60% of cases, then this research technique is effective and is characterized by high accuracy.

The effectiveness of using technical analysis in trading strategies can be analyzed by using historical data on price changes for the studied asset by using MS Excel software or specialized services. The implementation of such an analysis is called back testing. Back testing makes it clear that a trading strategy based on the historical profitability of an asset would give a financial result. In back testing we will analyze two strategies: a passive investment strategy Buy & Hold and an active investment strategy using two averages with periods of 50 and 200 days. The Buy & Hold strategy ("Buy and Hold") provides for the purchase of an asset with the purpose of its long-term maintenance with the expectation of a price growth. The strategy is based on the idea that the global long-term trend is always ascending.

As a rule, when using this strategy, it is not planned to fix losses in case the market moves down. In this case the expected return on capital is usually ten or even hundred percent. The basis of an average flow strategy is to receive a buy signal in case the price of the asset starts to rise and a sell signal when the price starts to decline. In long-term investment strategies, average sized flows with a period of 50 and 200 days are most often used for this purpose.

Taking into consideration the fluctuations in exchange rates that were in this period, the yield of 13.49% is leveled by a decrease in the purchasing power of the national currency in relation to the US dollar. At the beginning of 2007, the USD/UAH exchange rate was 5.05, at the end of October 2018, the USD/UAH exchange rate was 28.05 [13] and the USD/GEL exchange rate was 1.65, at the end of October 2018, the USD/GEL exchange rate was 2.80 [14]. Taking into consideration the precarious political and economic stability in Ukraine or Georgia and the lack of a stable exchange rate, long-term investment in the Georgian and Ukrainian stock markets is not attractive enough for foreign investors.

The main objective of an active investment is to reduce the risks associated with the fall in the value of the asset and obtain a higher return in the long term compared to the average return of the stock market. The use of technical analysis, namely secondary flow with a period of 50 and 200 days allows you to avoid losses arising from the use of Buy & Hold strategies, as with medium flowing there is a possibility to receive a signal about a reduced value of the asset and to limit losses by selling the asset before its price will continue to decline. Fig. 1 shows the price chart of SPY with the use of average flows.

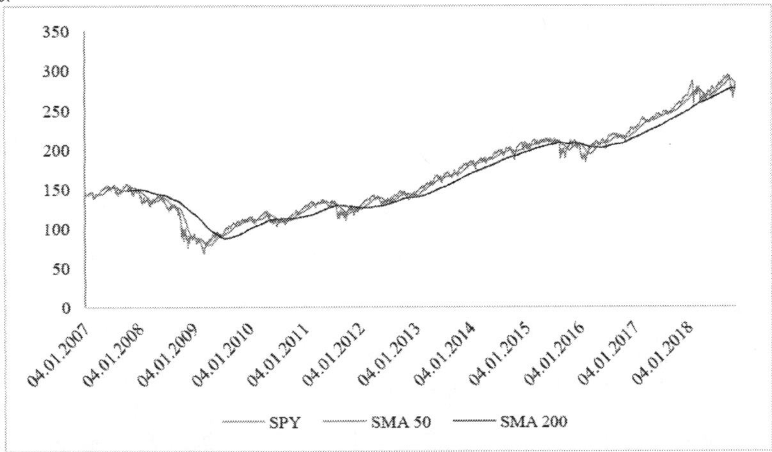

Fig. 1. ETF SPY Chart period from 2007 to 2018 using the average flow.
Source: Authors based on historical data [15].

According to Fig. 1 it can be seen that in the period from 2007 to 2018 5 signals were received to buy an asset and 5 signals to close a position. For a more detailed analysis of the strategy using flow averages, consider the chart of its profitability compared to the Buy & Hold strategy in Fig. 2.

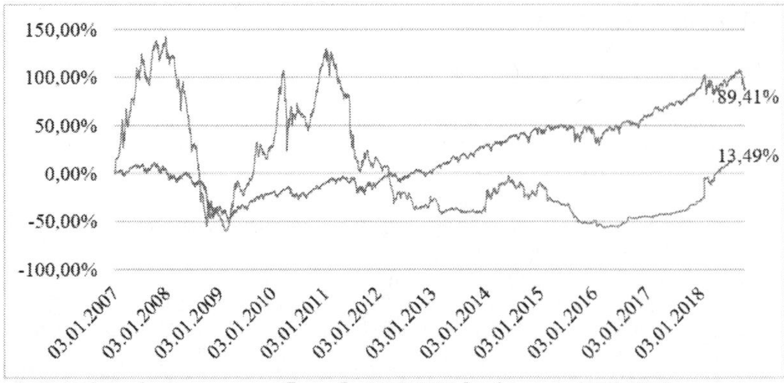

Fig. 2. Yield chart strategies using two average flows for ETF SPY for the years 2007-2018.
Source: Authors based on historical data [15].

Having considered Fig. 2 it can be seen that the use of medium flow reduces the risk of investments. So, compared to the Buy & Hold strategy, the maximum decrease in asset value was only - 8.62%, and the final profit due to the use of technical analysis was 101.53%. Compared to the Buy & Hold strategy, this is 12.12% higher.

Thus, the use of technical analysis in the future allows you to increase the profitability of trading strategies meanwhile reducing market risks associated with a decrease in the value of the asset. Fundamental and technical analyses have their advantages and disadvantages. Therefore, in order to conduct analysis of the economy of the country as a whole or of an individual company, it is necessary to carry out both fundamental and technical analysis. This will provide the greatest economic effect in the implementation of exchange operations.

4 Conclusions

The modern world stock market today is in a state of continuous transformation of the organizational structure of exchanges, exchange trading technology, exchange instruments, types of basic assets. From year to year, there is a tendency to increase trading volumes on stock exchange markets, especially Asian exchange markets, occupying the largest share in the world stock exchange trading. An objective implementation of fundamental analysis implies a transparent market economy, perfect legislation, political and economic stability of the country. When trading in any market, including the stock market, systematicity is important. For maximum profitability and stability, it is necessary to use a combination of technical and fundamental analysis. In order to improve technical analysis and increase the accuracy of forecasts, it is necessary to go through a complex process and the result can fluctuate only in a limited framework of the possibilities of technical analysis. The study of the dynamics of world markets is associated with the processing of large information arrays of various data. These difficulties can be overcome through the use of computer technology.

References

1. G. Abuselidze, A. Slobodianyk, Investment of the financial instruments and their influence on the exchange stock market development. In A. Auzina (Ed.), *Proceedings of the International Scientific Conference Economic Science for Rural Development*, **52**, 203-210 (2019).
2. G. Abuselidze, Georgia's capital market: Functioning problems and development directions in association with European Union. *Journal of Applied Economic Sciences*, **13**(7), 1929-1938 (2018).
3. S.V. Arzhenovsky, I.N. Molchanov, *Statistical forecasting methods* (RSEU, Rostov-on-Don, 2001). [in Rus.].
4. E.H. Nyman, *Small encyclopedia of trader* (Kiev Logos, Kiev, 1971).
5. A.N. Sokhatskaya, *Exchange business* (carte Blanche, Moscow, 2009).
6. V.V. Acharya, T.C. Johnson, Insider trading in credit derivatives. *Journal of Financial Economics*, **84**(1), 110-141 (2007).
7. D. Easley, N.M. Kiefer, M. O'Hara, J.B. Paperman, Liquidity, information, and infrequently traded stocks. *The Journal of Finance*, **51**(4), 1405-1436 (1996). DOI: 10.2307/2329399.
8. J.C. Bettis, J.L. Coles, M.L. Lemmon, Corporate policies restricting trading by insiders. *Journal of Financial Economics*, **57**(2), 191-220 (2000). DOI: 10.1016/S0304-405X(00)00055-6.
9. P. Brockman, D.Y. Chung, Investor protection and firm liquidity. *The Journal of Finance*, **58**(2), 921-937 (2003).
10. R.P.H. Fishe, M.A. Robe, The impact of illegal insider trading in dealer and specialist markets: Evidence from a natural experiment. *Journal of Financial Economics*, **71**(3), 461-488 (2004). DOI: 10.1016/S0304-405X(03)00188-0.
11. U. Bhattacharya, H. Daouk, The world price of insider trading. *The Journal of Finance*, **57**(1), 75-108 (2002).

12. J. Du, S.-J. Wei, Does insider trading raise market volatility? *The Economic Journal*, **114**(498), 916-942 (2004). DOI: 10.1111/j.1468-0297.2004.00249.x.
13. National Bank of Ukraine, *Official exchange rate of Hryvnia* (2019). URL: https://bank.gov.ua/control/en/. Accessed: 15.08.2019.
14. National Bank of Georgia, *Official exchange rate of LARI* (2019). URL: https://www.nbg.gov.ge/index.php?m=582&lng=eng. Accessed: 15.08.2019.
15. Nasdaq, *SPY S & P 500 Index* (2019). URL: https://new.nasdaq.com/market-activity/funds-and-etfs/spy. Accessed: 15.08.2019.

Management of Wage-To-Labor Productivity Correlation as a Factor of Samara Region Economic Development

I.V. Bogatyreva[1,*], *N.V.* Kozhukhova[1], and *D.A.* Acopyan[1]
* Corresponding author: scorpiony70@mail.ru.

[1] Samara State University of Economics, Samara, Russia

Abstract. The paper considers the problems in management of the wage-to-labor productivity correlation as the most important element in the strategic development of Samara region economy. The rational correlation between labor productivity and wages growth affects production cost reduction, and increases profitability and production efficiency. This explains the relevance of the study on the wage-to-labor productivity correlation in the digital economy. The authors present the results of the research on the dynamics of labor productivity and wages in Samara region in 2010 – 2017. They calculated the correlation figures of average wages growth in Samara region economy, described the model and developed the mechanism to manage the wage-to-labor productivity correlation.

Keywords: labor productivity, own-product real wage, wages, lead factor of wages.

1 Introduction

The problem of the wage-to-labor productivity correlation is an important economic issue. Change regularities in labor productivity and wages reflect many aspects of industrial relations in any society. The economic meaning of the priority development in productivity, in comparison with wages, on the one hand, reduces the unit costs of wages for production turnout and increases the surplus product share, which is spent on the production expansion and social needs, and, on the other hand, strengthens the material base of wages growth.

The importance of the problem to regulate the wage-to-labor productivity correlation in the digital economy becomes essential for each enterprise, region and country. The possibility of expanded reproduction, balance in the consumer goods market and product coverage of growing cash income depends on the efficient solution of this problem. In this regard, the study is extremely relevant. It is particularly relevant for Russia due to a number of serious problems that the country has faced in recent decades: increasing pressure from foreign competitors, dependence on raw materials exports and world prices on them, low living standards, etc.

2 Problem Statement

For a long period of time, the scientific community and politicians of the leading world countries are interested in the problem of the wage-to-labor productivity correlation [1, 2]. It has acquired particular relevance today in the context of the digital economy development. Economic progress in the modern world is based on labor productivity growth. It is labor productivity, more than any other factor, that determines the level and quality of life in the whole society, and is the best indicator of economic efficiency in the long term. The wage-to-labor productivity correlation reflects the ratio between production, distribution and consumption. Consequently, the value of product, produced by workers, must be greater than the value of products they consume. Without implementation of this principle, it is impossible to constantly expand and improve production. Objective economic laws show that there is a direct correlation between labor productivity and wages growth [3]. According to the results of some research studies, this correlation is bilateral [4].

In the economic and managerial literature, the issue of growth rate of labor productivity in excess of wages is widely discussed. It is impossible not to agree with the statement of some authors that labor productivity growth should be accompanied by a proportional increase in wages, because it affects increase in profits and economic efficiency [5]. However, in practice, the wage-to-labor productivity correlation often weakens due to the impact of external factors on the growth of these indicators [6]. Such factors may include inefficient utilization and distribution of labor and capital [7], form of ownership and size of exports [8]. Internal factors, such as rationalization of working time [9], implementation of digital technologies and efficient management of labor regulation, [10] also influence the optimal ratio of labor productivity and wages growth. Thus, the search for the optimal wage-to-labor productivity correlation is an important economic task.

3 Research Questions

The low level of labor productivity in enterprises affects the overall efficiency of the production system, because labor productivity is a criterion for efficiency of labor utilization in a production activity [11]. For this reason, for a long time, the subject of research of Russian and foreign scientists is productivity growth, payroll management as one of the effective tools to increase labor activity of workers to achieve the necessary level of their labor productivity, and rational wage-to-labor productivity correlation. However, discussions on these issues are still ongoing. Many scientists are interested in the following questions: is there a close wage-to-labor productivity correlation [12], what is the optimal ratio between labor productivity and wages, what model of wage-to-labor productivity correlation can be used for the further development of the Russian economy, what factors influence labor productivity and wages growth [13], and how to effectively manage the growth of labor productivity and wages with their ratio at the level of enterprise, region and country.

4 Purpose of the Study

The objectives of the study, the results of which are reflected in this paper, are to analyze and evaluate the wage-to-labor productivity correlation at the regional level and develop a mechanism to manage it. To achieve the objectives of this research, the following tasks were accomplished: the dynamics of labor productivity and wages in Samara region was analyzed, the correlation level between labor productivity and wages was assessed, the models of correlation between labor productivity and wages growth were described, and the mechanism to manage the wage-to-labor productivity correlation was elaborated.

5 Research Methods

The study is based on systematic and analytical methods of quantitative and qualitative analysis. Monitoring of data collection, study of statistical information on the dynamics of labor productivity and wages in the economy of Samara region allowed the authors to obtain reliable results of the research and formulate reasonable conclusions.

Analysis and evaluation of the wage-to-labor productivity correlation involves the study of the dynamics in the indicators under investigation, which is characterized by the rates of their growth and gain. To study the dynamics of wages and labor productivity in the economy of Samara region for a six-year period, the authors of the paper calculated the index for 2012-2017.

The calculated labor productivity rate is the volume of gross regional product per one employed in the economy of Samara region.

To examine in depth the wage-to-labor productivity correlation, the following indicators were calculated: own-product real wage, lead factor in nominal and real wages and increment rate of average wages for every one percent increase in productivity.

6 Findings

6.1 Labor productivity and wages in Samara region economy

The object of the study is Samara region, which belongs to one of the key industrial regions of the Russian Federation. Opportunities for productivity growth are determined by the developed regional production infrastructure, which is characterized by industry diversity, high level of labor potential, high opportunities for the development of transport communications and many other advantages in the regional economic and social life.

The study of trends in labor productivity and wages growth in Samara region economy over an eight-year period allows to identify the expansion and contraction in the dynamics of these key indicators. Our calculations showed that although during the period under review labor productivity in the region has a positive trend (growth to 23.3%), in recent years there has been a decline in the growth rate of this index. This situation was affected by decrease in the growth rate of labor force in the region and gross value added in 2015-2016. In 2017, the situation changed to the better: productivity growth was 3.3% due to increase in gross product by 0.8% (Table 1).

Table 1. Labor productivity and wages in Samara region economy in 2010-2017, %.

Index	2010	2011	2012	2013	2014	2015	2016	2017	2017 to 2010
Gross regional product	108.5	105.9	105.0	104.2	101.9	97.0	97.8	100.8	122.5
Number of employed persons	100.2	100.5	101.5	100.2	100.8	99.3	99.3	97.6	99.4
Labor productivity	108.3	105.4	103.4	104.0	101.1	97.7	98.5	103.3	123.3
Average monthly nominal wages	110.5	112.9	111.8	112.8	110.3	103.7	105.4	107.8	204.5
Real wages	104.3	104.4	107.5	106.3	102.4	89.9	98.8	103.7	117.4

Source: Authors.

Wages are the main source of income for Russians. In recent years, its share in total cash income has been fluctuated at the level of 65-70%. In Samara region, the population has a slightly different structure of monetary income: wages accounts for about 38.8% of all monetary income.

The analysis of the dynamics in nominal and real wages is of particular interest. The results of our research in Samara region indicate an annual growth of average monthly nominal wages by a mean of 9.4%. The fluctuant trend is present in the dynamics of average real wages: it reached the highest growth in 2012 (7.5%) and the negative growth in 2015 – 2016 (10.1% and 1.2%, respectively). The reason for the heterogeneous dynamics in nominal and real wages lies in the decrease in the purchasing power of nominal wages: the prices of goods and services grow faster than its size.

6.2 Growth rates in labor productivity and wages in Samara region economy

Growth rates in labor productivity and wages require a special attention on the part of employers and the state. According to statistics, over a long period of time, wages growth has significantly surpassed labor productivity growth both in the whole country and in particular regions. Our calculations show that the same situation is observed in Samara region. In 2010-2017 in the region, with labor productivity growth by 23.3%, average gross nominal wages increased by 2 times, and average real wages increased almost by 1.2 times (Fig. 1).

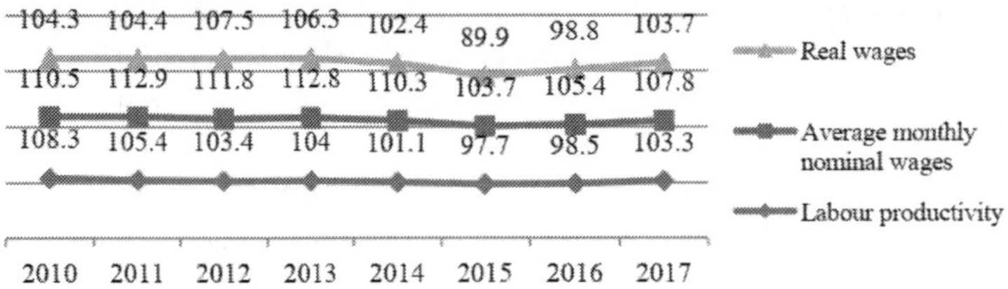

Fig. 1. Dynamics of labor productivity and wages in the Samara region economy in 2010-2017.
Source: Authors

Consequently, the lead factor of nominal wages (60.4%) is firmly low and own-product real wage is high (165.7%) (Table 2).

Table 2. Calculation of index of wage-to-labor productivity in Samara region economy in 2010-2017, %

Index	2010	2011	2012	2013	2014	2015	2016	2017	2017 to 2010
Lead factor:									
- nominal wages	98.0	93.4	92.5	92.2	91.7	94.2	93.5	95.8	60.4
- real wages	103.8	100.9	96.2	97.8	98.7	108.7	99.7	99.6	104.9
One-product real wage:									
- nominal	102.0	107.1	108.1	108.5	109.1	106.1	107.0	104.4	165.7
- real	96.3	99.1	103.9	102.2	101.3	92.0	100.3	100.4	95.1

Source: Authors.

The analysis of index of the wage-to-labor productivity correlation showed that the change in the unit costs of nominal wages per unit of production output, throughout the period under review, remains quite high (from 102% in 2010 to 104.4% in 2017). Consequentially, it affects the cost of production.

The ratio of the growth rate of average wages is characterized by another indicator. It is the increment rate of average wages growth for each percentage of labor productivity growth (Table 3).

Table 3. Calculation of the increment rate of average wages for each percentage of labor productivity growth in Samara region, %

Index	2010	2011	2012	2013	2014	2017
Average nominal wages growth	10.5	12.9	11.8	12.8	10.3	7.8
Average real wages growth	4.3	4.4	7.5	6.3	2.4	3.7
Labor productivity growth	8.3	5.4	3.4	4	1.1	3.3
Increment rate of average wages for each percentage of labor productivity growth	1.26	2.39	3.5	3.2	9.36	2.36
Increment rate of average real wages for each percentage of labor productivity growth	0.5	0.8	2.2	1.58	2.2	1.1

Source: Authors.

The results of the study show that from 2010 to 2014 there was an increase in growth of nominal and real wages for each percentage of labor productivity growth (7.5 and 4.4 times, respectively). In 2017, the situation changed. There was a decrease in growth of both nominal and real wages for each percentage of labor productivity growth. This is due to a decrease in the level of population's income including wages because of the United States and the European Union sanctions. High rates of wages growth have a positive impact on Samara region economy and contribute to improve the living standards. However, in the absence of a corresponding increase in labor productivity, the positive dynamics in wages increase comes to nothing. The increase in one-product real wage leads to increase in prices for goods and services, and the strengthening of inflationary development. This affects the economic and political climate of particular regions and the country [14].

At the legislative level, the government of the Russian Federation take steps to solve the problems to accelerate labor productivity growth in the country. Since 2017 the National Program to improve labor productivity: Priority Program Passport "Improving labor productivity and support for employment" in key regions, including Samara region, has been carrying out.

6.3 Models and mechanism to manage the wage-to-labor productivity correlation

In world practice, three models of correlation between labor productivity and wages growth gained popularity:

- classical model includes providing of direct proportional dependence between wages and labor productivity: an increase in labor productivity by 10% leads to increase in wages by 10%,
- inflation model is based on faster wage growth in comparison with labor productivity growth: wages are growing at a faster rate than labor productivity,
- stimulating model is based on the lag of wages growth behind labor productivity growth.

The classical model of the wage-to-labor productivity correlation is the most suitable for the economic development, but in practice, it is extremely rare. According to the results of our study, the Russian economy is based on the inflationary model of the wage-to-labor productivity correlation. It has negative consequences: lack of correlation between labor costs and payment, discouragement of labor productivity increase, reduction of wages purchasing power, etc.

The authors developed the mechanism to manage the wage-to-labor productivity correlation, which includes 4 stages (Fig. 2).

7 Conclusion

Today, ensuring the optimal wage-to-labor productivity correlation is one of the primary tasks of the Russian economy. The main directions to increase the growth level of labor productivity and achieve the optimal correlation between the labor productivity and wages growth in Samara region are: reducing the cost of financial resources, improving product competitiveness, reducing the credit and tax burden, increasing the innovation level and demand by stimulating consumers' purchasing power, strengthening workers' labor activity, increasing minimum earnings to the level of minimum restorative consumer budget [15], and implementation of scientifically grounded payroll management in enterprises.

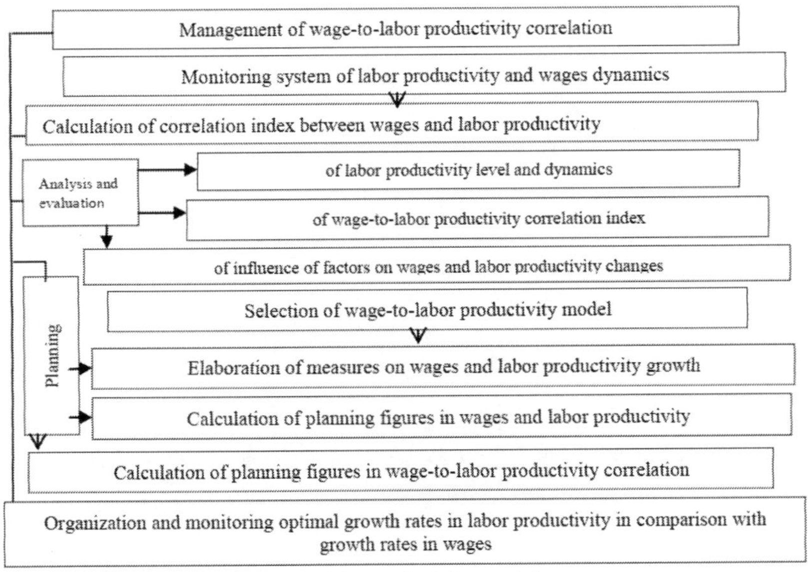

Fig. 2. Mechanism to manage the wage-to-labor productivity correlation.
Source: Authors.

References

1. J. Pehkonen, Wages and productivity growth in the Nordic countries. *European Economic Review*, **39**(6), 1181-1196 (1995). DOI: 10.1016/0014-2921(94)00017-T.
2. H. Bester, E. Petrakis, Wages and productivity growth in a dynamic monopoly. *International Journal of Industrial Organization*, **22**(1), 83-100 (2004). DOI: 10.1016/S0167-7187(03)00093-6.
3. R. Riley, C. Rosazza Bondibene, Raising the standard: Minimum wages and firm productivity. *Labor Economics*, **44**, 27-50 (2017). DOI: 10.1016/j.labeco.2016.11.010.
4. M. Millea, S.M. Fues Jr., Does pay affect productivity or react to it? Examination of U.S. manufacturing. *The Quarterly Review of Economics and Finance*, **45**(4-5), 796-807 (2005). DOI: 10.1016/j.qref.2004.06.004.
5. A.V. Kovalenko, E.S. Kalinskaya, I.V. Geleta, Areas of productivity growth. *Sustainable Development Economics*, **3**(19), 99-104 (2014). [in Rus.].
6. M. Fafchamps, S. El Hamine, Firm productivity, wages, and agglomeration externalities. *Research in Economics*, **71**(2), 291-305 (2017). DOI: 10.1016/j.rie.2016.12.003.
7. K. Nakamura, S. Kaihatsu, T. Yagi, Productivity improvement and economic growth: Lessons from Japan. *Economic Analysis and Policy*, **62**, 57-79 (2019). DOI: 10.1016/j.eap.2018.11.002.
8. J. Zhang, X. Liu, The evolving pattern of the wage – labor productivity nexus in China: Evidence from manufacturing firm-level data. *Economic Systems*, **37**(3), 354-368 (2013). DOI: 10.1016/j.ecosys.2013.05.001.
9. I.V. Bogatyreva, L.A. Ilyukhina, M.V. Simonova, N.V. Kozhukhova, Estimation of the efficiency of working time usage as a factor of sustainable increase of labor productivity. In: V. Mantulenko (Ed.), *Proceedings of the 17th International Scientific Conference "Problems of Enterprise Development: Theory and Practice", SHS Web of Conferences*, **62** (06002). Les Ulis: EDP Sciences (2019).
10. V.A. Schekoldin, I.V. Bogatyreva, L.A. Ilyukhina, V.M. Kornev, The development of IT-technologies in labor standardization and quality assessment of standards: Challenges and ways of solution in Russia. *Helix*, **8**(5), 3615-3628 (2018). DOI: 10.29042/2018-3615-3628.
11. I. Bogatyreva, M. Simonova, E. Privorotskaya, Current state of labor productivity in the economy of developed countries. In A. Zheltenkov (Ed.), *Proceedings of Topical Problems of Architecture, Civil Engineering and Environmental Economics (TPACEE 2018), E3S Web of Conferences*, **91**(08022). Les Ulis: EDP Sciences (2019).
12. A.A. Popkov, About the relationship of labor productivity growth and wage growth. *Modern Science*, **4**, 61-64 (2015). [in Rus.].
13. V.V. Aranzhin, The relationship of wages and productivity: Trends in the conditions of economy digitization. *Russian Journal of Labor Economics*, **6**(1), 523-534 (2019). [in Rus.].
14. I.V. Bogatyreva, N.V. Kozhukhova, E.P. Zheleznikova, Analysis of the current state of labor productivity in Russia. *Russian Journal of Labor Economics*, **5**(3), 683-698 (2018). [in Rus.].
15. V.A. Vaisburd, M.V. Simonova, I.V. Bogatyreva, E.G. Vanina, E.P. Zheleznikova, Productivity of labor and salaries in Russia: Problems and solutions. *International Journal of Economics and Financial Issues*, **6**(S5), 157-165 (2016).

Structure of the Performance Management at the Regional Level

M.V. Simonova [1]*, *N.N.* Gunko[1], *S.A.* Kolesnikov[2], and *A.N.* Parshin [2],
*Corresponding author: m.simonova@mail.ru.

[1] Samara State University of Economics, Samara, Russia
[2] Samara State Technical University, Samara, Russia

Abstract. In the article authors studied a problem of a practical use of scientific research on performance management on the regional level, suggested a mechanism of an institutional approach to solve the problem through the creation of a vertically integrated system with government and professional organizations in the region, which allows organizing cooperation on different management and economic levels. To justify the creation of the system, analysis of labor productivity growth in the regions of the Russian Federation had been carried out. In the system of productive factors social and economic drivers had been allocated, which can lead to dramatic improvements in productivity in the short term, without any use of considerable investment. Their implementation, however, requires well-developed institutional support. Functional and substantial structure of the regional center for competency of productivity had been developed. It is made for the scientific, methodological and organizational support of the project to improve productivity at the regional's workplaces. The need of creation of harmonized methodology to increase productivity, qualification and methodological process support on the principles of project management had been demonstrated.

Keywords: labor productivity, performance management, social and economic factors of productivity improvement, production efficiency.

1 Introduction

Implementation of National projects in Russia sets high measurements of economic growth, not less than 5% of gross domestic product annually, which depends to a large extent on the labor productivity growth rates. The possibility of productivity growth defines by regional specificities of productive infrastructure [1], which differs with industry variety, level of labor potential, development opportunities of transport communications and many others aspects of economic and social life of provinces that need creation of regionals management models. At the same time, country has huge reserves to increase productivity, based on available resources and comparing them with other counties. Reserves to increase productivity consists both through developments of advanced technologies [2] and through well-known technological solutions, without significant investment in fixed capital, but under the organization of effective cooperation of different governmental and managerial structures, which require scientific and methodological support.

2 Problem Statement

Potential, however, that available in each region, underutilized. In the last years there is a decline in development rates in many areas of the economy, including labor productivity growth, which averages no more than 1,3 % a year. With that rate of growth, it is unable to set the economies of regions on the necessary level and represents as an extremely low figure (Fig. 1).

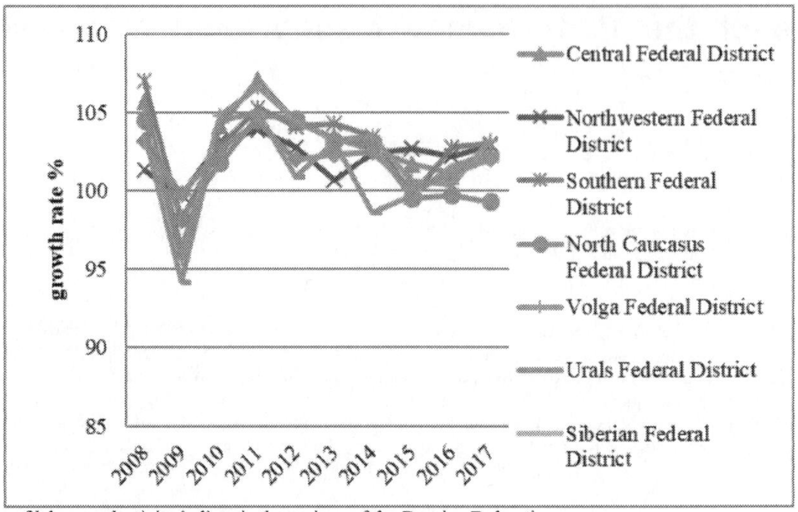

Fig. 1. Dynamics of labor productivity indices in the regions of the Russian Federation.
Source: Authors, based on the data of Rosstat [3].

Analysis of labor productivity trends in Federal areas of Russia shows both some differences and some general trends to a minor increase, rate of which cannot provide necessary growth of economic indicators.

3 Research Questions

The objective of significant labor productivity growth in the country has number of restrictions including financial nature. Methods and tools to increase productivity are sufficiently well studied by Soviet's scientists of Labor Economics School [4]. They would include technological upgrading of manufacture [5], when due to mechanization and automatization, number of personnel decrease, but the reduction of time to produce products happens, improvement in the quality and enhancing the competitiveness of products [6]. At the same time, in most cases, there are capital and other restrictions for quick investment under the existing need in improving productivity of production [7]. The lack of financial resources holds upgrading of basic means of production, which is the main engine of labor productivity growth [8]. However, labor productivity growth can be achieved not just because of technological innovation, but because of others, low-costed, but nonetheless important methods, which can give significant impact on existing equipment. Low productivity at many Russian enterprises is not only because of the high level of obsolescence of equipment, but also because of high level of different losses, for example number of defects, downtime of equipment, low quality of products, inefficient labor management, low motivation of personnel, and other factors that collectively frame the organizational culture. Without activation of these factors, productivity can be even lower on new equipment rather than on old one [9]. In so doing, awareness among enterprises of the potential improvement in productivity, not by investment methods in the short term, remains extremely weak. There is also a weak chance to make reform by themselves because of staff's workload by current responsibilities and lack of professionals with needed qualification.

4 Purpose of the Study

The relevance of the research topic is determined by the fact that among methods to increase productivity, there are very little number of non-investment matters to improve productivity, which can lead to an economic effect. To a large extent, the above social and economic factors [10] depend on combination of managerial tools and methods, applied at the enterprise, but at the same time most of them may be harmonized with certain scientific basis, that can serve as a basis to apply in any enterprise. The application of a harmonized methodology does not require significant investment, however, a need for an organizational and management reforms, painful for each enterprise. The efficiency of such activities can be significantly enhanced by skilled methodological support from specialized regional structures.

5 Research Methods

In the production process it is necessary to adherence to the technological parameters of their production, which allow to apply low-investment methods to increase productivity. They are determined by next factors – work organization, production loss, capital productivity and motivation of staff [11]. In order to achieve these goals its needed to: set short-, middle- and long-term priorities and goals to increase productivity; optimization of organizational structure through the reduction of duplicated and irrelevant functions; reconciliation and redistribution of responsibilities; outsourcing of functions and others; modification in staff motivation system, which aimed to achieve set goals on increasing productivity; reduction in non-productive losses of working time; establishment of workload standards and other methods [12]. Such instruments also include some action in social, economic and organizational factors of productivity. First group of factors designed to effectively manage human resources of an enterprise, second group - for improving the organization of work. To improve the productivity by traditional methods, managerial and motivational actions are needed, which enable to structure and synchronize operations of numerous participations by methods of project management.

The operational level to increase productivity is an area with the most quantifiable indicators and possibilities to achieve measurable results. Such results are combination of several parameters: value added and earnings by one worker, rate of use of the basic means of production.

6 Findings

The Federal Priority Program entitled "Productivity gains and employment reduction" [13], which is currently in force, does not fully reveal effectiveness of laid down activities in terms of focused priorities. One of the problems, that holding back implementation of the Program, is lack of the necessary skills of personnel of enterprises, which improve productivity by systematic and organizational support from government organizations and consulting centers. However, their actions are usually not harmonized and in the conflict between one another. Improving productivity in a short term by traditional technologies, despite significantly lower capital intensity in comparison with creation of new products and methods, still require some inputs: organizational and managerial. To do so in each region there is a need in The Project Offices' creation that manage improving of productivity [14] or The Competence Center, which allows to combine theoretical and methodological development with the requirements of each enterprise. The most appropriate way is to allocate specific structural divisions that will manage increasing productivity in created Research-Educational Centers (REC) (Fig.2).

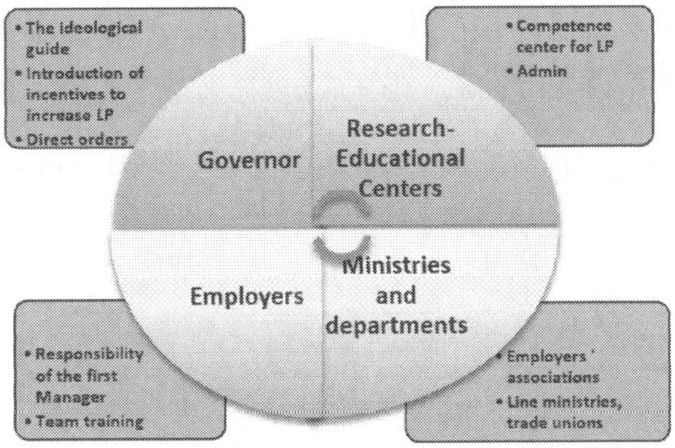

Fig. 2. Interrelations of the regional competence center on labor productivity (LP). Source: Authors.

Submission to the Governor is needed to confirm the considerable value of targets to increase productivity for the region, coordination of goals and tasks to gain better productivity with regards to regional development strategy. The direct supervision of The Competence Center must be guided by REC, but only in the areas of administrative functions, which defined by regulations and REC Council. The actual need of administrative management is determinate by institutional functions, which REC should be vested with. For example, development and functioning of the productivity service market, creation of infrastructural facilities, supporting of legislative initiatives, etc. The proposed structure and course of The Competence Center actions may look as follows (Fig.3).

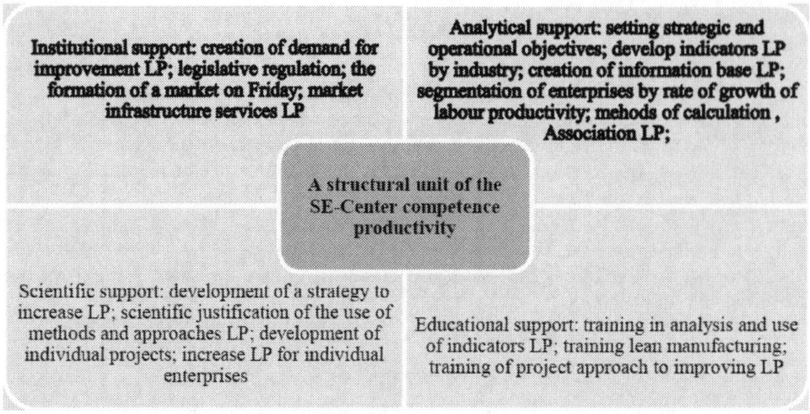

Fig. 3. The proposed functional structure of the competence center of labor productivity (LP).
Source: Authors.

Development of methodologies to increase productivity is one of the central issues, because using exactly proven methodologies can show algorithm of well-understood actions with predictable results, which are aimed at the achievement of short-, medium- and long-term goals. When planning efficiency of the system as a whole, its needed to use project approach, which allows to arrange priorities for time and accountability. Such attitude becomes applicable in various systems, including performance management.

7 Conclusion

Improvement of productivity requires a systemic, long-term effort on all economic and managerial levels. Studies confirm that it is possible to coordinate various organizational and economic institutions to achieve a common goal: increasing of productivity. Addressing those issues in a short term have been identified social and economic factors, which help to increase productivity through low-cost measures. However, it requires methodological and organizational support, which should be launched at institutional level in each region. In studies was created regional system between governmental and managerial structures, and functional system - The Competence Center - to improve productivity, which allows to organize the process of productivity increasing on principles of project management. In the future, for further research, it entails the development of project model to increase productivity and management tool, both on a regional level and level of an enterprise.

References

1. A.M. Jarkas, C.G. Bitar, Factors affecting construction labor productivity in Kuwait. *Journal of Construction Engineering and Management*, **138**(7), 811-820 (2014). DOI: 10.1061/(ASCE)CO.1943-7862.0000501.
2. G. Crespi, P. Zuniga, Innovation and productivity: Evidence from six Latin American countries. *World Development*, **40**(2), 273-290 (2012). DOI: 10.1016/j.worlddev.2011.07.010.
3. Federal service of state statistics, Region of Russia, *Socio-economic indicators. Stat. proceedings of the Rosstat*, Moscow (2018). URL: http://www.gks.ru/free_doc/doc_2018/region/reg-pok18.pdf. Accessed: 05.09.2019.
4. T.O. Selezneva, Economic essence and value of labor productivity. *Scientific Discussion: Economics and Management*, **5**, 91-101 (2015). [in Rus.].
5. C. Syverson, What determines productivity? *Journal of Economic Literature*, **49**(2), 326-365 (2011). DOI: 10.1257/jel.49.2.326.
6. N.N. Shash, A.I. Borodin, Indicators and methods of measuring labor productivity and the possibility of their application in enterprises. *Scientific Notes of Petrozavodsk State University*, **3-1**(148), 96-101 (2015). [in Rus.].
7. E.G. Repina, M.V. Simonova, E.I. Sukhanova, The development of a forecast model of labor productivity management at industrial enterprises. In Y. Silin, Y. Animitsa, E. Dvoryadkina, V. Blaginin (Eds.), *Proceedings of the 2nd International Scientific Conference on New Industrialization: Global, National, Regional Dimension* (SICNI 2018). *Advances in Social Science, Education and Humanities Research*, **240** (pp. 96-99). Atlantis Press: Paris (2018). DOI: 10.2991/sicni-18.2019.20.
8. V.A. Vaisburd, M.V. Simonova, I.V. Bogatyreva, E.G. Vanina, E.P. Zheleznikova, Productivity of labor and salaries in Russia: Problems and solutions. *International Journal of Economics and Financial Issues*, **6**(5S), 157-165 (2016).

9. V.A. Lobanova, N.V. Trofimova, Factor analysis of social labor productivity in the Volga federal district. *Bulletin of Belgorod State Technological University named after V.G. Shukhov*, **5**, 251-255 (2015). [in Rus.].
10. T.I. Ovchinnikova, Y.A. Salikov, A.V. Markov, Productivity and technological development aggregate factors of competitiveness of regional enterprises. *Bulletin of Voronezh State University of Engineering Technologies*, **1**(63), 236-243 (2015). [in Rus.].
11. S.E. Black, L.M. Lync. How to compete: The impact of workplace practices and information technology on productivity. *Review of Economics and Statistics*, **83**(3), 434-445 (2001).
12. A.A. Fedchenko, Methodological approaches to the study of labor productivity. *Russian Journal of Labor Economics*, **3**(1), 41-62 (2016). DOI: 10.18334/et.3.1.35153. [in Rus.].
13. Official website of the Government of the Russian Federation, *Passport priority program "increase productivity and employment support»* (2017). URL: http://government.ru/news/29354/. Accessed: 01.09.2019.
14. F. Nasirzadeh, P. Nojedehi, Dynamic modeling of labor productivity in construction projects. *International Journal of Project Management*, **31**(6), 903-911 (2013). DOI: 10.1016/j.ijproman.2012.11.003.

Sustainable Development Concepts as a Response to Expectations of Modern Business Management

D.V. Filatova[1,2], and *R.V.* Fedorenko[3,*]
* Corresponding author: fedorenko083@yandex.ru.

[1] CHArt, EPHE, Paris, France
[2]Belarusian State University of Informatics and Radioelectronics, Minsk, Belarus
[3]Samara State University of Economics, Samara, Russian Federation

Abstract. The interactions among individual economic agents require some regulations to hedge the risk of business activity. This paper examines the general equilibrium theory, backgrounds of system analysis and optimal control to show that these regulations present the mechanism of balanced management of economic processes at various economic level (enterprise, region, country). The study shows that sustainable development is self-regulatory and adaptive effort to remain on the market under the condition of reliable cooperation. Furthermore, by investigating the hierarchy of the business management systems, the conceptual and mathematical models of the mechanism for sustainable development are built and illustrated.

Keywords: sustainable development, management, region, integration.

1 Introduction

The ongoing process of globalization demands searching for new, effective methods of business management. Independent actions undertaken by business entities in order to reduce costs or improve their profitability often prove to be insufficient. Nowadays, establishing cooperation with other enterprises or even regions is a way of increasing their strength and achieving market success. The legal framework of establishing and operating agreements showed that those vertical agreements these days are one of the most beneficial form of cooperation between partners. An opportunity for better coordination between the involved entities is their main advantage, which results in improving the economic efficiency of business activities. The European Commission presents a similar viewpoint, claiming that these agreements may contribute in particular to lowering the transaction and distribution costs as well as to optimizing the levels of sales and investment of cooperating parties [1]. Thus further attention will be focused only on enterprises or regions cooperating on vertical principles (they will be referred to as "a group"). In order for the undertaken cooperation to satisfy all the participants of the agreement and to provide the intended benefits over a longer period of time, its development must be balanced. This means considering the interests of all parties involved. Various methods are known to reconcile the interests of cooperating enterprises. Determining their so-called sustainable development is one of these methods. There exist various definitions of sustainable development [2, 3]. Most often it is defined as ***"development that meets the prevailing needs of society, without diminishing opportunities for development of future generations"*** [4]. This definition is too broad and does not make it possible to implement strict methods of managing complex economic structure. It is then necessary to indicate the determinants of such development. Therefore the aim of the paper is to determine the mechanism of sustainable business management of a group as a complex economic entity.

2 Methods

The study will be conducted using theoretical and empirical methods. The general equilibrium theory, the concepts of system analysis and optimal control will be applied to achieve the objective of the paper. This will allow us to indicate the determinants of sustainable business management, build a dynamic mathematical cooperation model in the form of optimal control problem.

3 Concepts of sustainable management

3.1. Determinants and definition

Let us consider a business situation of sustainable management of an enterprise (it can be also a region). The solution to a problem can be obtained by means of the system analysis. That is to say, the enterprise (on micro-level) or region (on macro-level), as the object of investigation, has to be discussed as a complex system [5].

Intuitively, the complex system can be defined as a hierarchical multilevel structure of interconnected homogeneous or heterogeneous elements which constitute an ordered and single whole. The elements of the system constitute the internal environment (each enterprise is a subsystem), the system itself is surrounded by an uncertain external environment (enterprise faces competition, whose behavior is difficult to forecast). Apart from that, the system must have the following features, including: integrality and emergence, organization (having structure and functions), functionality (showing certain properties when affecting the external environment and pursuing a goal), development (irreversible, targeted, regular change of the internal environment, which results in a new quality), self-regulation (striving to maintain the state of internal balance despite the changes in the external environment), endurance (suppressing harmful effects of the external environment), adaptation (ability to change behavior in order to maintain or improve the existing features or acquire new ones in an unspecified environment), reliability (ability to maintain structure despite the incapacity for work of distinct elements of the system).

The determinants of group's sustainability can be distinguished when analyzing the above-mentioned properties. They include: self-regulation, endurance, adaptation, reliability. These properties indicate the possibility of achieving business objectives not only for a group but also for its each member, as well as long-term cooperation regardless of the harmful activities of the competition. Therefore group sustainability will be understood as such ***self-regulatory and adaptive efforts to remain on the market under the condition of reliable cooperation, which results in development aimed at achieving business objectives***.

3.2. Decomposition and integration of the business management systems

Achieving sustainability requires the implementation of appropriate methods of group management or coordination. It should be mentioned that within the framework of vertical agreements the interests of each party are equally important, hierarchy of enterprises (regions) and favoring the interests of one partner is not allowed. The management method is to be formulated as a joint program, e.g. manufacturing. This requires a quick exchange of the most important information that is usually subject of a cooperation agreement. This exchange is achieved through appropriate decomposition of the management systems of each enterprise and further integration of selected subsystems into a coherent group management (sometimes only informative). The following items are characteristic of each enterprise: hierarchy of management (management levels - strategic, tactical or operational management), and functions or subsystems of management (planning, controlling, motivation, organization, coordination) [6]. The five management subsystems can be considered as the decomposition of management system of each enterprise. Integration demands analyzing the functions of particular subsystems and understanding their hierarchy [7].

Let us consider this problem exclusively from the perspective of information flows. The elements of the planning subsystem include information system containing a database along with the database management system, the procedures for collecting and analyzing the information acquired and the system of interrelated plan of delivery, production, marketing, etc. at each management tier.

The formation of an integrated planning subsystem consists of four stages. The first stage consists in developing the parameters of the information system and thus determining the required amount of information, its scope, direction, content and specificity level, as well as defining the methods of its storage and access. The second and third stage consists respectively in developing the procedures of collecting and analyzing information. The last, fourth stage consist in developing procedures of devising plans, developing entry forms, time frames and appointing individuals responsible for formation, coordination and approval. The elements of the control subsystem as part of the information system include: a system of plans, with a system of target indicators, a system of standards for the operations of entities; "management indicators". On the other hand the procedures of subsystem control involve: monitoring the current operations of an enterprise; assessment/measurement of the business entities perform assessment/measurement of the implementation of plans; transmission and dissemination of information. It should be noted that the parameters of the control subsystem are entirely dependent on the parameters of the planning subsystem, therefore the control subsystem should be formed after the planning subsystem has been created. The formation of this integrated subsystem involves defining a set of indicators and determining their quantitative values, as well as developing the procedures for monitoring, assessment/measurement, transmission and dissemination of information. The staff motivation subsystem depends on the two previous subsystems, and it is created to meet the strategic and operational goals of enterprises. The system of indicators used to assess the performance of various entities and the system of indicators of employee performance are the elements of the information system here. The procedures include assessing and motivating employees. The elements of the subsystem of forming the organizational structure include the description of the functions performed in the enterprise, powers and responsibilities of the executive entities and the organization scheme of the enterprise. The organizational structure itself is created in several stages, namely: defining executive functions in the enterprise and executive entities implementing the functions, distribution of the functions by links; organizational analysis; defining professional responsibilities, devising the "Statute of the organizational structure". The key elements of the coordination subsystem, whose formation is based on the principles of operation of the four previous subsystems, include responsibility and authority understood as a limited right to use the organization's resources and perform delegated tasks. Management procedures imply clear decision-making methods [8].

It results, inter alia, from the fact that the powers are limited to plans, procedures, rules and verbal orders of managers, as well as environmental factors. In the case of the organizational structure formation subsystem the integration involves the organization of the strategic management process, organization and formalization of the management process as well as the implementation of new management procedures. After a detailed analysis of each management subsystem, allowing for the management tier, subsystems can be incorporated into an integrated management system (IMS) which fully reflects the needs of all stakeholders: employees, customers and the public.

This integration enables effective (and consequently, sustainable) management of the region, enterprise, projects, processes, etc. as it is easy to track the performance of individual tasks and information flows [9] The advantage of the IMS is not only simplicity, transparency, uniformity and explicitness, but also the capability of parallel existence with other business management systems. One can assume that IMS is one of the solutions to sustainable group management. This means that enterprises cooperating in a group can agree on selected model (a kind of mechanism) to ensure sustainable development of the group without specifying activities that will lead to achieving business objectives. Summing up, the integration of the group management systems can be reduced to the integration of planning subsystems (at the strategic level) on the basis of the selected model using the determinants of sustainable development.

4 The mechanism for sustainable management

Let us consider a certain product (stock) whose overall production size at time $t \in [t_0, t_1] \subset R_+$ is a function $x:[t_0, t_1] \to R_+$. Denote by $\theta_1 \in R_+$ the intrinsic growth rate and by $\theta_2 \in R_+$ the warehouse optimal capacity. Here we assume that the production growth rate takes the form of an ordinary differential equation

$$\frac{dx(t)}{dt} = f(x(t)), X(t_0) = x_0,\ t \in [t_0, t_1], \tag{1}$$

where x_0 is the initial value of the stock and the C^2 function $f: R_+ \to R_+$ depends on θ_1 and θ_2 and meets the following conditions, namely: f is concave (c1), $f(0) = 0$ (c2), and $f(\theta_2) = 0$ (c3). There is no loss of generality to assume that f is a C^2 function. Combining (c1) with (c2) and (c3) we get that the function f has a unique maximum, say x_{max}, as soon as f is not the null function on the set $(0, \theta_2)$. The choice of f function as a limited growth one

$$f_0(x) = \theta_1 x\left(1 - \frac{x}{\theta_2}\right),\ x \ge 0, \tag{2}$$

gives the following solution of (1)

$$x(t) = \frac{\theta_2}{1 + a \cdot \exp(-\theta_1 t)},\ t \in [t_0, t_1], \tag{3}$$

where $a = \frac{\theta_2 - x_0}{x_0}$ is a real number. The solution (3) has the following explanation, namely for any $t \in [t_0, t_1]$ whenever $x(t) < \theta_2$ the overall production will increase, if $x(t) > \theta_2$, it will decrease, and if $x(t) = \theta_2$, it will stay in a stable equilibrium. This conclusion can be used as a measure of the sustained management. We assume that the product is continuously consumed at a rate $u(t)$, $u:[t_0, t_1] \to R_+$. This allows on the rewriting (1) as follows:

$$\frac{dx(t)}{dt} = f(x(t)) - u(t),\ x(t_0) = x_0,\ t \in [t_0, t_1], \tag{4}$$

where the C^2 function f fulfills the conditions (c1) – (c3).

Let us study the situation when the function u is constant. As it is possible to see for any $t \in [t_0, t_1]$ whenever $0 < u < f(x_{max})$ the curve of the function f and the strait line $y = u$ intersect at the points (x_1, u) and (x_2, u), where $0 < x_1 < x_{max} < x_2 < \theta_2$. So, for constant consumption rate u whenever $u < \max f(X)$, the equation (4) has one stable and one unstable equilibrium, however, whenever $u > \max f(X)$, there are no equilibria, if $u = \max f(X)$ a single semi-stable equilibrium exists at $X = x_{MSY}$. The point x_{MSY} is referred to maximum sustainable yield and can be used as a performance indicator. Applying the same reasoning, we can show that the sustainable consumption is given by

$$u(t) = \begin{cases} u_{max} & if\ x(t) > x_{MSY}, \\ f(x_{MSY}) & if\ x(t) = x_{MSY}, \\ 0 & if\ x(t) < x_{MSY}. \end{cases}$$

In the sequel we will focus on the following situation. The profit of economic activity related to the product is some function $\pi(x, t)$. Taking into account the value of time as the standard economic mechanism, we introduce the discount factor:

$$\rho(t) = \exp\left(-\int_{t_0}^{t} \delta(\tau) d\tau\right),$$

where $\delta(\cdot)\in \mathrm{R}_+$ is a time-varying discount rate. The optimal sustainable consumption problem can be formulated as the following:

$$\max_u \int_{t_0}^{t_1} \rho(t)\pi(t)u(t)dt\,, \quad (5)$$

subjected to (4) and

$$0 \le u(t) \le u_{\max}\,. \quad (6)$$

The task (4) – (6) is a singular optimal control problem, which solution is well-known and can be obtained by the Pontryagin's Maximum Principle. Therefore, referring to [10] and omitting technical details, we write the Hamiltonian

$$\mathrm{H} = \rho(t)\pi(x,t)u(t)+\psi(t)\big(f(x)-u(t)\big),$$

where $\psi(t)$ is the adjoint variable. Solving the task (4) – (6) we get for the consumption the bang-bang control

$$u(t)=\begin{cases} u_{\max} & \text{if } \psi(t)<\rho(t)\pi(x(t),t), \\ 0 & \text{if } \psi(t)>\rho(t)\pi(x(t),t), \end{cases}$$

and the singular control $u(t)=u^S(t)=\left(x^*(t)\right)-\frac{dx^*}{dt}$ for $\psi(t)=\rho(t)\pi(x(t),t)$ over $t\in[t_0,t_1]$, where the singular path $x^*(t)$ of (4) refers to the optimal sustainable production. As it is possible to notice from Fig. 1, the initial production is reduced by bag-bang control to the singular path $x^*(t)$, which can be calculated from the current changes of the economic parameters. To conclude, this solution is the mechanism for the sustainable management of the production-consumption relations.

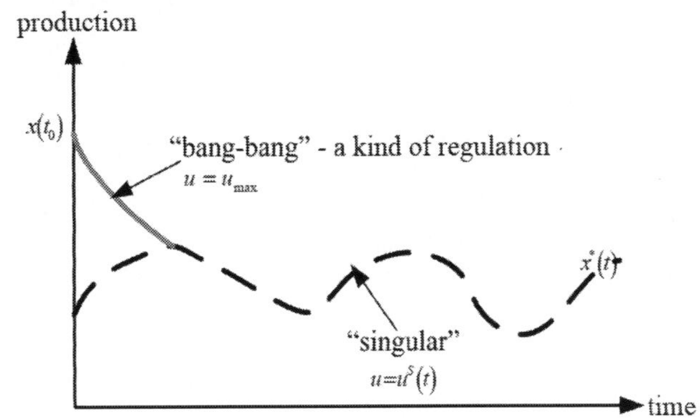

Fig. 1. The mechanism of sustainable management of the production process
Source: Authors.

5 Conclusion

The ongoing globalization processes and the complexity of the interactions among individual economic agents require some regulations in order to achieve a kind of sustainable development. These regulations are considered as the mechanism of balanced management of economic processes at the level of enterprises and regions. The proposed balanced management mechanism has the potential for implementation at various economic levels from micro to mega one [11]. Obviously, the application of this mechanism for the decision-making will allow an enterprise or a region to maximize profits from economic activity by successfully integrating into the existing demand-supply economic system of a higher level. In some future research, the proposed approach can be considered as one of the tools for developing strategies for international economic cooperation at the level of states or interstate associations.

Acknowledgment

The reported study was funded by RFBR and FRLC according to the research project № 19-510-23001.

References

1. W.V. van Themaat, B. Reuder, *European competition law. A case commentary* (Elgar Commentaries series). 2nd Edition (Edward Elgar Pub, New York, 2018).
2. J. Markard, R. Raven, B. Truffer, Sustainability transitions: An emerging field of research and its prospects. *Research Policy*, **41**(6), 955-967 (2012). DOI: 10.1016/j.respol.2012.02.013.
3. B.B. Hughes, P.D. Johnston, Sustainable futures: Policies for global development. *Futures*, **37**(8), 813-831 (2005). DOI: 10.1016/j.futures.2005.01.017.
4. R.R. Shaker, The spatial distribution of development in Europe and its underlying sustainability correlations. *Applied Geography*, **63**, 304-314 (2015). DOI: 10.1016/j.apgeog.2015.07.009.
5. G.P. Farnell, A.J. Saddington, L.J. Lacey, A new systems engineering structured assurance methodology for complex systems. *Reliability Engineering & System Safety*, **183**, 298-310 (2019). DOI: 10.1016/j.ress.2018.11.024.
6. A.C.-L. Chian, *Complex systems approach to economic dynamics* (Springer, Berlin, 2007).
7. T. Kaizoji, Towards economics as a new complex system. *The European Physical Journal Special Topics*, **225**(17-18), 3225-3230 (2016). DOI: 10.1140/epjst/e2016-60161-x.
8. A. Bergek, S. Jacobsson, B. Carlsson, S. Lindmark, A. Rickne, Analyzing the functional dynamics of technological innovation systems: A scheme of analysis. *Research Policy*, **37**(3), 407-429 (2008). DOI: 10.1016/j.respol.2007.12.003.
9. C. Binz, B. Truffer, Global innovation systems – A conceptual framework for innovation dynamics in transnational contexts. *Research Policy*, **46**(7), 1284-1298 (2017). DOI: 10.1016/j.respol.2017.05.012.
10. A.V. Dmitruk, On the development of Pontryagin's maximum principle in the works of A.Ya. Dubovitskii and A.A. Milyutin. *Control and Cybernetics*, **38**(4A), 923-957 (2009).
11. S. Nambisan, M. Wright, M. Feldman, The digital transformation of innovation and entrepreneurship: Progress, challenges and key themes. *Research Policy*, **48**(8), 103773 (2019). DOI: 10.1016/j.respol.2019.03.018.

Social and Economic Factors of Sports Activity in Russian Regions

T.V. Kramin[1,4*], *M.R.* Miftahov[2,4], *W.*Andreff[3], and *S.B.* Eroshkina[4]
*Corresponding author: kramint@mail.ru.

[1] Kazan Innovative University named after V. G. Timiryasov, Kazan, Russia,
[2] Ministry of Sports of the Republic of Tatarstan, Kazan, Russia,
[3] French Ministry of Sports, Paris, France
[4] Federal Research Center of Physical Culture and Sports, Moscow, Russia

Abstract. In this study, the key factors of sports activity in the Russian regions were identified and substantiated. The regional panel data for the period 2007–2017 was used as data for the study. It is shown that traditional material factors are highly significant in models for assessing the level of sports activity in the Russian regions. These include, in particular, the following indicators: the one-time carrying capacity of sports facilities, the per capita expenditures for physical culture and sports development, the number of sports facilities per 100 thousand people, the number of full-time physical culture and sports personnel per capita. Consequently, forecasting based on them will give a stable and reliable result.
Along with the evidence of the significance of the material factors of sports activity, a justification of the influence of a number of intangible, institutional factors is given. Such factors include the awareness about the benefits of systematic physical education and sports, institutions of public encouragement and support for those involved in sports and physical culture, etc.

Keywords: factors of sports activity, physical activity, physical culture, sports, institutions, digital economy.

1 Introduction

Development of mass sport and physical culture of the population is an important component of social-economic development of any country. One of the key indicators of mass sport development is the share of population systematically engaged in physical culture and sport.

According to the data of Sport and Active Society Commission under the International Olympic Committee, in 2008 in some countries the share of people regularly engaged in physical culture and sport was 60–70%. These countries include Brazil, Iceland, Columbia, Finland, France, Sweden, and Japan. In most of the economically developed countries this indicator is over 40%.

In compliance with the “Strategy of physical culture and sport development in the Russian Federation up to 2020”, in 2020 the share of citizens systematically engaged in physical culture and sport must be not less than 40%.

In particular, a report by E.P. Kakorina [1] stated that the most disadvantageous period for sport development in the RSFSR and Russia was 1975-1995. In that period, the analyzed indicator of sport activity decreased 5 times (from 37.5% to 7.1%). Since 1995, the situation was gradually recovering.

In this context, it is important to note the systemic understanding of physical culture and sport development in the USSR demonstrated by foreign scholars [2], who highly evaluated the USSR public institutions in this sphere. Physical activity of the population depends on many individual material and nonmaterial factors, including gender, age, social environment, income level, educational level, etc. [3].

The main obstacles for doing physical culture and sport are the lack of spare time and motivation. According to P. Edwards and A. Tsouros, overcoming these barriers plays the key role in solving the problem of physical culture and sport activation [4]. In should be noted that most of the conclusion of foreign authors cited in the present work are confirmed by the Russian researchers (see below).

A research by A. Bauman et al. showed a lower physical activity in women compared to men, as well as heterogeneous influence of educational level [5].

The social-psychological factors of sport activity are:

1) the social-economic status of a person, which determines the degree of sport infrastructure accessibility; this is especially important for low income groups [6,7];
2) the family traditions, the social environment, support of colleagues and friends [8, 9, 10];
3) self-confidence, ability to get pleasure and benefits from physical culture and sport [11, 12];
4) the established way of living and habits [13].

Physical activity is promoted by the development of city environment, while its influence can be both positive and negative. L. Frank et al. note that the availability of convenient walking sites with quality surface stimulates regular walks

[14]. Researchers [15, 16] showed that the increase of such zones leads to the growth of physical activity. Others emphasize that physical activity decreases due to urbanization processes and broader use of automobile transportation [17, 18].

Notably, according to the data of All-Russia Center for Public Opinion Research, the reasons impeding engagement on sport in Russia in 2008 were: "lack of spare time" (39%), "lack of desire" (41%), "state of health (20%) [1].

It should be noted that these data are close to the results of a more recent poll [19]. The above data allow defining the main directions of researching the factors of sport activity of the Russian population.

The physical activity of the population is one of the most important components of the social life of any country. Development of mass sport and physical culture of the population is an important component of social-economic development of any country. One of the key indicators of mass sport development is the share of population systematically engaged in physical culture and sport. According to the data of Sport and Active Society Commission under the International Olympic Committee, in 2008 in some countries the share of people regularly engaged in physical culture and sport was 60–70%. Presently this indicator in Russia does not reach 40%.

In Russia, a lot is being done to increase the sports activity of the population - new sports facilities are being built and existing one are reconstructed, national and regional sports events are being held with the assistance of regional and federal authorities. To increase the effectiveness of this work, it is necessary to identify the most important factors of the sports activity of the population. Their identification and justification will allow for the implementation of a more balanced and productive regional policy in the field of physical education and mass sport.

The objective of this research is to reveal and quantitatively estimate the influence of the main factors of sport activity in order to develop efficient regional policy in the sphere of physical culture and mass sport. In particular, it is essential to reveal the main barriers and limitations, as well as the underused reserves and potentials of increasing sport activity. The regional agenda will help to reveal the best practices in managing the system of physical culture and mass sport in the Russian regions and spread them in other subjects of the Russian Federation.

2 Methods

The main tool of this research is econometric modeling. We formulate and test a number of hypotheses to identify factors of sports activity of the population in the Russian regions based on a literature review based on the official regional statistics of Russia.

The research data are regional panel data for Russia in the period 2007-2017 [20]:

1. The share of population systematically engaged in physical culture and sport in the total number of the population (per cent). In the constructed models, this indicator is denoted N1 with the two last digits of a calendar year. For example, N115 stands for the indicator for the year 2015;

2. The share of students systematically engaged in physical culture and sport in the total number of student (standed for N2).

Taking into account the above said, the indicators of potentially significant factors of sport activity (the explanatory variables) are: the number of sport facilities (stadiums, swimming pools, other constructions) per 100 thousand of population (units, standed for FPC); one-time capacity of sport facilities (%, standed for EP); expenditures for the development of physical culture and sport per capita (a year, thousand rubles, standed for ZP); the number of permanent staff members in physical culture and sport per capita by the subjects of the Russian Federation (standed for WPC).

The overall objective of the modeling is to reveal and substantiated the factors (and respective indicators) facilitating the growth of sport activity in the Russian regions and municipalities and the degree of their impact.

3 Results

Analysis of data was carried out on the basis of formulating and checking the following hypotheses:

1. The level and dynamic of sport activity of the population in the Russian regions (indicator N1 in the models) depends on the level and dynamic of one or several explanatory variables, listed above.

2. The level and dynamic of sport activity of the students in the Russian regions (indicator N2 in the models) depends on the level and dynamic of one or several explanatory variables, listed above.

The results of testing the hypotheses are shown in Table 1. It presents the models of estimating the factors of sport activity of the population in the Russian regions. The dependent variables are N1 and N2; modeling was carried out with panel data; fixed spacing effects based on Hausman test results were applied. The models are well specified: the coefficients at all explanatory variables are highly significant. The models possess good explanatory power: values of R^2 show that the models explain over 90% of differences in the values of dependent variable in the Russian regions. Adding the lag variables allowed overcoming serial autocorrelation, which is explicitly reflected in the values of Durbin-Watson statistics.

Table 1. Models for estimating the factors of sport activity in the Russian regions
Dependent variables: N1, N2. Fixed effects, period 2011-2016. Regions included: 82
Panel least-squares method, total number of panel observations: 492

Dependent variable	N1	N2
Variable\ Model number	1	2
C	-0.096899 (0.0000)	-0.064320 (0.0544)
EP	0.024714 (0.0008)	0.071416 (0.0001)
ZP	0.001529 (0.0000)	0.002149 (0.0023)
ZP(-1)	0.001206 (0.0000)	
FPC	0.069401 (0.0000)	0.071016 (0.0003)
WPC	14.61097 (0.0028)	
WPC(-1)		29.83800 (0.0076)
N1(-1)	0.690421 (0.0000)	
N2(-1)		0.750549 (0.0000)
R^2	0.947856	0.919645
F-statistics	84.41084	53.89678
Probability by F-statistics	0.000000	0.000000
Durbin-Watson statistics	1.923263	2.331804

Source: Authors.
Note: specification of (-1) in denotations of variables shows a time lag of one period (a year). Figures in brackets are null-hypothesis probabilities by t-statistics for coefficients of independent variables.

Table 1 shows that many hypotheses of groups 1 and 2, described above, were conformed. Comparing models 1 and 2 in Table 1, it should be noted that in general the impact of the analyzed factors on the indicators of sport activity N1 and N2 are similar. One should note the high dependence of the current values of dependent variables on their previous year values. Thus, in the studied period, stable trajectories of sport activity development are formed in the Russian regions. The indicators "One-time capacity of sport facilities" (EP), expenditures for the development of physical culture and sport per capita (ZP), the number of sport facilities per 100 thousand of population (FPC), and the number of permanent staff members in physical culture and sport per capita by the subjects of the Russian Federation (WPC) are significant in the models explaining the differences in the Russian regions by the level of their sport activity.

Thus, the material factors, such as availability of sport facilities, the level of per capita expenditures for the development of physical culture and sport in the Russian regions, and others, significantly influence the sport activity of their population. The model specification proves the necessity to include lag explanatory variables into them, which demonstrates their so called delayed (in time) effect on the explained variable.

Despite the high quality of the built models, judging by the R^2 value, there are also factors of sport activity not taken into account in these models. Besides, the dynamics of sport activity in the Russian regions takes place along distinctly trackable trajectories, which is displayed, in particular, by autocorrelation of time series of the relevant variables. There are also various trajectories of sport activity indicators in different Russian regions. These facts testify to the necessity of taking into account the non-material, institutional factors of sport activity, as well as of revealing and registering the best practices in the Russian regions.

4 Conclusion

The present research identifies and substantiates the key factors of sport activity in the Russian regions. It should be noted that the presented models of estimating the factors of sport activity are well specified: the coefficients at all explanatory variables, included in them, are highly significant; the use of lag variables enabled to actually exclude autocorrelation of the model errors, which shown by the values of Durbin-Watson statistics. The models possess good explanatory power (R^2 of the models exceeds 90%). Consequently, forecasting based on them will yield a reliable result.

In future research, we plan to continue the study of the above issues, in particular, in the sphere of identifying the new institutional factors of sport activity, and to formulate new proposals for taking these factors into account when forming the regional policy in the field of physical culture and sports and developing new institutions.

References

1. E.P. Kakorina, Attitude of the population to engagement in physical culture and sport. *Healthcare in Russia*, **11**, 137-140 (2011). [in Rus.].
2. O'Mahony, *Sport in the USSR: Physical culture – Visual culture* (University of Chicago Press, Chicago, 2006).
3. T.S. Solov'eva, The level of physical activity and motivation of urban population for physical exercises and sports. *Issues of Territorial Development*, **3**(83), 119-136 (2016). [in Rus.].
4. P. Edwards, A. Tsouros, *Promoting physical activity and active living in urban environments. The role of local governments. The solid facts* (World Health Organization, Geneva, 2006).
5. A.E. Bauman, R.S. Reis, J.F. Sallis, J.C. Wells, R.J. Loos, B.W. Martin, Correlates of physical activity: Why are some people physically active and others not? *Lancet*, **380** (9838), 258-271 (2012). DOI: 10.1016/S0140-6736(12)60735-1.
6. L.H. McNeill, M.W. Kreuter, S.V. Subramanian, Social environment and physical activity: A review of concepts and evidence. *Social Science and Medicine*, **63**(4), 1011-1022. (2006).
7. M. Braubach, M.E. Héroux, N. Korol, E. Paunovic, I.A. Zastenskaya, The value of housing conditions and urban environment for the health. *Hygiene and Sanitary*, **1**, 9-15 (2014). [in Rus.].
8. K. Young, M. Atkinson (Eds.), Qualitative research on sport and physical culture. *Research in the Sociology of Sport*, **6** (Emerald Group Publishing Limited, Bingley, 2012). DOI: 10.1108/S1476-2854(2012)0000006015.
9. S.L. Gustafson, R.E. Rhodes, Parental correlates of physical activity in children and early adolescents. *Sports Medicine*, **36**(1), 79-97 (2006).
10. C.A. Yao, R.E. Rhodes, Parental correlates in child and adolescent physical activity: A meta-analysis. *International Journal of Behavioral Nutrition and Physical Activity*, **12**(10) (2015). DOI: 10.1186/s12966-015-0163-y.
11. C. Currie, Ch. Roberts, A. Morgan, R. Smith, W. Settertobulte, O. Samdal, V. Barnekow Rasmussen (Eds.), *Young people's health in context. Health Behaviour in School-aged Children (HBSC) study: International report from the 2001/2002 survey* (World Health Organization, Copenhagen, 2004).
12. F.J. Van Lenthe, J. Brug, J.P. Mackenbach, Neighbourhood inequalities in physical inactivity: The role of neighbourhood attractiveness, proximity to local facilities and safety in the Netherlands. *Social Science and Medicine*, **60**(4), 763-775 (2005).
13. D.A. Kurakov, N.L. Maltseva, Physical inactivity is a result of urbanization or laziness? (Materials of medicosociological study). *Sociology of a City*, **3**, 74-81 (2014). [in Rus.].
14. L.D. Frank, J.F. Sallis, T.L. Conway, J.E. Chapman, B.E. Saelens, W. Bachman, Many paths from land use to health: Associations between neighborhood walkability and active transportation, body mass index, and air quality. *Journal of the American Planning Association*, **72**(1), 75-87 (2006).
15. L. Frank, S. Kavage, T. Litman, *Promoting public health through Smart Growth: Building healthier communities through transportation and land use policies and practices* (2006). URL: https://www.vtpi.org/sgbc_health.pdf. Accessed: 17.09.2019.
16. A.T. Kaczynski, K.A. Henderson, Parks and recreation settings and active living: A review of associations with physical activity function and intensity. *Journal of Physical Activity and Health*, **5**(4), 619-632 (2008).
17. B.E. Saelens, J.F. Sallis, J.B. Black, D. Chen, Neighbourhood-based differences in physical activity: An environment scale evaluation. *American Journal of Public Health*, **93**, 1552-1558 (2003). DOI: 10.2105/AJPH.93.9.1552.
18. R.L. Mackett, B. Brown, *Transport, physical activity and health: Present knowledge and the way ahead* (Centre for Transport Studies University College London, London, UK, 2011).
19. N.V. Zybunovskaya, A.N. Pokida, Spare time and leisure as conditions for forming healthy lifestyle of the Russian population. *Izvestiya of Saratov University. Series: Sociology. Politology*, **11**(2), 37-41 (2011).
20. Federal Agency of State Statistics, *Annual report "Regions of Russia. Social-economic indicators"* (2018). URL: https://www.gks.ru/free_doc/doc_2018/region/reg-pok18.pdf. Accessed: 17.09.2019. [in Rus.].

Intermodal Line between Ploce and Brcko: Future Perspectives for the Pan-European Transport

N. Palic[1], *E.* Velinov[2,3], and *V.* Vassilev[4]
*Corresponding author: emil.velinov@savs.cz.

[1]CEPS College "Center for business Studies", Kiseljak, Bosna and Hercegovina
[2]Skoda Auto University, Mlada Boleslav, Czech Republic
[3]RISEBA University of Applied Science, Riga, Latvia
[4]University of Transport, Todor Kableshkov, Sofia, Bulgaria

Abstract. The aim of this paper is to propose the intermodal logistics Corridor Vc, relying on the European Union transport policy, which will be in the function of regional development. The main goal is to analyze a potential route of freight intermodal transport between the ports of Ploče and Brčko in order to popularize this intermodal line at the international level. Among the most important activities in achieving this goal is the integration in the Euro corridor system. This integration is of crucial importance for companies to successfully execute their supply chain processes both domestically and internationally. The paper is based on the analysis of the current socio-economic situation and the condition of the intermodal transport system in Bosnia and Herzegovina (BiH), as well as existing strategic studies and development plans. This paper sheds a light on the creation of a strong integrated intermodal freight transport network in BiH, that will be a quality alternative to other international intermodal flows from the aspect of shorter transport time, lower transport costs and more favorable ecological aspects of transportation.

Keywords: intermodal, transport network, Bosna and Hercegovina.

1 Introduction

Bosnia and Herzegovina (BiH) is a land of extraordinary resources - both natural and human, but the synergy effect of their interaction is missing. There is a lack of cross-sectoral coordination in all key policy areas - regulatory, investment, fiscal, social, spatial, environmental and so on. Logic states that the regional organization at the national level and the local unit level is based on the recognition of spatial specifics, their development potential and functional integration. In addition to the uneven regional development, the problem for BiH is also the trend of mass emigration of the population to Western European countries. The response to these problems is to be found in the development of a quality and attractive intermodal logistics corridor, which will popularize the entire BiH area, stop the outbreak of population and thus contribute to better regional development.

Revitalization of key elements of the intermodal logistics corridor will contribute to the establishment of a competitive intermodal logistics route (Port Ploče – Corridor Vc - Port Brčko) at international level, thereby enhancing regional development in BiH and beyond. To prove this, it is necessary to evaluate the potentials of intermodal logistic Corridor Vc. During the valorization, a large source of secondary data was used to enhance the validity of the analysis that will be offered in the work: White Paper 2011 Strategy, TEN-T Network Development Strategy, Construction project "Highway in Corridor Vc", Intermodal Transport Study in Bosnia and Herzegovina, Strategy for Development of Air Traffic Management System in Bosnia and Herzegovina, Transport Master Plan Study in BiH (JICA, 2001), Project B.E.P.O. Neretva, Intermodal Transport System "Intercontainer", BiH Transport Strategy 2016-2030, projects with EU funds, papers from international scientific conferences etc.

First, the analysis of socio-economic data along the route of the Corridor Vc was carried out. The geopolitical importance of the Corridor Vc was highlighted, where an influential corridor zone was presented and a Framework plan for optimization of the intermodal logistics corridor was offered throughout BiH. In the last chapter, the valorization of the intermodal logistics corridor Vc was done by comparing the potential revitalized route through BiH with already existing international intermodal routes. The key criteria to be taken into consideration are transport time, transport costs and ecological aspects of the transport process. If these criteria prove to be more efficient, it would be justified to invest in key elements of the intermodal logistics corridor. The revival of this framework would contribute to regional development in BiH and beyond.

2 Analysis of socio-economic data along the route of corridor Vc and cohesion potential

During the analysis socio-economic socio-economic data it is especially important that the geo-traffic position of the specific area is valorized in the context of its territorial and functional affiliation to a wider region at the transregional

 https://doi.org/10.1051/shsconf/20197104010

and national level, and following the elaboration of development projections ensure integration into the competitive European macro-region.

Table 1: Share of urban and rural populations by European regions / countries

Region/State	Urban population	Share (%)	Rural population	Share (%)
Europe	533.295.008	72,8	199.463.538	27,2
East Europe	200.938.427	68,9	90.546.160	31,1
North Europe	1.827.955	79,1	20.691.418	20,9
West Europe	149.930.742	79,5	38.656.465	20,5
South Europe	104.208.508	67,8	49.569.495	32,2
Bosnia and Herzegovina	1.827.955	48,6	1.931.678	51,4
Region of Corridor Vc*	846.407	40,51	1.243.160	59,49

Source: Own elaboration based on Geohive."Urban/rural division of countries for the years 2015 and 2025." Geohive. Accessed, January 7, 2017 [1]. *Authors calculations on the basis of NASBIH (http://nasbih.com/), June, 2016

By analyzing the statistical data of the urban-rural population structure in Europe, there are significant differences between BiH and the corresponding territorial unit of Southern Europe, and even more so in relation to the urban-rural structure of the populations of the regions of Northern and Western Europe.

BiH has two urban centers with more than 100.000 inhabitants - Sarajevo and Banja Luka. However, when we look at the aspect of the municipality, then we can mention at least four more centers here: Tuzla, Zenica, Bijeljina and Grad Mostar, which together with the surrounding area and the formerly mentioned urban centers form agglomerates with about 1 million inhabitants or 24% of the total population (Figure 1).

For our research it is especially important to carry out an analysis of the area through which passes the route of Corridor Vc: Herzegovina-Neretva Canton (HNK 207), Sarajevo Canton (KS 209), Zenica-Doboj Canton (ZDK 204), Doboj Region (RD 212) and Brčko District (BD „A"), and areas of significant value: Central Bosnia Canton (SBK 206) and Banja Luka Region (RBL 211). This analysis will be done by applying the criterion of population share in the new typology of spatial units NUTS3 (Table 2).

For comparison, in Europe, 40.3% of the population lives in urban areas, 35.6% in transition regions and 24.1% of the population in rural areas (EU27, 2011.). By analyzing the statistical data, a significant deviation of the urban-rural distribution of the population of this part of BiH can be determined in relation to the average of the European Union – over twice as small a population belongs to the urban category, and more than twice the category of rural population. Sarajevo Canton only belongs to the category of mostly urban areas (82.9%), while the HNK and the Banja Luka Region can be sorted into the category of transition areas.

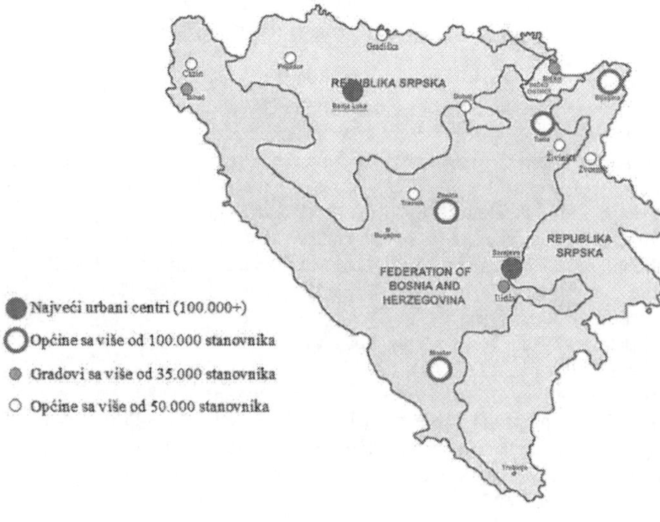

Fig. 1: Urban agglomerations and cities in BiH (left); Distribution of Regions / Cantons in BiH (right)
Source: Authors, based on data provided by the Agency for Statistics of Bosnia and Herzegovina, Population, Household and Housing Census in Bosnia and Herzegovina, 2013, Sarajevo, June, 2016 [2]

In the southern area of BiH, namely in the HNK, population density is 53.69 inhabitants per square kilometer, and in the central part – Canton Sarajevo, ZDK and SBK, population density is 140.43 inhabitants per square kilometer area and in the north – Doboj Region, Banja Luka and Brčko District, population density is 59.98 inhabitants per square kilometer area. It can be concluded that the population density is twice as high in the central part of BiH, which is also the focal point of development. These indicators show that it is necessary to work on the popularization of other parts of the region in order to evenly distribute the development and thereby stimulate migration of the population and reduce the emigration, which could be greatly contributed by the implementation of the Corridor Vc. These data are indicative in the planning and programming of national development strategies, especially in the bases for redefining the regional organization of BiH and regional territorial cooperation.

Table 2: The share of urban and rural areas in BiH in the area of Corridor Vc according to the new NUTS3 typology (2013)

Canton/Region	Total population	Predominantly urban	Share (%)	Intermediate	Share (%)	Predominantly urban	Share (%)
HNK	236.278	0	0	118129	49,99	118149	50,01
SBK	273.149	0	0	49.298	18	223.851	82
KS	438.443	363.314	82,9	0	0	75.129	17,1
ZDK	385.067	0	0	115.134	29,9	269.933	70,1
RD	226.523	0	0	0	0	226.523	100
RBL	437.079	0	0	200.532	45,9	236.547	54,1
BD	93.028	0	0	0	0	93.028	100
TOTAL	2.089.567	363.314	17,38	483.093	23,12	1.243.160	59,49

Source: Authors

Analyzing socio-economic indicators, there are significant differences in the status of the regional units of BiH that will be covered by the Corridor Vc in the rates of unemployment and the level of income. For example, the areas within the Vc corridor (HNK, ZDK, SBK and RD) reach a share of 40 - 50% of unemployment (The unemployment rate was determined on the basis of data on the number of unemployed by municipalities in relation to the labor force (employed + unemployed). Unemployment rate indicators by BiH regional units to a large extent coincides with the indicator of the level of income of the employed population, which has the smallest size in the observed area (ZDK, SBK, RD and RBL).

According to the recent census of 2013, BiH has 14,07% of the population without any and with incomplete primary education; 21.4% of the population has completed elementary school; 51.1% secondary school; 3.2% vocational school; and 9.6% higher education schools/universities.

The region of Central and Northern BiH as a predominantly rural area (SBK, ZDK, RD, BD) has an extremely unfavorable status, expressed in the size of the low educated population, reaching 75% in the total population over the age of 15. This data is also associated with an indicator of the age structure of the population, from which it is evident that the area of Central and Northern BiH has a median age of 35-44 years. This tells us that this is the area with the predominant population that is in full work force.

3 Analysis of Intermodal Logistic corridor Vc

The main function of the transport system is to insure area integration and social cohesion, as well as economic (market) integration. Competitive transport systems have a fundamental importance for Europe's competitiveness in the world, for economic growth, for job creation and for the quality of everyday life of people. While transport improvements have the potential to yield significant economic benefits, the regional distribution of these benefits is dependent on the level of transport costs. In order to keep low transport costs, we need to have an operational and well-connected intermodal corridor. That kind of corridor will be suggested in this paper as an alternative to already existing international routes. The main idea is that goods from Asia flowa through Suez and continues from the Port of Ploce across the Corridor Vc by rail to Port Brcko and further to Port Vukovar, where the Danube and other goods routes go to Central and Western Europe, eg Rotterdam, Frankfurt, Praha, Wien, Bratislava, Budapest etc.

In Bosnia and Herzegovina (BiH), the main route of intermodal transport corridor should be acccros the Corridor Vc within the Trans-European-Network which runs in the middle part of the country passing through the most favorable conditions – the valleys of the Bosna and Neretva rivers, main railway network and the capital city Sarajevo. Corridor Vc has a strategic position in the European transport system and an ideal geotraffic location, making the connections between the countries of Central Europe, starting from Budapest to the south, through Sarajevo, and to the Adriatic Sea. Because of the proximity of the river Sava and the Danube it allows further connection with the Black Sea, but also with Turkey, providing a good basis for the development of intermodality in the South East European region. By completing all the components of the Vc corridor, it will represent the link between North, Central and Eastern Europe with Southern Europe, which is why the corridor has a great value in the economic and traffic integration processes of the Central European Space. However, the key importance for the development of economy and intermodality in this region has the entry point of the intermodal corridor that accepts goods from overseas countries (Asia and Africa), which is Port Ploče.

 https://doi.org/10.1051/shsconf/20197104010

3.1. Influential (gravitational) corridor zone

Given the spatial and temporal availability, where an average of 30 minutes is needed to overcome the distance of 30 km, an influential (gravitational) zone of Corridor Vc (Figure 3 – in the middle) has been determined. [3]

This zone occupies less than 20% of BiH territory, where 56% of the population is inhabited, and 59% of the total number of employees, accounting for almost 63% of the state's GDP. It covers parts of the HNK, SBK, Sarajevo Canton and ZDK, Doboj Region i Brčko District. In addition, 61 municipalities are in this zone. It runs through four of the six largest centers in BiH, including one of the two urban centers (Sarajevo, the capital). There are many important business facilities where is also located the largest road and rail node.

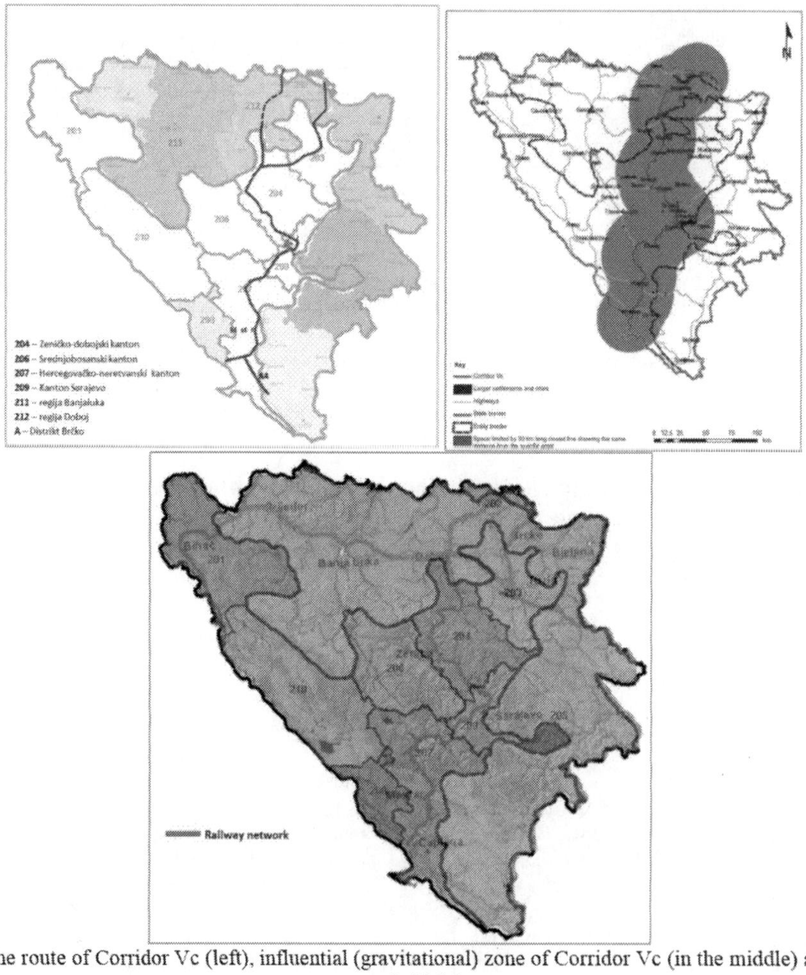

Figure 3: The route of Corridor Vc (left), influential (gravitational) zone of Corridor Vc (in the middle) and railroad network (right)

Source: Authors.

3.2. Strategic plan for optimization of the intermodal logistics corridor Vc

In order for this intermodal corridor to experience the maximum potential in the regional development of BiH, it is necessary to optimize the specific points of the logistic intermodal corridor. The main goal is to take over part of goods flows from Asia and India to the central and western part of Europe, which is passing through the Suez Canal. The framework plan for optimizing the intermodal logistics corridor is given in Table 3. Listed activities are already part of existing development strategies mentioned in the introduction. Activities are mostly based on modernization of intermodal lines (especially the railway network and inland waterways) and nodes (Port Ploče, Sarajevo and Port Brčko to Vukovar). In order to attract the interest of freight forwarders who manage international goods flows to streamline the goods flow through the intermodal corridor through BiH, it is most important to adequately equip the entry point of Corridor Vc, which in this case is Port Ploče.

Table 3: Framework plan for optimization of the intermodal logistics corridor 2016-2030.

Activities in the period 2016-2030.	**2016-2020**	**2020-2025**	**2025-2030**
Modernization of Port Ploče	■		
Modernization of railroads in BiH	■	■	
Modernization of Port Brčko	■	■	
Adaptation of main intermodal nodes in BiH		■	■
Establishment of the operators and application of ICT			■
Taking over part of the goods flows from Asia and India			■

Source: Authors

Linking Port Ploče with the quality transport infrastructure of Corridor Vc with Corridor VII is a prerequisite for the extension of the Port Ploče gravitational area. In order to have that quality transport infrastructure we need to invest in the railway system from Ploče to Vukovar and to the Sava-Danube canal to connect the European Union with the most acceptable modes of transport from the point of view of costs, energy efficiency, ecology and safety: railroads and inland waterways.

By constructing a multipurpose channel Danube - Sava and Pan - European Corridor Vc (Budapest - Osijek - Sarajevo - Ploče), Port Ploče will gain importance in combined and intermodal transport with all the Danubian countries as well as the western part of Europe through the Rhine - Main - Danube channel. The central Adriatic sea link and natural extension of the Vc corridor will trough Port Ploče connect the southern region of Italy with the countries of Central and Eastern Europe to the most economically acceptable way, with better conditions for achieving intermodality in freight transport [4].

Project Implementation of the Park B.E.P.O. Neretva that was designed by the Port Ploče Administration and the Shipping Institute of Zagreb could also contribute to the strengthening of Port Ploče. It is a business zone located in the Neretva valley with an ideal geostrategic location through which passes the Corridor Vc. One of the first strategic goals of this project is to initiate a sustainable economic development of the region within the radius of about 50 km on a geographical area that belongs to the Republic of Croatia and BiH [5].

However, the entry point in the intermodal corridor is not sufficient to redirect the flow of goods through the Corridor Vc. It is necessary to have a connected complete chain, which in this case can be achieved by modernizing the routes through BiH. The first thing to do is adjust the railway network according to the South-east Europe Transport Observatory SEETO standards, and thus join the paths of „Intercontainer", one of the most important systems of intermodal transport in South East Europe. It is also necessary to adapt certain intermodal nodes through BiH, thinking primarily on the largest urban center of BiH, which is the capital city Sarajevo. Sarajevo is the optimum choice for building a modern logistics center from more influential factors, for example due to the highest volume of goods flow in the country in this part, the largest population in this geographical area, the existence of an international airport nearby, the vicinity of the Ploče - Doboj railway line and the influential corridor zone Vc passing through the industrial zone of the city [4]. Contribution to the development of the region and the popularization of this corridor would be improved by enabling the Tuzla airport with greater capacity for freight transport. Tuzla is also interesting because of the fact that the in Tuzla's railway junction faces the lines from Serbia over Zvornik and from Doboj with the connection to Corridor Vc. This position on the railway network offers great opportunities in the area of rail transport. This also applies to Banja Luka.

Fig. 4: Influential zone of Park B.E.P.O. Neretva and potential intermodal nodes through BiH and region
Source: Own elaboration with the support of Google Maps API

The exit point from BiH and the entry into EU countries is Port Brčko, which in addition to the fact that it is located on Corridor Vc, can become a consolidation site for container flows that are coming from the East to BiH and further to the whole region [7]. Thus, Port Brčko has the potential to become the main container center in South East Europe [8].

The establishment of operators with strong ICT support would provide a rational division of labor market, concentration of commodity work and warehouse subsystems, coordination and cooperation of work by selecting the best carrier of transport in the macro distribution and concentration of activities with the unique service of urban areas and industrial zones in micro distribution, with the compatibility of ICT technology in all parts of the transport chain [9]. The last precondition is the establishment of a favorable system of the Schengen Border regime and the functional transboundary movement of people and goods, which would remove all barriers and open the door to trade flows through this corridor.

4. Valorization of Intermodal logistic corridor Vc

Valorization of the proposed intermodal logistics corridor Vc will be presented through comparison of the potential route through BiH with already existing international routes (Suez-Rotterdam-Frankfurt, Suez-Trieste-Frankfurt-Rotterdam, Suez-Constanta-Wien-Frankfurt-Rotterdam).

Since economic criteria (the power of receptive markets, transport costs, time savings, etc.) and environmental protection criteria (the release of poisonous gases into the atmosphere) are now almost decisive in making entrepreneurial and consumer decisions, data on distances, transport costs and emission of harmful substances are taken into considiration. Data are collected and processed for a number of possible travel routes for several cities in Central, North and Western Europe, which are inevitable in making decisions for every reasonable seller and buyer.[10]

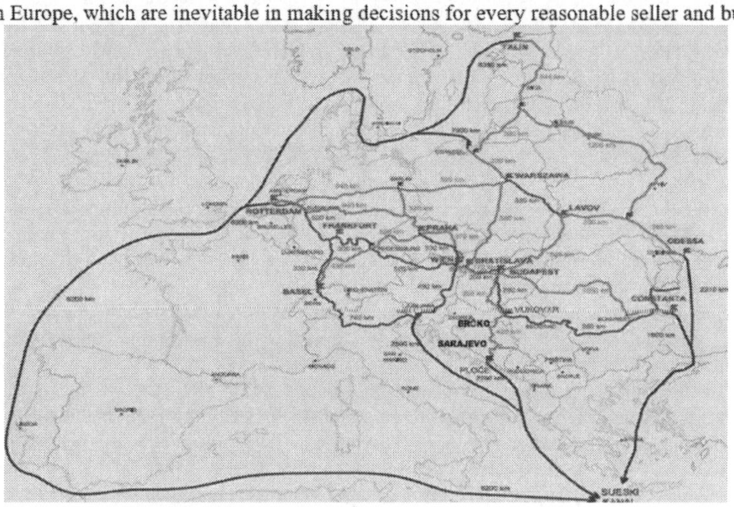

Fig. 5: Display of the distances of certain cities of Europe from Suez
Source: Authors.

Calculations of route distance, done according to the Figure 5 are shown on Figure 6 (left). In order to calculate transport costs we have used the recommendations about energy consumption costs by different modes (the value of the transport cost of 1000 tons of goods). The results are also shown on Figure 6 (right). More detailed analysis is shown in the ANEX 1. It can be seen that directing goods flows through the intermodal logistics corridor Vc will shorten transport time and also reduce transport costs. We can conclude that the revitalization of the intermodal logistics Corridor Vc, made according to the proposed Framework Plan (Table 3), is much more acceptable than lot of other existing routes, and therefore justified.

Using the ship (across the sea and river) and the railway transport in relation to road transport, more than noticeable efficiency in environmental protection is achieved, which is in accordance with one of ten goals of White Paper transport policy: „Thirty per cent of road freight over 300 km should shift to other modes such as rail or waterborne transport by 2030, and more than 50% by 2050, facilitated by efficient and green freight corridor" [11].

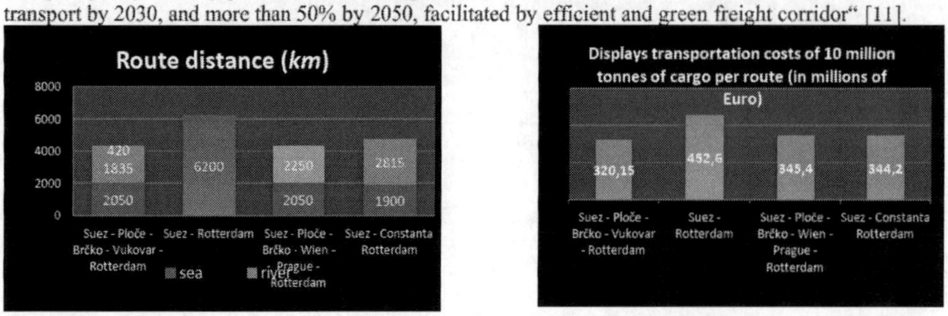

Fig. 6: Route distance (upper); Transportation costs of 10 million tonnes of cargo per route (lower) Source: Aurhors.

5. Conclusion

The planning of an intermodal logistic corridor with supporting operational logistical and distribution terminals, investment prioritization of the use of existing corridor capacities of the regional transport network and affirmation of the transport function of natural resources and ecologically acceptable transport options are necessary frameworks of transport policy in relation to the requirements of sustainable development – economic growth, ecological equilibrium and social progress [12]. In the context of regional development, it is necessary to re-evaluate the pan-European transport corridor Vc and, through the combined instruments of cohesion policy, to devise a unique development strategy for the participating regions, taking into account the set thematic objectives and investment priorities of the common strategic framework, the national development strategy, the partnership agreement and operational programs and territorial cooperation program. In this spirit, the Framework Plan for the Optimization of the Intermodal Logistic Corridor Vc has been proposed, whose chronologically stated activities are already part of existing development strategies. By valuing and comparing with other alternative goods flows, it has been established that this intermodal logistic corridor would have great advantages over existing routes, reflected in shorter transport times, lower transport costs and environmentally acceptable results. We can conclude that revival of the suggested framework of intermodal logistic Corridor Vc has the potential to contribute regional development in BiH and beyond.

References

1. Geohive, *Urban / rural division of countries for the years 2015 and 2025* (2017). URL: http://www.geohive.com/earth/pop_urban.aspx. Accessed: 07.01.2017.
2. Agency for Statistics of Bosnia and Herzegovina, *Population, Household and Housing Census in Bosnia and Herzegovina, 2013, Sarajevo, June*. URL: http://www.popis.gov.ba/popis2013/mapa/?lang=eng. Accessed: 20.09.2019.
3. M. Đuro, Corridor, VC as a factor of integration of Bosnia and Herzegovina into the European Union. *Zbornik Radova: Geografski Institut "Jovan Cviji ć"*, **62**(1), 89-101 (2012).
4. M. Martinovic, *Povezivanje riječnog i pomorskog prometa u europskim lukama. Diss. University of Split*. Master's thesis. University of Split (2017).
5. Foreign Investment Promotion Agency of Bosnia and Herzegovina, *HIPP Group plans to build a plant in the Valley of Neretva* (2014). URL: http://www.fipa.gov.ba/novosti/vijesti/default.aspx?id=5824&langTag=en-US. Accessed: 20.09.2019.
6. H. Pilko, D. Majcan, Ž. Šarić, G. Zovak, Turbo roundabouts: Case study of Croatia. In I. Bajor, A. Vidovič (Eds.) *Proceedings of the International Scientific Conference "Science and Traffic Development" (ZIRP 2018)* (pp. 221-229). Zagreb: Faculty of Transport and Traffic Sciences University of Zagreb (2018).
7. S. Steiner, Air transport development potential in surroundings of Port Ploče. In S. Steiner, V. Koroman, J. Božičević, J.Z. Blašković, M. Nikšić, N. Brnjac (Eds.), *Valorisation of the Intermodal Logistics Corridor Ploče-Mostar-Sarajevo-Vukovar (Middle Adriatic - Danube Region)* (pp. 63-74). Zagreb: Croatian Academy of Sciences and Arts (2014).
8. T. Rožić, M. Petrović, D.Ogrizović, Container transport flows as a prerequisite for determination of inland terminal location. *Pomorstvo*, 28(1), 3-9 (2014).
9. V. Roso, N. Brnjac, B. Abramovic, Inland intermodal terminals location criteria evaluation: The case of Croatia. *Transportation Journal*, 54(4), 496-515 (2015).
10. M. Muschkiet, Binnenschiffgüterverkehr. In U. Clausen, C. Geiger (Eds.) *Verkehrs-und Transportlogistik* (pp. 179-202). Berlin, Heidelberg: Springer (2013).
11. P. Golinska, M. Hajdul, European Union policy for sustainable transport system: Challenges and limitations. *Sustainable Transport*, 3-19. Berlin, Heidelberg: Springer (2012).
12. V. Vassilev, E. Velinov, Monitoring and evaluation of quality in rail transport services: An element of quality management. *Serbian Journal of Engineering Management*, **2**(1), 1-7 (2017).

Corporate Governance Modernization: Legal Trends and Challenges

M.A. Tokmakov[1,*]
*Corresponding author: maxim.tokmakov@gmail.com.

[1]Samara State University of Economics, Samara, Russia.

Abstract. Development of digital technology opens up new opportunities for corporate governance. At the same time, modern law faces a difficult task – to find a balance between creating conditions for development of technologies including by means of non-interference, and providing the stakeholders in corporate governance with proper legal guarantees. This paper considers the impact of some of the most significant digital technologies on corporate governance, such as distributed ledger technology, smart contracts and artificial intelligence. There are certain legal trends and challenges arising from such innovations including the pursuance of sociability, peer-to-peer and decentralization of corporate governance which in many cases is associated with abolishing of bodies (of a part of bodies) for a corporation management, or transferring their powers (a part of powers) to the corporation members and/or to a computer program (artificial intelligence, algorithm, smart contract). Besides, the paper considers occurrences of new subject of corporate relations – crypto-assets (tokens) holders as well as the possibility for recognition of the legal personality of computer programs, in particular, decentralized autonomous organizations and artificial intelligence.

Keywords: corporate governance, distributed ledger technology (DLT), tokenization, crypto-assets holders, artificial intelligence, decentralized autonomous organizations (DAO).

1 Introduction

Digitalization processes taking place in the world affect even the most stable modern society institutions formed over decades and therefore resistant to external effects. Such a traditional area of public relations as corporate governance is not an exception. As a result, the outlines of a completely new, requiring a deep rethinking of the corporate governance paradigm are outlined, based on openness, peer-to-peer and decentralization.

However, modernization of corporate governance challenges the legal system that supports it. Law traditionally tends (at least, it has such a goal) to stability and constancy trying to ensure stability of regulated relations and therefore inevitably lags behind in comparison to technologies. Ultimately, law shall not change as quickly as technology. The law task, on the one hand, is not to interfere with technological progress (legal barriers or legal uncertainty often impede this), and on the other hand, it is necessary to build the necessary behavior models in changing conditions preventing negative consequences from the use of new technologies. At the same time, it makes sense to talk about the need to change the legal regulation only if it is impossible to solve the indicated problems. Altering law in order to please hype or political trends just because technology has altered is not only unreasonable, but also pointless. Forms change, but not the content.

It is traditionally assumed that relations in the area of corporate governance constitute the subject of corporate law (or corporate legislation at least). Without focusing on the problem, this approach will be followed in the paper. Therefore, modern corporate law has a difficult task to find a balance between creating conditions for development of technologies including by means of non-interference, and providing the stakeholders in corporate governance with proper legal guarantees.

Is corporate law ready to introduce new technologies in corporate governance? May these technologies be implemented without legal alterations, or will such changes be required (first of all, radical ones, and not the targeted ones)? And finally, will all the elements of corporate governance, and together with them the corresponding institutions of corporate law, "survive" the transformation? Searching for answers to these questions formed the goal of the present analysis.

The paper considers the effect of some of the most visible (breakthrough) technologies on corporate governance, such as distributed ledger technology, smart contracts, and artificial intelligence. There are also some legal trends caused by the use of such innovations and some challenges that can potentially arise before law as a result for introduction of researched digital technologies into corporate governance.

2 Methods

The present analysis was carried out without reference to law and order of any particular state since the technologies are transnational in nature and exert influence on corporate governance around the world. Moreover, the problems identified in the paper are staged and appear to be relevant for majority of jurisdictions. In this regard, the basis of the analysis, primarily, consisted of methods for scientific generalization, modeling and forecasting.

3 Results

3.1 Using distributed ledger technology (DLT) in corporate governance

3.1.1 Holding meetings

Distributed ledger technology (DLT) offers many opportunities for corporate governance. A lot of papers are devoted to the use of DLT in corporate governance both in Russia and abroad [1-4]. Most often, researchers note a high potential for using DLT in holding general meetings of shareholders due to which reliability, transparency and accuracy of vote counting are achieved. Options for using DLT while holding board meetings are more rarely considered.

Generally, a fair conclusion of such studies is that holding meetings by means of DLT are a form of electronic (virtual) voting known and permitted (with some variations) in majority of law and orders. In essence, DLT does not introduce new content but provides new qualitative characteristics for voting: transparency, reliability, cost reduction, etc. In this regard, restructuring regulation of such relations is not breakthrough in nature and is easily ensured by targeted amendments to national legislation.

At the same time, majority of researchers overlook the fact that capabilities of DLT and its decentralized applications (Dapps) go far beyond usual accounting of transactions (votes). Using DTL, it is possible to create not just a voting system but a full-fledged platform for remote interaction of corporation members with each other and members of corporate bodies in real-time mode, in which transparency and reliability are provided by DLT. Such platform will not only allow, as noted, to hold a general meeting of corporation members much more often, and also not to limit the general meeting to a time frame [2] but will transfer all corporate governance online and make it look like a social network. This means that all the corporate bodies of the corporation including the general meeting of shareholders and the board of directors, will become permanent since any issue can be brought up for discussion and voting within a short time.

The role of corporation members in implementing such model will certainly increase. The general meeting of members may become the leading corporate body making decisions on a significantly wider range of issues. This will entail redistribution of powers in favor of the general meeting of members from the board of directors and even the executive body of the corporation. Potentially, such a redistribution may lead to restructuring the system of corporate bodies and to abolition of some of them.

The described scenario does not seem so implausible considering the global trend in desire to ensure maximum involvement of corporation members in management. At the same time, implementing the indicated model will require serious changes from law – it is necessary to review the structure and functions of corporate bodies, in particular, the role of the general meeting of corporation members as well as provide ample opportunities for formation of competencies of various corporate bodies.

At the same time, the redistribution of powers in favor of the general meeting means imposing additional responsibility on the members for managing the corporation. Furthermore, recognition of corporation members as members of the corporate body and charging them with liability for damnification to both the corporation itself and other stakeholders is not new to most jurisdictions and will not cause serious problems.

3.1.2. Tokenization of corporation assets

Many researchers also associate potential of distributed ledger technology in corporate governance with creation of a system for accounting shares and rights to them by issuing appropriate cryptographic tokens (crypto-assets) [1,3,4,5] which ensures transparency of share ownership and helps on reducing costs for tracking their beneficial owners.

Without denying viability of using DTL in this direction, however, it can be noted that tokenization of stocks (shares) of the corporation is nothing more than the issue of securities or, as noted by the Russian lawyers [6], derivative financial instruments similar to the content-wise existing ones, and therefore shall not encounter special legislative encumbrances.

In view of this analysis, possibility of creating cryptographic assets (tokens) that content-wise are different from known financial instruments is of great interest. Modern trends in attracting investments show desire of corporations to limit investors in profit participation rights by creating different classes of shares for different types of shareholders (founders, premier investors, posterior investors, etc.), as well as by issuing several voting shares or non-voting shares [1], or desire to refuse to grant investors the status of a shareholder completely by means of issuance of other financial instruments (primarily bonds) or the use of alternative methods to attract investment opportunities (crowdfunding, ICO, etc.).

Along with that, many tokens to be issued as part of ICO do not provide for any claim to their issuer, so their holders may not be qualified as holders of securities or creditors of the corporation. Nevertheless, on the one hand, their property interest is inextricably linked with the corporation activities, on which the value of the token depends, and on the other hand, holders of tokens with high liquidity in the secondary market are capable of exerting serious economic and reputational pressure on the corporation.

In this situation, the holders of various crypto-assets issued by the corporation and not providing for the right to participate may be qualified as corporate stakeholders. Herewith, the legal status for the crypto-assets holders is uncertain and largely depends on the token type, the legal regime of which is still not defined or contradictory in many countries.

From the point of view of corporate governance, the law has only to figure out what corporate rights and obligations the crypto-assets holders issued by a corporation should have. It appears that in some cases the interest of the token holder is more likely by nature to be the interest of the shareholder than the creditor of the corporation or another stakeholder. Law needs to consolidate the opportunities corresponding to this interest as well as form an interaction system between token holders and the corporation.

3.2 Using artificial intelligence in corporate governance

Artificial intelligence (AI) technologies have long occupied a niche in various areas of life and, in particular, in the corporate sector. AI can answer calls, perform functions of a customer support service, set reminders, draw up work schedules and meetings, prepare reports, etc., including performing more complex tasks: medical diagnostics, property management and financial consultations, legal expertise and training for sales.

At the same time, in recent years, using artificial intelligence in making managerial decisions has been increasing significantly in many business areas, especially in financial ones. M. Petrin gives the following examples: in 2014, Deep Knowledge Ventures, a Hong Kong-based venture company, integrated the VITAL computer program which is able to provide investment advice to the board of directors; in 2016, the Finnish IT company Tieto informed the public that it had appointed artificial intelligence as a member of the managerial group of the new data management unit; in 2018, California-based software provider SalesForce announced that it was using an artificial intelligence device called the Einstein in its weekly staff meetings and asked Einstein to comment on the proposals being discussed [7].

The above examples confirm that in cases of decision-making based on numerous and complex data sets, computer programs are increasingly surpassing people, especially if AI allows these programs to constantly improve their respective capabilities [8].

Today, using AI is possible only as an assistant consultant and should not encounter any special obstacles from law, since the final decision (or control over the decision) is made by a human being. However, many researchers are confident that in the near future, the auxiliary role of AI in managing a corporation will be transformed into a leading one: the role of AI director (robo-director). The robo-director can work around the clock, process any information available to him, call and use this information almost instantly and perform its functions without payment.

Using AI calls into question existence of many corporate bodies, especially supervisory bodies: board of directors, audit committee, etc. "honesty" and "impartiality" defined by the code (algorithm) are also unquestionable. On the one hand, this solves many problems with corporate liability of managers, on the other hand, it again implies shifting responsibility to its members who decide to "appoint" AI as a manager. The literature already actively discusses the problems of liability for harm arising from using robots [9], there are also attempts to justify the legal capacity and legal personality of AI [10]. Achievements of such research can be crucial for corporate law.

For further development of AI technology in corporate governance, first of all, it is necessary to allow (allow the possibility) the assignment of the responsibilities of the corporate bodies to AI (robot, algorithm, program), which in fact means their abolition, or - to recognize AI as legal personality which will allow AI to become a member of the corporate bodies. Herewith, both of these approaches should be accompanied by a solution to the question for distribution of liability for AI actions.

3.3 Decentralized autonomous organizations

One of the trends for digitalization process going on around the world is craving for transfer the majority of social relations from the real world to the cyberspace. On the one hand, this process is accompanied by "digitization" of existing public institutions and creation of their "digital counterparts," on the other hand, emergence in the virtual space of new, previously non-existent elements generated by technology development.

These were "incorporeal creatures floating in cyberspace without a visible form" [11] – decentralized autonomous organizations (DAO) which are a complex system of smart contracts that operate on the basis of the block-chain which makes the corporate hierarchy unnecessary, creating a peer-to-peer communication system between all members. The traditional model of corporate governance is largely based on the fiduciary responsibilities of corporate managers – due to the fact that the efforts and focus of work processes are important in the DAO structure, management oversight and imposition of legal duties on management are less necessary because there are few or no supervisory bodies [12].

The uniqueness of DAO technology provides for the possibility to create an organization type that can exist exclusively in cyberspace which complicates the legal impact on it. DAO "encodes", i.e. it creates its own rules to which it is subject, and by default, legal rules do not have a legal point of intervention [13].

Potential recognition of legal personality of DAO in any law and order will challenge the prevailing legal entity theories. Consequently, much attention is paid in modern literature to determining the legal form of such organizations, as well as to issues for regulation of their activities [13-15].

Despite DAO virtuality, the creators and participants are quite real, which means both the need to protect their rights and interests, and a possible "point of intervention" by means of which law can "affect" DAO. The bitter experience of the embarrassing high-profile failure of the first DAO in 2016 – DAO project – showed the need to develop a model of legal regulation of relations for participation in such organizations.

4 Conclusion

Development of digital technology opens up new opportunities for corporate governance. At the same time, introduction of majority of them does not require significant alterations in the legal system. Along with that, some technologies establish conditions for a fundamental restructuring of corporate governance which does not fit into the existing legal models and requires their adaptation.

The main trends are aimed at establishing openness, peer-to-peer and decentralized corporate governance which in many cases is associated with abolishing of bodies (a part of bodies) for a corporation management, or transferring their powers (part of powers) to the corporation members and (or) to a computer program (artificial intelligence, algorithm, smart contract).

Modernization of corporate governance generates new subjects of corporate relations – crypto-assets (tokens) holders whose legal position requires the determination, and once again raises the question of the possibility for recognition of legal personality of computer programs, particularly decentralized autonomous organizations and artificial intelligence.

Along with that, this allows to talk about the need to rethink many of the provisions of modern corporate (and not only) law to ensure further technological development of corporate governance.

References

1. S. Blemus, D. Guegan, *Initial crypto-asset offerings (ICOs), tokenization and corporate governance* (2019).URL: https://ssrn.com/abstract=3350771. Accessed: 24.09.2019.
2. A. Lafarre, Ch. Van der Elst, Blockchain technology for corporate governance and shareholder activism. *European Corporate Governance Institute (ECGI) - Law Working Paper*, **390** (2018). URL: https://ssrn.com/abstract=3135209. Accessed: 24.09.2019.
3. L.V. Sannikova, Blockchain in corporate governance: Challenges and opportunities. *Law and Economics*, **4**(374), 27-36 (2019). [in Rus.].
4. D. Yermack, Corporate governance and blockchains. *Review of Finance*, **21**(1), 7-31 (2017).
5. G.S. Geis, Traceable shares and corporate law. *Northwestern University Law Review*, **113**(2), 227-278 (2018).
6. A.I. Savelyev, Some risks related to tokenization and blockchainisation of civil-law relations. *Statute*, **2**, 36-51 (2018). [in Rus.].
7. M. Petrin, Corporate management in the age of AI. *UCL Working Paper Series*, **3** (2019). URL: https://ssrn.com/abstract=3346722. Accessed: 24.09.2019.
8. F. Möslein, *Robots in the boardroom: Artificial intelligence and corporate law*. In W. Barfield, U. Pagallo (Eds.), *Research Handbook on the Law of Artificial Intelligence* (pp.649-670). Cheltenham, UK: Edward Elgar (2018).
9. S.P. Bortnikov, Robots liability or liability for products? In S. Ashmarina, A. Mesquita, M. Vochozka M. (Eds.), *Digital Transformation of the Economy: Challenges, Trends and New Opportunities. Advances in Intelligent Systems and Computing*, **908** (pp. 32-41). Cham: Springer (2020).
10. P.M. Morhat, *Legal personality of artificial intelligence in the field of intellectual property law: Civil law problems*. Doctor Thesis. Moscow: Russian State Academy of Intellectual Property (2018).
11. D.J. Shakow, The Tao of the DAO: Taxing an entity that lives on a blockchain. *Tax Notes*, **160**(7), 929-934 (2018).
12. W.A. Kaal, Blockchain solutions for agency problems in corporate governance. In Kashi R. Balachandran (Ed.), *Economic Information to Facilitate Decision Making*. World Scientific Publishers (2019). URL: https://ssrn.com/abstract=3373393. Accessed: 24.09.2019.
13. U.R. Rodrigues, Law and the blockchain. *Iowa Law Review*, **104**(2), 679-729 (2018).
14. A.J. Kolber, Not-so-smart blockchain contracts and artificial responsibility. *Stanford Technology Law Review*, **21**(2), 198-234 (2018).
15. C.L. Reyes, If Rockefeller were a coder. *George Washington Law Review*, **87**(2), 373-429 (2019).

Blockchain Technology as Efficiency Improvement Tool for the Agricultural Sector

A.E. Bogomolov[1], *L.E.* Popok[1*], *D.N.* Savinskaya[1], and *E.B.* Tyunin[1]
*Corresponding author: lpopok@gmail.com.

[1] Kuban State Agrarian University (named after I. T. Trubilin), Krasnodar, Russia

Abstract. Blockchain technology is primarily associated with cryptocurrency, but the potential of this technology is enormous and can be a base not only in finance, but also in other sectors of the economy. This article discusses application prospects of blockchain technologies in agricultural production. Particular attention is paid to the possibility of the formation of new, more efficient business models at the farm level. Given seven possible areas of using blockchain technology in agricultural production, affecting both small farmers and international relations in the field of food security of the member countries of the Eurasian Economic Community. Also in this article the first successful examples of implementation experiences blockchain technology in agriculture are described.

Keywords: Blockchain, information technology, business model, agriculture, farming, automation of data collection, data digitization.

1 Introduction

The market of agricultural products plays an important role in the welfare of the countries of the Eurasian Economic Community. Food exchange problems, transparent supply schemes, customs clearance and fair regulation of prices for Eurasian Economic Community always been acute.

In agriculture blockchain can become a unique system, which is a source of information about the product status, resources and contracts - from small farmers to large multinational agribusiness corporations. Blockchain technology can radically change our understanding in three areas of the agricultural market: management, financing and building supply chains and mediation.

Blockchain saves time in the value chain of agricultural products, providing a single source of data on production and minimizing the burden in collecting information. Blockchain itself without increasing income producers, but by providing a unique technological infrastructure, it leads to the fact that agriculture is more profitable in today's competition.

2 Problem Statement

Agricultural enterprise - is a complex ecosystem with a variety of elements, whose activity is tied to compliance with schedules, results of other elements. Such systems are characterized mainly seasonal structure funding. Products produced by agricultural producers become elements of the other, larger, logistic system with a lot of participants: distributors, grocery chains, processing enterprises, retailers and consumers. These participants at any time, information on the whereabouts of the goods shipped to food, to which specific farm grown coffee beans, etc.

The trend of development of agriculture using blockchain promises to simplify each stage of food cultivation and distribution. Blockchain provides all parties involved with a single source of accurate information for the agricultural supply chain. In world and Russian practice, there are already examples of implementation of all types of blockchain in agriculture.

3 Research Questions

The ultimate goal of blockchain in agriculture is the digitization and automation of data collection for product quality control. When the manufacturer needs information, it can get this inforamtion almost instantly and provide accurate irrigation, based on the information on rainfall, making conventional fertilizers depending on the soil characteristics and the use of targeted measures to combat pests in response to an outbreak.

On a global scale with the help of blockchain technology Sweden, Ukraine and the United Arab Emirates plan to keep the land registry, and the Indian government is already struggling with a land fraud using blockchain [1]. There are opinions that soon blockchain technology will become as familiar as the data transmission protocol TCP / IP, which in the early days of its existence, also caused a lot of controversy and has been used in the narrow closed networks. In any case, progress is unstoppable and blockchain gradually gaining confidence. The possibility of using blockchain limitless, its functionality depends only on the imagination of developers. The most important thing - to learn to apply the basic principles of its operation and advantageous to apply the benefits of using this system in each case [2].

4 Research Methods

To consider application prospects of blockchain technologies in agricultural production and to describe the first successful examples of implementation experiences blockchain technology in agriculture, the authors used general scientific methods: methods of critical analysis, methods of comparison and generalization, expert assessments, logical constructions and observation.

5 Findings

5.1 Settlement System

In households with a pronounced seasonality (e.g. cultivation and sale of seedlings of fruit, conifers and ornamental trees) sales take a few months of the year. Rest of the time - laying plants and processes for the care of plants. At the same time the funds for the maintenance of the functioning of the economy are needed constantly. The blockchain technology can help to develop a system of payments to customers, which will allow to optimize cash flow. For example, payment on the fact of receiving the products can be replaced with a down payment before setting the seedlings. For a customer, this calculation may be even more favorable, because the seller will be able to offer a discount or installment plan, calculated on the basis of the customer's capabilities and in accordance with the agreement between the parties to the transaction [3].

Saplings of fruit trees, especially mature trees for landscape design - the purchase of an expensive and hardly belongs to the category of spontaneous. Everyone understands that the trees need a few years to find the right size and shape. The customer can choose from the range of his desired varieties of trees and watch them grow, periodically pays advance payments.

Such an approach may be implemented without the blockchain using, but it significantly increases the ease accounting transactions and customer guarantees calm. Of the participants in the transaction excludes lawyers who helped make the treaty. Entry of an order placed in blockchain, see all its members, to deceive any party to the transaction will fail. The system completely eliminates this development.

5.2 Customer interaction system

Blockchain allows you to build a fundamentally new scheme of interaction between farmers and consumers of farm products. Such a scheme may be based on equal footing registered in blockchain. Share is a certain amount of conventional hard currency, and it may acquire anyone. Binding conditional exchange is carried out to the set produced farming products. Share owner can pre-select a set of products and get it in its natural form, or elect to receive a dividend from the sale of products on the open market. Farmer sends the collected funds for production and further provides unitholders latest environmentally friendly products at a reasonable price. Thus, a new production unit facilities that can provide the other shareholders.

The same approach can be implemented in the case where a farmer has a very small farm and has no plans to expand it for various reasons (lack of territory, labor or free time). Suppose his farm consists of a pair of cows, two dozen chickens, a dozen geese, ducks and a few rabbits and twenty acres of land under potato. Instead of a few food baskets, so the farmer can stay on the same and act the same scheme. Farmer registers blockchain share-purse in an amount that believes it is possible, and sell them for cash. For example, the cost of such a share can be $ 500. On this amount the farmer provides its customers with food basket, which includes: three hundred eggs, five liters of milk, three poultry carcasses, two rabbits and twenty kilograms of potatoes.

5.3 Sale of assets of listed cryptocurrency

As an example, consider the production of alcoholic beverages. For this particular winery and distillery, popular manufacturers crafting a beer does not require any changes to its production technology. The only difference is that each bottle or keg manufacturer supplies the individual certificate and Wallpapers label with a unique QR-code, photographs and sends in a cellar for storage or ripening. Further, the manufacturer makes an entry in blockchain thus placing an asset (a bottle or cask alcohol) as an asset on cryptocurrency exchange at an initial (start) price.

Derivative financial instruments are based on cryptocurrency is now very much in demand, and traders will use alcohol cryptoasset as speculative derivative. Thus, traders will trade commitments on the transfer of alcohol to mature as a kind of futures contract for the supply of, for example, in the future wines already present in the fixed price. Such a contract will allow the buyer to insure against excessive prices, but will cause additional costs if commodity prices decline. But, firstly, the price of rare and unique alcohol almost never fall, and manufacturers such scheme implementation is advantageous because it is guaranteed by laying in the price of all production costs.

According to such a scheme can produce assets breeding birds or small animals. For example, a farmer makes an entry in blockchain and release assets for the purchase of geese in 2021. The money raised from the sale of assets, the farmer buys breeding eggs, feed, materials, building goose house, and by the end of the agreed term already has the output of commodity geese.

5.4 Delivery Tracking

In addition to the successful solution of the problems of agricultural producers "on the ground", blockchain technology may be less useful on a global scale.

To this day a huge problem for the world's agro-industrial complex is to track the delivery and payment products. Large multinational corporations sell thousands of tons of products in a geographically fragmented markets. In most cases of placing and receiving orders is coordinated by a third party (agent), functioning locally. Respectively, and the seller, and the buyer must be some agent that acts as a guarantor of delivery of goods, performs its testing and tracking, recommends the most advantageous payment schemes.

Through the use of this complex system of blockchain agents at the regional level can be simplified to a single distributed registry, allow buyers to products contact the supplier directly without any loss of time and money on fees. This will eliminate many of the "paper" processes, because of which the settlements on the supply can be stretched for a few weeks. The use of such a system would allow farmers to receive a large portion of the proceeds from the sale of the harvest.

Louis Dreyfus company recently conducted by blockchain first trading in agricultural commodities. They are sold to the Chinese government 60 thousand tons of US soybeans. According to the people who own the information about the transaction, the using of blockchain reduces total logistics by 80% - the whole operation took only a week. [4]

Moreover, the delivery of operations are often associated with complex agreements specifying conditions of delivery and payment. They are ideal candidates for management using smart contracts with blockchain. Because smart contracts are increasingly being used in agriculture, the farmer can sell the products directly to restaurants or even individuals without intermediaries.

5.5 Products origin and supply chain

Popular in recent years, organic products are increasingly finding themselves not as organic as it says the seller. Publication of fake organic products are increasingly appearing in the press. Of course, this upsets consumers, to spread a lot of money for the 'organic' vegetables or meat, but more importantly, it devalues the efforts of farmers producing these eco-friendly products.

Keeping a register of proven organic products in blockchain - a great tool confirm the authenticity of the product supply chain. Manufacturers can use such a registry to add their products to the history of each phase of cultivation and processing of the product. The final buyer with the mobile app in the store will be able to ensure the ecological purity of the product and to follow all the way from the field to the counter.

When tracing through blockchain if the vendor claims that coffee beans come from ecologically clean mountainous plantations in Jamaica, any customer can easily verify these statements. Agriculture based on blockchain allows to trace the path of beans from the farm to the coffee shop with your phone.

From a regulatory perspective, it would make the work of state bodies in the field of agriculture is much easier. If you find stocks of contaminated food will be easier to track where it came from, and to establish the related manufacturer. As a result, regulators can quickly isolate the diseases associated with food. [4]

5.6 Decentralization of power of international companies

Globally, in modern agriculture is dominated by international corporations. They are often the major resellers in the market, always dictate their prices and indirectly indicate which manufacturers products will be in great demand in the next season.

The implementation of blockchain can turn the existing system in the opposite direction - small businesses and agricultural cooperatives will have a dominant position in the market. The idea is the implementation of the above schemes farm products through the issuance blockchain-shares. Consumers buy a share, and farmers supply their products throughout the growing season. This approach carries with two distinct advantages:

- first, farmers receive funding at the beginning of the season, and invest in the production and stabilization of its seasonal cash flow;
- secondly, consumers are confident in the reliability of its farmers and get the products at the best price according to the season.

Agriculture based on blockchain can solve some problems of management, distribution and equity related to the management of the agricultural enterprise, sponsored by the community. Due to the distribution of shares on the basis of smart contract, community-supported agriculture can significantly improve the efficiency, directly linking farmers with consumers.

5.7 Adjustment of prices

In a situation where commodity markets controlled by large companies, they can hardly be called transparent. Prices of products into separate groups may vary greatly depending on a combination of many factors: the demand, weather, regional scale, with regional mediators, etc. [5].

The using of Blockchain technology in agriculture can significantly improve the transparency of processes of pricing for the products of mass demand for the end-consumer, and for all members of the agrarian market.

Market based on blockchain allows buyers and sellers to compare the conditions offered in the negotiations with information on such transactions, committed in the past or from other market participants. Farmers around the world will be able to determine how much it actually costs their harvest, and to make a sale in real time.

6 Conclusion

The modern development of the world food market is defined by two main groups of factors:

– population growth, and as a consequence, the growth in agricultural production needs;

– concern for the environment, and a marked increase in demand for organic products with transparent and "fair" supply chain.

Thus, in the foreseeable future to agriculture will be in dire need to address two issues: the organization of efficient logistics and finance purchases of food and quality control to guarantee the detection of fraud. Implementation of blockchain technology - the best way to solve these problems.

A single transparent storage of the data supply chain can effectively organize the relationships between suppliers, transporters, consumers and processing industry. The real experience of blockchain in farming activities already allows to simplify the marketing of products and to eliminate intermediaries in the distribution chain.

Organization of certification of production and processing technologies using blockchain will effectively protect bona fide suppliers, thus substantially reduce the passage of certification procedures costs (compared to conventional systems).

Before agriculture Eurasian Economic Community countries worth an ambitious task of export development, active promotion on the world market of high-quality, organic products. Simplify its decision to allow the use of blockchain, thus ensuring the quality and origin of the buyer of goods, simplifying procedures for veterinary and phytosanitary certification.

References

1 A. Genkin, A. Mikheev, Blockchain: How it works and what awaits us tomorrow (Alpina Publisher, Moscow, 2018). [in Rus.].

2 T.A. Nedogonova, L.E. Popok, Prospects of application blockchain. In E.V. Popova, D.A. Zamotaylova, S.A. Kurnosov, R.U. Rakhmetova, A.F. Rogachev, et al. (Eds.), *Proceedings of the X International Student Forum "Information Society: Current State and Development Prospects"* (pp. 180-182). Krasnodar: KubGAU (2017).

3 Cryptor, *Blockchain and agriculture. Interviews with the world's first blockchain farmers* (2018). URL: https://cryptor.net/persona-grata/blokcheyn-i-selskoe-hozyaystvo-intervyu-s-pervym-v-mire-blokcheyn-fermerom. Accessed: 03.06.2019. [in Rus.].

4 International Independent Institute of Agrarian Policy, *Blockchain technology in agriculture* (2017). URL: http://xn--80aplem.xn--p1ai/analytics/Tehnologia-blokcejn-v-selskom-hozajstve/. Accessed: 03.06.2019. [in Rus.].

5 A. Evstigneeva, FTS plans to cancel declarations. The ability to work without workflow can be a revolution in terms of simplifying business. *Izvestiya* (2016). URL: https://iz.ru/news/636682. Accessed: 03.06.2019. [in Rus.].

Motives for Early-Stage Entrepreneurship in Modern Economies

I.S. Pinkovetskaia[1], *T.V.* Gromova[2], and *I.N.* Nikitina[2,*]
*Corresponding author: i.n.nikitina@gmail.com.

[1] Ulyanovsk State University, Ulyanovsk, Russia
[2] Samara State University of Economics, Samara, Russia

Abstract. The relevance of the study is due to the important role played by small and individual entrepreneurship in the vast majority of modern national economies. The aim of the study is to assess the indicators describing the ratio of voluntary, highly demanding and necessity-driven entrepreneurs in the total number of start-ups. The results of the Global Entrepreneurship Monitor project for 2018 were used as input data. Modeling of empirical data and evaluation of differentiation of the considered parameters was based on the development of normal distribution density functions. According to the results of modeling average values and change intervals of specific indicators were established. The originality of the work is associated with the assessment of the share of voluntary, highly demanding and necessity-driven entrepreneurs in the total number of early-stage entrepreneurs in 48 countries. The novelty of the study is due to the assessment of indicators characterizing the current motivation of entrepreneurs, as well as the ratio of the above mentioned indicators. The results of the study can be used in the formation of an entrepreneurial climate aimed at the development of a highly demanding business.

Keywords: entrepreneurship, motivation, countries, voluntary entrepreneurs, necessity-driven entrepreneurs, highly demanding entrepreneurs.

1 Introduction

Entrepreneurship plays an important role in modern economy. It provides a significant increase in the production of goods, operations and services in the vast majority of countries, creates new jobs, increases the level of competitiveness and innovation. Theoretical work and empirical research confirm that entrepreneurship is a key factor in the development of national economies. Thus, the paper [1] emphasizes that earlier industrial policy was aimed at the creation and growth of state-owned companies, and now such policy should be based on partnership between entrepreneurs and the state. Moreover, the relationship between entrepreneurs and the state depends on the level of economic development of a particular country. The article [2] considers how internal and external factors affect the development of small and medium enterprises.

Surveys of Global Entrepreneurship Monitor projects that have been conducted over the past 20 years are of great importance in the study of modern entrepreneurship. In 2018, a survey of working-age adults produced data that characterize indicators of entrepreneurial activity in 48 countries, including data on motivation of early-stage entrepreneurs [3]. Such entrepreneurs are adults (18-64 years) who at the time of the survey are in the process of starting a business or owning a business for less than 3.5 years.

The current methodology divides early-stage entrepreneurs into three main types based on their motives. The first type includes voluntary entrepreneurs. They are adult citizens who use the existing conditions and opportunities to benefit from creating their own business compared to employment. These advantages include the improvement of their financial situation through their own work, independence and autonomy in activities, improving social status, the possibility of self-realization and creative activity. Voluntary entrepreneurs include also the so-called highly demanding entrepreneurs who are oriented towards greater independence and the desire to earn more money than hired labor. They form the second type of motivation of entrepreneurs. The third type includes necessity-driven entrepreneurs (entrepreneurs by necessity). They include adult citizens who do not have any income and therefore start business because of the lack of other opportunities for making money. Motivation of entrepreneurs determines the features of the approach to entrepreneurship. Therefore, many of the necessity-driven entrepreneurs, with the appearance of employment opportunities for hired work prefer to stop business. Their business has a low survival rate. Accordingly, the short-term nature of their activities causes disinclination to invest in their business development. This is the most significant difference between voluntary (including highly demanding) entrepreneurs and necessity-driven entrepreneurs. Voluntary entrepreneurs focus on the development and improvement of business, getting the maximum positive effect from it. The data from the Global Entrepreneurship Monitor project show that voluntary entrepreneurs are more likely to explore new markets, create innovative products and services, and plan to increase the jobs. The share of voluntary, highly demanding and necessity-driven entrepreneurs in the total number of early-stage entrepreneurs is used as indicators of motivation.

Over the past decade, a number of studies have been conducted by foreign and domestic scientists related to the study of the motivation of entrepreneurs. The most interesting among them are the following studies. The article [4] analyzes data from a survey of entrepreneurs in the United States and 32 countries in Europe and Asia. Significant differences between voluntary and necessity-driven business owners in terms of socio-economic characteristics, personality and perception of entrepreneurship support are shown. In [5] it is proved that the higher degree of economic development of countries, measured by the gross domestic product per capita, positively correlates with the level of voluntary motivation of entrepreneurial activity. The article [6] examines the relationship between the development of the theory of entrepreneurship, and the motives of people making efforts to start their own business. The results of the study [7] show that many of those who at the initial stage were necessity-driven entrepreneurs in the UK, at the next stages of the development of their business changed motivation and moved to voluntary entrepreneurs group. The work [8] demonstrates that entrepreneurs are characterized by a higher level of satisfaction with their activities than employees.

The article [9] concludes that potential entrepreneurs in Russia are aimed at self-assertion. However, doing business to increase self-esteem and recognition is difficult because public authorities do not provide support for entrepreneurial activity of the population. The analysis of the published works shows that the studies have not paid sufficient attention to the comprehensive assessment of the motivation indicators of early-stage entrepreneurs in modern economy.

2 Methods

The article is devoted to the analysis of levels of motivation for entrepreneurship in the economy of modern countries. In this case, the data of the Global Entrepreneurship Monitor project for 2018 are used.

The study examined the following indicators characterizing the participation of adults in entrepreneurial activity at its initial stage in 48 countries:

- the share of voluntary entrepreneurs in the total number of early-stage entrepreneurs;
- the share of highly demanding entrepreneurs in the total number of early-stage entrepreneurs;
- the share of necessity-driven entrepreneurs in the total number of early-stage entrepreneurs.

In addition, the following indicators that describe the prevailing relationships of the above mentioned motives for entrepreneurship were evaluated:

- the ratio of the share of voluntary entrepreneurs to the share of necessity-driven entrepreneurs in each country;
- the ratio of the share of highly demanding entrepreneurs to the share of necessity-driven entrepreneurs in each country.

Assessment of the above indicators was based on the construction of density functions of the normal distribution. The construction of such functions, as shown by studies conducted by the authors, allows obtaining unbiased characteristics of the studied economic processes. The methodology of using density functions of normal distribution for estimation of specific and relative indices is given in the article [10].

Spatial data describing proportions and ratios for each of the 48 countries were considered as input data, respectively.

The study tested the following two hypotheses:

- indicators characterizing the motivation for early-stage entrepreneurship have a significant differentiation across countries;
- in most countries, voluntary and highly demanding entrepreneurs predominate in the early-stages of entrepreneurial activity.

3 Results

3.1 Results of a computational experiment

This article presents the models developed by the authors. The development of these models was based on information collected during the survey on the Global Entrepreneurship Monitor project in 2018. As models, we have developed dependencies characterizing the normal distribution of the studied parameters. Such functions (y) describing the share of entrepreneurs belonging to each of the above mentioned types of entrepreneurs (x, %) in the total number of early-stage entrepreneurs are given below:

- the share of voluntary entrepreneurs, %

$$y_1(x_1) = \frac{390.00}{9.82 \times \sqrt{2\pi}} \cdot e^{\frac{-(x_1-71.69)^2}{2\times9.82\times9.82}} \quad ; \quad (1)$$

- the share of highly demanding entrepreneurs, %

$$y_2(x_2) = \frac{357.02}{11.89 \times \sqrt{2\pi}} \cdot e^{\frac{-(x_2-46.93)^2}{2\times11.89\times11.89}} \quad ; \quad (2)$$

- the share of necessity-driven entrepreneurs, %

$$y_3(x_3) = \frac{377.14}{10.36 \times \sqrt{2\pi}} \cdot e^{\frac{-(x_3-23.14)^2}{2\times 10.36\times 10.36}} \quad ; \qquad (3)$$

In addition, normal distribution functions (y), describing the relationship of the established indicators (x) in each country were developed. They are listed below:

- the ratio of the share of voluntary entrepreneurs to the share of necessity-driven entrepreneurs

$$y_4(x_4) = \frac{88.01}{2.16 \times \sqrt{2\pi}} \cdot e^{\frac{-(x_4-3.85)^2}{2\times 2.16\times 2.16}} \quad ; \qquad (4)$$

- the ratio of the share of highly demanding entrepreneurs to the share of necessity-driven entrepreneurs

$$y_5(x_5) = \frac{51.57}{1.50 \times \sqrt{2\pi}} \cdot e^{-\frac{(x_5-2.48)^2}{2\times 1.50\times 1.50}} \quad . \qquad (5)$$

The quality of the developed models was assessed using three tests. The computational experiment showed that estimated values by Kolmogorov-Smirnov test range from 0.041 to 0.110. These values are significantly less than the tabulated value of 0.152 (significance level 0.05). The estimated values by Pearson test range from 0.038 to 4.334 that is less than the tabulated value (9.49). The estimated values by Shapiro-Vilka test exceed the tabulated value 0.93 (significance level 0.01). Econometric analysis of these 3 tests showed high quality of each of the functions (1) -(5).

3.2 Assessment of early-stage entrepreneurs' motivation

Using density functions of the normal distribution (1) - (5), characteristics were obtained that show the existing patterns of prevailing motivation for early entrepreneurial activity in the economics of different countries in modern conditions.

The main indicators of functions (1) - (5) are given in the table. The average values are shown in column 2. Column 3 of the table shows the change interval of the indicators under consideration typical for most (68%) countries.

Table 1. Indicators characterizing motivation for early-stage entrepreneurship.

Indicators	**Average value**	**Change interval of values typical for most countries**
1	2	3
Share of voluntary entrepreneurs,%	71.69	61.87-81.51
Share of highly demanding entrepreneurs,%	46.93	35.04-57.82
Share of necessity-driven entrepreneurs, %	23.14	12.78-33.50
Ratio of the share of voluntary entrepreneurs to the share of necessity-driven entrepreneurs	3.85	1.69-6.01
Ratio of the share of highly demanding entrepreneurs to the share of necessity-driven entrepreneurs	2.48	0.98-3.98

Source: Authors.

The data in column 2 allows concluding that in 2018 the average country-specific share of voluntary entrepreneurs amounted to almost 72%. Therefore, voluntary motivation was prevailing. For most countries, this indicator ranged from 62% to 81% (column 3). The indicators that went beyond the upper boundary of this change interval were noted in Poland, Switzerland, Colombia, Panama, Cyprus, Great Britain and Greece. The indicators that went beyond the lower boundary of this change interval were observed in Brazil, Iran, Angola, Russia, Egypt and India. However, only in Egypt and India the figure was less than 50 per cent (47.5 per cent and 43.2 per cent, respectively).

The average share of highly demanding entrepreneurs in the countries under consideration was about 47 per cent. That is, almost half of all early-stage entrepreneurs in the analyzed countries regarded themselves as highly demanding. For most countries, the figure ranged from 35 per cent to 58 per cent. The indicators above the upper boundary of the interval shown in column 3 of table 1 occurred in 2018 in Panama, the Netherlands, Switzerland, South Korea, Chile, Cyprus, France, Thailand. Lower values of the share of highly demanding entrepreneurs from 25.4% to 31.6% were observed in India, China, Egypt, Bulgaria, Turkey, Italy and Russia.

The average share of necessity-driven entrepreneurs across 48 countries, in the total number of early-stage entrepreneurs was 23%. At the same time, in most countries this indicator ranged from 12.8% to 33.5%. Values above the upper boundary of this range were noted in Egypt, India, Russia, Angola, Guatemala, Brazil, Iran, Lebanon. Values beyond the lower end of the range (namely, from 12% to 7%) were observed in Luxembourg, Cyprus, Italy, Sweden, the Netherlands, Poland, the USA, and Switzerland. It is interesting to note that in countries such as Russia, Japan, Germany, and Italy, a low level of early-stage entrepreneurship and a low share of entrepreneurs who ceased operations were simultaneously observed.

The ratio of voluntary to necessity-driven entrepreneurs averaged 3.8 in all the countries under consideration. The trend of exceeding the share of voluntary entrepreneurs over necessity-driven ones was observed in 46 countries. In one country only (India) the figure was 0.9. Values below the lower end of the range shown in column 3 of table 1, but more than 1.0 occurred in Guatemala, Brazil, Angola, Russia and Egypt. The highest values of the ratio of the share of voluntary and necessity-driven entrepreneurs (from 11.8 to 6.5) were observed in Switzerland, Poland, the USA, the Netherlands, Sweden, Cyprus, Italy, Colombia, Luxembourg, Panama and the UK.

The ratio of the share of highly demanding entrepreneurs to the share of necessity-driven entrepreneurs was 2.5 on average in all the countries under consideration. The trend of exceeding the share of highly demanding entrepreneurs over necessity-driven ones was observed in 43 countries. In five countries, the figure was less than 1.0. The highest values of the ratio of the share of highly demanding and necessity-driven entrepreneurs (from 9.1 to 4.4) were observed in economically developed countries, such as Switzerland, Poland, the USA, the Netherlands, Sweden, Cyprus, Italy, Luxembourg, Panama and the UK.

As follows from the data of column 3, in 2018 there was a significant differentiation of all five analyzed indicators by countries. This is due to social, economic, demographic, historical features of the development of specific countries.

In general, the above analysis confirmed the correctness of the hypotheses about the significant differentiation of the considered indicators by countries and the excess of the share of both voluntary and highly demanding entrepreneurs over the share of necessity-driven entrepreneurs in most countries.

4 Conclusion

The scientific novelty of the research findings includes the following:

- the indicators describing the existing levels of motivation for early-stage entrepreneurship, using the density functions of the normal distribution are assessed;
- the indicators reflecting the correlation of motivation levels are evaluated;
- high quality of approximation of initial data by functions (1) - (5) is proved;
- average values of the considered indicators for 48 countries included in the Global Entrepreneurship Monitor project are established;
- the prevalence in most countries of voluntary and highly demanding entrepreneurs over necessity-driven is confirmed;
- significant differentiation of indicators by countries is proved;
- the countries with high and low levels of the considered indicators are presented.

The models developed in the research process can be applied by government bodies in substantiating strategic documents for the development and improvement of entrepreneurship.

References

1. Z.J. Acs, W. Naudé, *Entrepreneurship, stages of development, and industrialization*. WIDER Working Paper 2011/080 (UNU-WIDER, Helsinki, 2011).
2. P.D. Gupta, S. Guha, S.S. Krishnaswami, Firm growth and its determinants. *Journal of Innovation and Entrepreneurship,* **2,** 15 (2013).
3. N. Bosma, D. Kelley, *Global Entrepreneurship Monitor. 2018-2019 Global Report*. Global Entrepreneurship Research Association (GERA) (2019). URL: https://www.gemconsortium.org/report. Accessed: 25.09.2019.
4. P. van der Zwan, R. Thurik, I. Verheul, J. Hessels, Factors influencing the entrepreneurial engagement of opportunity and necessity entrepreneurs. *Eurasian Business Review*, **6**(3), 273-295 (2016).
5. M. Benz, B.S. Frey, The value of doing what you like: Evidence from the self-employed in 23 countries. *Journal of Economic Behavior & Organization*, **68**, 445-455 (2008).
6. G. Segal, D. Borgia, J. Schoenfeld, The motivation to become an entrepreneur. *International Journal of Entrepreneurial Behaviour & Research*, **11**(1), 42-57 (2005).
7. R.L. De Silva, Business start-up and growth motives of entrepreneurs: A case in Bradford, United Kingdom. *Manchester Business School Working Paper,* **597** (2010). URL: https://www.econstor.eu/bitstream/10419/50712/1/656920246.pdf. Accessed:09.09.2019.
8. D. Bobera, B. Lekovic, N. Berber, The relationship between motives of entrepreneurial behavior and venture maturity. *Industrija*, **45**(1), 81-98 (2017).
9. N.V. Ket'ko, O.E. Akimova, Researching the motivational preferences of entrepreneurs for Russian small business development. *Economic Analysis: Theory and Practice,* **9**, 27-38 (2016). [in Rus.].
10. I.S. Pinkovetskaia, I.N. Nikitina, T.V. Gromova, The role of small and medium entrepreneurship in the economy of Russia. *Montenegrin Journal of Economics,* **14** (3), 177-188 (2018).

Systematization of Methodical Approaches to the Analysis of Economic Entity Sustainable Development

O. Likhtarova[1,*]
*Corresponding author: lihtarova.olga@yandex.ru.

[1] Samara State Economic University, Samara, Russia

Abstract. The main approaches to the calculation and classification of indicators used for assessment of the economic entity sustainable development are considered in the study. Methodological advantages and shortcomings of approaches to the development of the system of indicators of an organization sustainable development within stakeholder approach are analyzed. The results of comparative analysis of the main international and Russian methodological approaches to the formation of the system of indicators for assessment of economic entity sustainable development are presented. The conclusion is drawn that in terms of stakeholder approach methodological approaches to the formation of integrated assessment of the organization sustainable development level on the basis of the formation of indexes and rating models are of great scientific and practical interest.

Keywords: stakeholder, sustainable development, indexes and rating of sustainable development, indexes of development.

1 Introduction

Today the methodology of the assessment of economic entities development on the basis of stakeholders' requirements is widely discussed in the researches of Russian and foreign scientists. Practical interest in the formation of a system of indicators of an economic entity sustainable development on the basis of stakeholders' requirements is caused by the fact that the balance of interests of an economic subject and interested parties are the leading factor of sustainable development of a social-economic system in general in the conditions of the developed market infrastructure [1].

2 Problem Statement

The identification of features and criteria of the analysis, the assessment and monitoring of activity of an economic entity within the concept of sustainable development demands a further specification in the form of the complete system of indicators for tasks of management [2, 3, 4]. The task of the analysis of the main approaches to the evaluating and monitoring of activity of an economic entity was set in the research.

Considering contents and features of evaluating an economic entity development on the basis of stakeholder approach, it should be noted that the important aspect of such an assessment is the dynamic nature of the parties' interest. In general, the degree of the parties' interest and also their concern about the activity of the organization is capable to change. So, at an initial stage only some individuals and legal entities who are stakeholders can express the concern about the activity of the organization [5]. Some time later more stakeholders including those of primary groups can show their concern. In process of other interested parties' more awareness (including through information publicity of problems and contradictions in mass media) all primary groups of stakeholders can show concern about activity of the organization [6].

3 Research Questions

The carried-out analysis of various scientific points of view on the quantity of stages in development of social interactions submitted R.W. Ackerman's works [7], R.A. Buchholz [8], S.C. Zyglidopoulos [9], R. Van Tulder and A. Van Der Zwald [10], S. Zadek [11], allowed to draw the following conclusions. The stakeholders' concern, the level of their interaction among themselves and the extent of intervention in the activity of the organization begins in the *insignificance period* (when only some individuals and legal entities are interested), develops through the period of *high awareness* of stakeholders (it is noted by the growth of attention from mass media, the population, public authorities and local self-government), to the stage of *the institutionalization of new management decisions and the practice* (when there are formal and informal changes in the management practice not only of the economic entity, but also in the institutional environment in general).

4 Purpose of the Study

The purpose of study is the development of the system of indicators for carrying out the analysis, the assessment and monitoring of an economic entity activity. For the achievement of a research goal the following complex of research problems was created:

- definition of the most optimum approaches to the analysis and the assessment of economic entity sustainable development;
- synthesis of the main scientific approaches to calculate and subsequently to group the indicators used for the analysis and assessment of economic entity sustainable development;
- systematization of methodological advantages and shortcomings of the main approaches to assessment of the organization sustainable development within stakeholder approach.

5 Research Methods

The research is based on the system approach assuming complex studying and structural and logical streamlining of a cluster as a complex economic system; methods of structural and comparative analysis were also applied. The solution of the given questions demanded the application of the following general theoretical research methods: abstraction and specification, analysis and synthesis, induction and deduction, comparison and opposition.

6 Findings

At the same time, there were several points of view on the methods of calculation and the subsequent grouping of the indicators used for assessment of an economic entity development in the economic literature. Yu.A. Sedykh developed a model of assessment of the industrial enterprises development which can be also applied within the stakeholder approach. The peculiarity of such a model is the possibility of evaluating the rationality of the choice of one of the integrated directions of an organization capital use (financial, production, human, etc.). This model of assessment consists of four key blocks of indicators which include:

- analysis of the potential of branch commodity markets, works and services;
- identification of the organization development barriers within the considered branch market;
- definition of key indicators of an economic entity development;
- assessment of risk level of conducting economic activity [12].

Russian economist E.V. Korobeynikova suggest carrying out the system assessment of an economic entity development on the basis of cooperation communications and interaction with products consumers. The ranging of development indicators considering their deviations from values, optimum for this industry is carried out within this approach to the assessment of the organization. Optimum values are defined on the basis of expert assessment. E.V. Korobeynikova offer the following groups of development indicators of an economic entity: indicators of economic development; indicators of social development; integrated indicators of development [13].

Methodological advantages and shortcomings of the following approaches to development of the system of indicators of the organization sustainable development within stakeholder approach were analysed in the research:

1. Project approach to the evaluating and monitoring of an economic entity activity within the realization of sustainable development strategy. Methodological feature of project approach is more expressed binding of indicators to specific goals and tasks of the organization and also indicators of their achievement [14].
2. The matrix approach to the formation of a system of indicators of effectiveness and efficiency of an economic entity activity within stakeholder approach which includes definition of strategic objectives achievement extent and problems of sustainable development of the organization on the basis of meeting requirements of interested parties [15].
3. Rating approach to the assessment and monitoring of an economic entity activity for the purpose of formation of an integrated indicator of an economic entity sustainable development. The most significant rating assessments for practice of management are the following: Bloomberg ESG Index, Dow Jones Sustainability Index, etc.
4. The approach on the basis of formation of sustainable development standards consisting in preparation the standardized form of the reporting assuming the disclosure of information on the indicators the organization stakeholders are interested in. It is possible to give Global Reporting Iniciative (GRI) as an example of this approach implementation to assess and monitor the activity of an economic entity.

Respectively, the application of each of the listed above approaches to the assessment of stability and balance of economic entity activity has a certain area and application conditions and has to be agreed with the purposes of the organization strategic development. Therefore, the analysis and critical judgment of the organization purposes and problems of sustainable development allow to choose the most optimum approach and on its basis to create analytical tools.

On the basis of the analysis of the main international and Russian methodological approaches to formation of a system of indicators for assessment of economic entities sustainable development table 1 was made which contains the comparative analysis of methodological advantages and shortcomings of techniques of disclosure of information on the organization sustainable development.

Table 1. Comparative analysis of methodological advantages and shortcomings of techniques of disclosure of information on the organization sustainable development.

Technique of sustainable development assessment	Methodological advantages	Methodological shortcomings
Technique «Bloomberg ESG Index»	- orientation to the information disclosure which stakeholders are interested to; - it is carried out in a section of different types of economic activity for the purpose of carrying out comparative analysis; - ability to integrate into analytical model of key performance indicators of the organization.	- covers the largest corporations presented at the global market; - does not correlate with the system of the Russian accounting and static account, regulations on accounting; - has high specificity of branch indicators.
Techniques «Dow Jones Sustainability Index», «Global 100 Index»	- combination of traditional indicators of the analysis with non-material factors of competitiveness; - it is carried out in a section of different types of economic activity for the purpose of carrying out comparative analysis.	- it is carried out on the basis of voluntary questioning of the organizations; - covers the largest corporations presented at the global market; - does not correlate with the system of the Russian accounting and static account, regulations on accounting.
Technique «Global Reporting Initiative» (GRI)	- broad coverage of indicators of the organization corporate responsibility; - it is focused on the public reporting before stakeholders; - it is constructed on standardization of data with the purpose of comparative analysis; - contains mechanisms of assessment of the organization strategic development and support of management decisions adoption.	it is suitable for sustainable development assessment of the large industrial enterprises; - does not correlate with the system of the Russian accounting and static account, regulations on accounting; - demands high level of training of economists. engaged in the reporting, rather labor-consuming.
Technique of the Russian Union of Industrialists and Entrepreneurs (RUIE)	- has high complementarity with a technique of "Global Reporting Initiative" (GRI); - contains the blocks of indicators differentiated on the main spheres of action (economic. social, ecological); - it is focused on the Russian system in accounting and statistical reporting; contains detailed methodical explanations for calculation of indicators in Russian conditions.	- does not include a number of important factors of sustainable development on the basis of interested parties' requirements: energy saving, labor laws of the person, innovative management, etc.; - the assessment objectivity is in many respects limited a set of standard data for the Russian organization; - the aspect of administrative innovations and standards of corporate responsibility is insufficiently disclosed.
Rating Era Interfax	- it is carried out in a section of different types of economic activity for the purpose of carrying out comparative analysis; - discloses aspects of power, technological and environmental efficiency of the organization; - it is focused on the Russian system of accounting and statistical reporting of the organization.	- it is carried out on a voluntary basis; - rather narrow set of indicators for monitoring and control; - it is not focused mainly on the concept of the company sustainable development.

Source: Authors.

7 Conclusion

Thus, the main international and Russian methodological approaches to formation of a system of indicators considered in table 1 for assessment of sustainable development have both the methodological advantages, and the shortcomings. Together within a research perspective in terms of stakeholder approach methodological approaches to formation of integrated assessment of level of the organization sustainable development on the basis of formation of indexes and rating models are of considerable interest.

The rating systems of an economic entity development assessment within stakeholder approach are of scientific and practical interest in terms of assessment of investment attractiveness of the organization and its long-term social and economic stability. Therefore Russian companies require an organic combination of the principles and approaches put in the system of the corporate reporting "Bloomberg ESG Index" and also a technique Interfax Era Rating which applies aspects of the power, technological and environmental efficiency of the organization. At the same time it is necessary to add to the given techniques a system of indicators of "Global Reporting Initiative" (GRI) which are applied within a technique as the most widely used in the international corporate governance practice and also the technique of the Russian Union of Industrialists and Entrepreneurs (RUIE) focused on the Russian system in accounting and statistical reporting of the organization.

References

1. A.Yu. Yakovleva, O.A. Pisarenko, Open programme stakeholders' expectations – Is it possible to manage? *Russian Journal of Project Management*, **6**(1), 15-24 (2017). [in Rus.].
2. M.H. Zaharia, F.A. Hodorogea, Research stakeholders identification using an mobile entity's framework. *Expert Systems with Applications*, **72**, 18-29 (2017).
3. R.E. Freeman, *Strategic management: A stakeholder approach* (Pitman, Boston, 1984).
4. S. Antognelli, M. Vizzari, Landscape liveability spatial assessment integrating ecosystem and urban services with their perceived importance by stakeholders. *Ecological Indicators*, **72**, 703-725 (2017).
5. V.A. Semidotsky, A.E. Akopyan, The stakeholder-focused methodical approach to strategic planning of development of industrial business. *Journal of Economy and Entrepreneurship*, **6**(95), 1026-1031 (2018). [in Rus.].
6. V.A. Korovina, Assessment of efficiency of the enterprise activity with use of a system of the balanced indicators. *Bulletin of South Ural State University*, **20**(120), 40-51 (2008). [in Rus.].
7. R.W. Ackerman, *The social challenge to business* (Harvard University Press, Cambridge, 1975).
8. R.A. Buchholz, *Public policy issues for management* (Prentice Hall, Upper Saddle River, New Jersey, 1991).
9. S.C. Zyglidopoulos, The issue life-cycle: Implications for reputation for social performance and organizational legitimacy. *Corporate Reputation Review*, **6**(1), 70-81 (2003).
10. R. van Tulder, A. van der Zward, *International business-society management: Linking corporate responsibility and globalization* (Routledge, London, 2006).
11. S. Zadek, The Path to corporate responsibility. *Harvard Business Review* (2004). URL: https://hbr.org/2004/12/the-path-to-corporate-responsibility. Accessed: 27.09.2019.
12. Yu.A. Sedykh, *Theoretical and methodical approaches to branch priorities assessment in business development: On the example of the small industrial enterprises of the Rostov region*. Ph.D. thesis. Rostov-on-Don: Rostov State University of Economics (2004). [in Rus.].
13. E.V. Korobeynikova, System assessment of efficiency of consumer cooperation activity. *Problems of Modern Economy*, **4**(32), 371-375 (2009). [in Rus.].
14. N.N. Kireev, Performance indicators of entrepreneurial activity in the region. *Regional Economy and Management: Online Scientific Journal*, **2**(26) (2011). URL: https://eee-region.ru/article/2605/. Accessed: 26.09.2019. [in Rus.].
15. T.N. Zhukova, Marketing aspects of strategic management of the organization's relationships with stakeholders, *Bulletin of Faculty of Management of Saint Petersburg State University of Economics*, **1-1**, 272-276 (2017). [in Rus.].

Mechanism to Form Labour Potential of Innovative Type: Logistic Aspect

I.A. Toymentseva[1]*, *N.P.* Karpova[1], *O.N.* Denisova[2], and *V.D.* Chichkina [2]
*Corresponding author: tia67@rambler.ru.

[1]Samara State University of Economics, Samara, Russia,
[2]Syzran Branch of Samara State Technical University, Syzran, Russia

Abstract: The role of labor potential in the current economy as the main factor of its development is considered with taking into account the influence of innovative trends in the world economy and the need to develop a competitive domestic economy in the near future. Nowadays, the problem solution to improve the quality parameters of labor potential under current circumstances is possible with taking into consideration the following trends: first, structural changes in the economy of the postindustrial type contribute to an increase in labor potential number in the non-productive sphere. Secondly, destructive changes in the socio-economic system, on the one hand, and the need for further development of the national economy in accordance with the competition laws in the global market, on the other hand, determine the change in the management system of its formation and development. Economic instability of our time especially concerns industrial enterprises that affects the reduction of enterprise competitiveness, and response flexibility of the logistics system of industrial enterprises to environmental changes. Consequently, enterprise management pays more attention to modern methods and ways to conduct activities based on the logistics principles.

Keywords: knowledge economy, effective employment forms, innovative development, labor potential, logistics, regional strategy.

1 Introduction

The relevance of the paper is not in doubt, because there are significant differences in the development level of productive power and living conditions of the population between the Russian regions, thus, in the qualitative characteristics of the labor potential. The labor potential theory is constantly developing and enriching, and the importance to improve its implementation mechanism under current circumstances is steadily increasing. Modern requirements to labor potential allow to develop a set of tools and methods to improve the mechanism of further development of labor potential, including the regional level.

At present, the problem of studying theoretical and methodological provisions that reflect the peculiarities of formation and use of labor potential in the regional economy is becomes actual under the following circumstances:

- multivariance of human behavior in the labor market that contributes to the expansion of the parameters range in labor potential, based on growth rate of regional development and economic situation,

- permanent necessity to constantly search for new forms and effective methods to form and use labor potential of the regional economy for the purpose to achieve multiplicative effect under the conditions of cyber multipolar development.

In order to achieve the goals, which are strategic in nature, it is necessary to build partnerships with all participants of the logistics system in the industrial enterprise [1, 2]. A well-managed supply chain develops strategic and operational objectives, comparing the participants' acts at each level [3, 4].

It is logistics use in strategic management of the industrial enterprise that leads to the optimization of both quantitative and qualitative parameters of flow processes. Thus, in order to increase competitive position in the market, it is necessary to use logistics to manage flow processes in the supply chain [5, 6]. Therefore, it is necessary to adopt the strategic management concept that is the main activity of leading companies.

2 Methods

To identify the enterprise efficiency sources in the supply chain and formation and development of labor potential, the following experimental and theoretical methods were used: system analysis, economic and statistical comparisons and groupings, forecasting and modeling methods, cognitive and normative approaches, as well as methods of demographic research. To determine strategic guidelines in the process of human resources management in the enterprise, a strategic management tool Balanced Scorecard, BSC, developed by R. S. Kaplan and D. P. Norton [7-10] was used. The analysis of the factors of innovative responsiveness is conditioned by the fact that the technological development level is proportional to the innovations scale implemented into production. The mechanism of innovative development of human potential is presented by Becker G.S. [11]. It is evident that the technological achievements of the last decades, responsible for high demands to the labor potential quality, lead to the reason and motivation transformation in labor activity [12-16]. It can be stated that in the future, there will be a labor activity replacement by a new innovative type of labor activity that is peculiar to the postindustrial civilization and based on the creative labor elements. The acceleration of transition to the innovative development contributes to increase in knowledge, experience, working conditions improvement, wage growth, life quality growth and, accordingly, to the need to form the employee's innovative genotype (figure 1, Appendix A.).

The main idea is that competences formation of the innovative person should be started before school, because the skill foundations for critical perception of information, creativity, ability to work in a team and socialization skills are laid at this stage. The further education should develop and complement these skills (and, if necessary, form them from the beginning level). Thus, the conditions to form employee's innovative genotype and have effects such as possibility to diversify the employment spheres, employment mobility extension and implementation of labor ambitions, are created.

3 Results

Current problems of economic environment regionalization, related to the sustainable socio-economic development of subfederal systems and their harmonious incorporating in the convertible meso-economic environment, lead to a new look at the problem of effective formation and development of labor potential on the territory. labor potential is a multi-level integral category that includes a set of possibilities, evaluation of able-bodied population, as well as social, moral and other characteristics that contribute to their implementation at the achieved level of productive forces and production relations. labor potential is closely related to the productive forces and objectively determined by both their development level and the development level of production relations in the society. Thus, the level of their development determines the essence, nature and quality of current labor potential. According to the authors, it is advisable to distinguish 3 types of labor potential that are differentiated by a number of characteristic features (table 1).

Table 1. Differentiation of labor potential types based on the goals of the multivariate strategy of regional development.

Item	Innovative and progressive	Stabilizing and inertial	Depressive and regressive
Basic immanent qualities	High degree of manageability due to the high level of education and availability of necessary professional competencies. High degree of adaptability to innovations over a long period of time.	Quite high degree of manageability. At the same time, the further development and use of labor potential do not provide the maintenance of sustainable socio-economic development of the territory in all directions, but stimulate the development of only certain areas characterized by high efficiency in the medium term.	There is not integrity and structure due to the insufficient level of professional and personal qualities. At the same time, there is a low level of labor activity that requires additional external influence.
Influence on territory development	It is innovative technologies donor. The formation and development experience of this type can be extended to other territories without significant changes. Within the existing regional socio-economic system, it is an incubator of innovative ideas and innovations.	It is an example of positive experience in innovation technologies that, however, cannot be extended due to specific features, as well as insufficient power to extrapolate stabilizing functions in the economy of other territorial formations.	The experience of applied management technologies cannot be extended to other territories due to low efficiency. It is the main consumer of other types experience due to insufficient mechanism of formation and development.
Method to manage labor potential	It is characterized by complexity and system approach that assumes high efficiency of applied methods of management.	It is inertially extended and characterized by a selective approach to labor potential formation with the aim to level certain parameters in accordance with the economic development strategy.	It is intrusive and suggests imperative impact from outside.
Educational status	High	Moderate	Low

Source: Authors.

Qualitative characteristics of labor resources formation and development depend on the level of economic and social development in the region. They are determined by the sectoral, professional and qualification structure of those employed in public production, standard of living, development of health and social security systems, education and training. Thus, to a certain extent, polarity and asymmetry as two complementary categories are, ex ante, peculiar to labor potential of any spatially localized system, due to various natural, functional and systemic qualities, uneven use and development, impact of socio-economic trends and different degrees of state and public institutions efficiency. The presence of permissible asymmetry in the labor potential development has a positive impact on the functioning of regional socio-economic systems, as it affects the spatial distribution of labor application and localization of specialization zones of labor activity.

In Russian regions, there is a real opportunity to develop the labor potential of innovative and progressive type, but some of its elements are presented in varying degrees in different federal districts and regions. The current employment state and the labor market in the regions of the Russian Federation shows quite significant differences in the opportunities to form and develop labor potential of innovative type. The presence of such differences is reflected both in the development speed of labor potential of innovative type and possibility to synchronize the processes of labor potential formation and development in different regions.

Industry and territory determine the specifics of both the reproduction process, as a whole, and labor potential functioning, in particular industries and regions. If the reproduction process is characterized by territorial distinctness, the industry in the spatial sense is very fragmented, and therefore, indirectly affects the process of labor potential formation and use through its enterprises located in the region. The industry is represented by a set of relatively separate enterprises, each located and operates within a specific geographical unit.

Analysis of strategic directions and specific forms to develop the labor potential management system should be correlated with the goals of regional economic policy in the labor resources, taking into account the need to develop basic regional clusters and the preservation of the regional specialization priority. The intensity to develop innovative labor potential depends on the technological development level in the region, adopted consumption patterns, flexibility, entrepreneurial behavior and culture, regional management quality that provides conditions for commercialization of innovative technologies, and effective employment.

Changes in public production affect the nature of human activity. They change not only the type of activity, but also the nature of its values, needs and preferences. Under the current production conditions, labor change is both natural need and a qualitatively new feature of modern management of the multi-structured system of labor potential. It is aimed at providing the necessary scientific and technical progress with employment mobility and changes in labor qualities.

Thus, labor potential in the modern economy reflects the labor opportunities and people's ambitions as read in conjunction with information, economic and social aspects of its use. The initial structural unit of the labor potential analysis is labor potential of innovative type that forms the basis for the formation of labor potential of higher structural levels. Its demonstration is effective innovation and entrepreneurial activity. In other words, the way to realize labor potential is full employment, that is, the maximum use of employee's labor abilities in producing economic benefits. The strengthening of the personal principle in the human needs and values structure, and the dominant of intellectual and social capital among the sources of growth of the material well-being identified the most important methodological postulate of the theory of innovative economy: a person becomes a subject, not a factor of the economic system.

The economic and social backgrounds for the labor potential realization are diverse and change throughout a person's life. Therefore, the mechanism of labor potential management can and should be continuously improved under the influence of innovative development in the regional economy.

Overcoming the economic crisis consequences and further economic growth cannot be achieved without a deep restructuring and innovative economic development, and a positive decision of population problems in the region. Therefore, for the regional economic growth, it is important to determine the place of the regional system, its financial basis for development, sources of funds and mechanisms for their attraction. All this is possible through studies of the regional economy as a whole system with its set of reproduction relationships. Such a mechanism should be based on the theory of the regional reproduction process that is the renewing circular movement of physical, financial and labor resources in regions. At the same time, resources should be directed both to the development of reproduction conditions and functioning of the reproduction process.

The paper reveals that the most important aspect to improve the mechanism of labor potential management is to focus on the effective use, systematic reproduction and continuous improvement of labor force (figure 2, Appendix B). We offer the following principles of the mechanism suggested to manage formation and development of regional labor potential on the basis of the logistic approach:

- the principle of improving the institutional environment quality that provides development and implementation of measures to improve the social infrastructure of the society, corresponding to the intellectual sector of the regional economy,
- the principle of cooperation between regional authorities, education and entrepreneurship that provides development of communication methods to create promising commercial technologies, new products (services), attraction of investment resources for research with the participation of all concerned parties (business, science, government and civil society), and improving the legal and regulatory setting in scientific, technological and innovative development,
- the principle of innovative entrepreneurship and education that prioritizes the education system and problem-oriented educational technologies aimed at teaching the skills, which are necessary for the innovative behavior formation, including analytical, critical and creative thinking, and contributing to the continuous self-learning, readiness for predictable risk and work in a highly competitive entrepreneurial environment,

- the principle of innovative education nature where competencies formation of the innovative person should be started before school, because the basics of skills for critical perception of information, the ability to nonstandard solutions, creativity, ingenuity, communication skills and diligence are laid both at this stage and in primary school,
- the principle of innovative activity motivation that is based on the use of all regional mechanisms (economic, financial, organizational, technological and information), contributing to competitiveness increase of individuals in the labor market.

The institutional basis for the development should be a network of territorial and industrial entities in the form of innovative educational zones and centers for the educational technologies transfer that are specialized in innovations promotion both at the level of particular industrial enterprises and administrative territories.

The cyclical pattern of the process to form labor potential of innovative type that is implemented in the course of employment provides, at each stage, knowledge increment, which is realized through production of new or improved products/services, creating a new technology of the labor process, combination of new factors and improvement of employees' intellectual and innovative potential.

Further evolution of the local society and regional economy involves strengthening of competitive advantages, based on the use of knowledge as a labor potential resource that are used for effective market conduct of all market participants (companies, organizations, sectors, fields of activity).

4 Discussion

The main purpose of the paper was to develop theoretical, methodological and methodical approaches to quality management of labor potential for effective human activity on the basis of modern production and educational environment formation, and the corresponding innovative development of information and socio-economic infrastructure in the regional society.

According to the scientists, innovation as a qualitative characteristic of labor potential reflects the basis of employees' adaptability to the requirements of the modern economy. It is aimed at changing the employees' behavior on the basis of new models development of labor behavior and motivated by the need to obtain a competitive advantage of the enterprise.

According to the analysts, the problem of labor potential innovation should be segmented into several levels. The first level should include the ability to generate new knowledge that implies a high degree of fundamental science development. The second level involves the fundamental knowledge transformation into innovative and applied researches and workings. The third level generates demand from the state and business for new technologies and developments through creation of institutional conditions that would stimulate and facilitate the know-how commercialization.

In the course of the research, it was revealed that labor potential, under the circumstances of economy of innovative type, is a difficult, dynamic and multifunctional category that characterizes professional, qualification, psychophysiological and personal characteristics of employees, defines a new quality of work (dynamic, information and creative intensive), turns into product, technological, managerial, organizational and social innovations at all levels of the economic system, and generates a synergy of social and economic effect at the individual, personal and social levels. Thus, it is established that the main law of labor potential development of innovative type is the emergence of innovative labor elements that provide improved adaptation of the labor force to the requirements of the economy of knowledge, and have both probabilistic and deterministic nature.

5 Conclusion

The condition for realization of actual labor potential is the interest in creativity and constant participation in innovative search. There are some distinctive features of a current-day employee. They are not only aspiration to increase knowledge, abilities, skills and personal development, but also urgent need to have all this. Creativity is a kind of conscious human activity. It corresponds to a highly organized type of workers, i.e. human capital.

The transformation of labor into creativity will be a fundamental factor that will ensure not only the human person's progress, but also sustainable economic growth. The indicated trend leads to the structuring of a new employment type according to the following areas: redistribution of social services and labor from the industrial sector to the professional one, improving the educational structure of employment, growth in the proportion of women in the employment structure, intellectualization of the job structure, creative community increase, dilution of the traditional sectoral structure of employment (influenced by the processes of diversification, integration and conglomeration), development of the cyber employment structure, blurring of lines between formal and informal employment of creative workers, emergence of many transitional forms of labor distribution (table 2).

Table 2. Classificational qualities of innovative employment forms.

Innovative employment forms	Qualities of innovative employment forms	Structuring of innovative type of employment	Trends in development of innovative type of employment	Result/ Institutional regulation entity
New technologies to search for a job /employee	Economic qualities (continuous updating of the structure and forms of employment, new labor quality, demand for new forms and types of employment and their effectiveness)	Redistribution of labor from the industrial sector to the professional and social services sector	Individualization	Increment of knowledge/ Intermediary institutions in the labor market
Innovative labor contracts	Organizational qualities (new labor and employment management, variety of integration and cooperation forms of innovative labor)	Development of the educational structure of employed people	Increase in creative staff shortage	Manufacture of new or improved product and services/ flexible labor laws
Freelancing, cyber employment, employment in virtual organizations	Institutional qualities (emergence of fundamentally new or improved employment institutions, etc.)	Development of the cyber employment structure, blurring the lines between formal and informal employment of creative workers	Virtualization	Development of a new technology of the labor process/ New educational standards and technologies
Directly innovative labor in organizations	Technical and technological qualities (intellectualization and computerization of workplaces, etc.)	Intellectualization of the professional structure and creative community growth	The emergence of innovative labor elements that provide adaptation improvement of employment to the requirements of economy of knowledge	Combination of new factors of the process/Institute of licensing
Polyvectorial and intensive labor mobility	Social and psychological qualities (employee's innovative genotype, motivation to innovative work, etc.)	Increase in the share of women in the employment structure	Reduction of the gender gap in total employment	Improvement of employee's intellectual and innovative potential/Institut e of social partnership

Source: Authors.

As a result, we have two multidirectional trends: the first is labor socialization that covers workers of creative labor and the second is labor desocialization that covers workers of formulaic labor. Polarization of social status provokes discrimination of routine workers, which means that the innovative solution of this conflict is possible only on the basis of new quality development of labor potential, taking into account the information and socio-economic components of labor under the circumstances of innovative economic development.

References

1. D. Kumar, Z. Rahman, Sustainability adoption through buyer supplier relationship across supply chain: A literature review and conceptual framework. *International Strategic Management Review*, **3**(1-2), 110-127 (2015).
2. D. Waters, S. Rinsler, Global logistics: New directions in supply chain management. 7th Edition (Kogan Page imited, London, 2014).
3. T. Lee, H. Nam, An empirical study on the impact of individual and organizational supply chain orientation on supply chain management. *The Asian Journal of Shipping and Logistics*, **32**(4), 249-255 (2016). DOI: 10.1016/j.ajsl.2016.12.009.
4. A.E. Saak, Traceability and reputation in supply chains. *International Journal of Production Economics*, **177**, 149-162 (2016). DOI: 10.1016/j.ijpe.2016.04.008.

5. G. Kumar, R.N. Banerjee, P.L Meena, K.K. Ganguly, Joint planning and problem solving roles in supply chain collaboration. *IIMB Management Review,* **29**(1), 45-57 (2017).
6. P. Banchuen, I. Sadler, H. Shee, Supply chain collaboration aligns order-winning strategy with business outcomes. *IIMB Management Review*, **29**(2), 109-121 (2017).
7. R.S. Kaplan, Conceptual foundations of the balanced scorecard. *Handbooks of Management Accounting Research,* **3**, 1253-1269 (2009).
8. R.S. Kaplan, D.P. Norton, *The balanced scorecard: Translating strategy into action* (Harvard Business Press, New York, 1996).
9. R.S. Kaplan, D.P. Norton, *Balanced scorecard. From strategy to action* (Olimp-Biznes, Moscow, 2017). [in Rus.].
10. R.S. Kaplan, D.P. Norton, *Strategy maps: Converting intangible assets into tangible outcomes* (Harvard Business School Press, Boston, 2004).
11. G.S. Becker, *Accounting for tastes* (Harvard University Press, Cambridge, 1998).
12. D. Bowersox, D. Closs, *Logistics. Integrated supply chain* (Olimp-Biznes, Moscow, 2017). [in Rus.].
13. V.I. Sergeev, *Corporate logistics. 300 answers to questions from professionals* (INFRA-M, Moscow, 2013). [in Rus.].
14. R.E. Lucas, On the mechanics of economic development. *Journal of Monetary Economics*, **22**(1), 3-42 (1988).
15. F. Machlup, *The production and distribution of knowledge in the United States* (Literary Licensing, LLC, Whitefish MT, 2012).
16. M. Porter, *Competitive advantage: How to achieve a high result and ensure its stability* (Alpina Publisher, Moscow, 2016). [in Rus.].

Appendix A

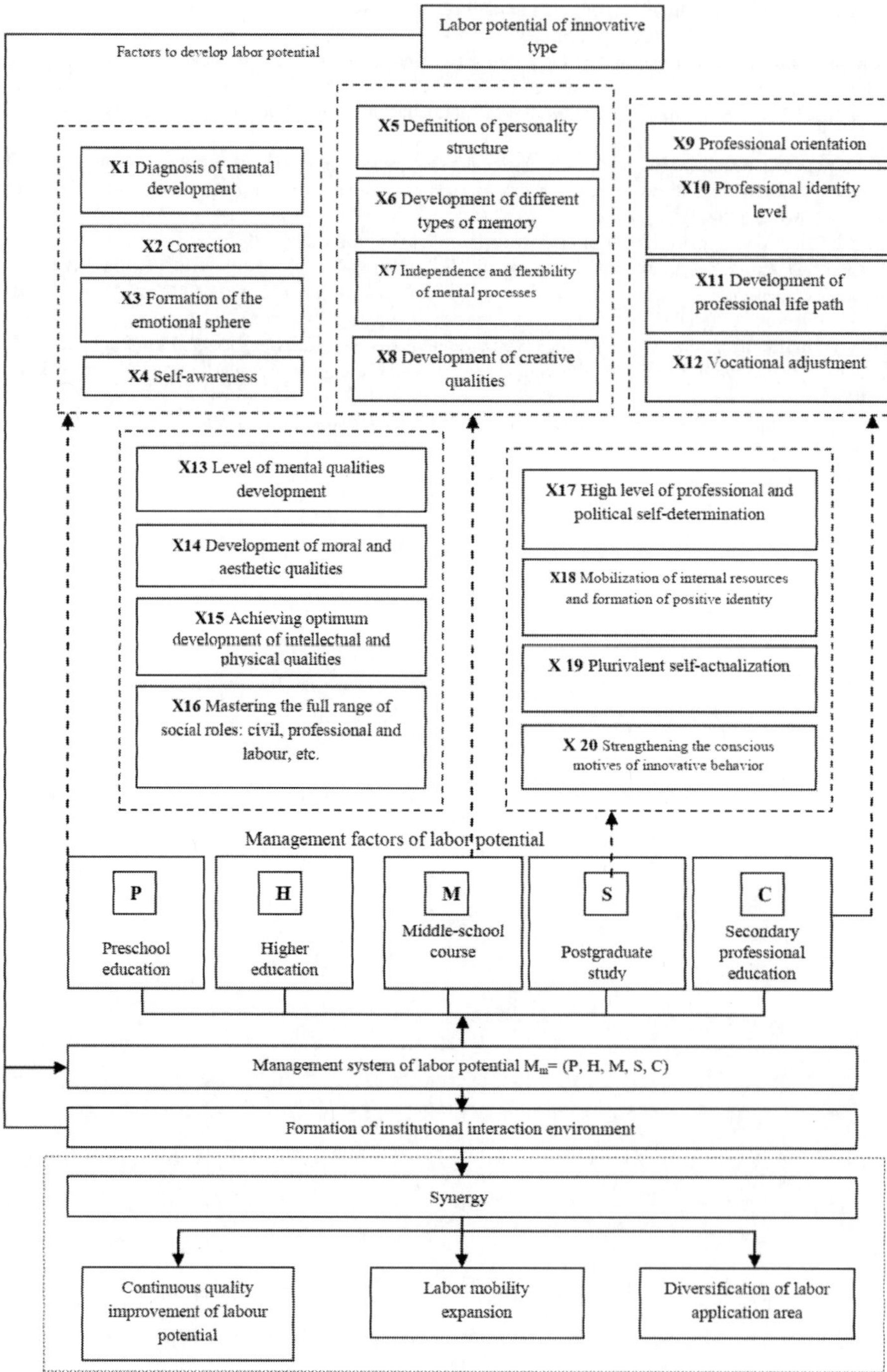

Fig. 1. Stages of the process to form the innovative type of labor potential
Source: Authors.

Appendix B

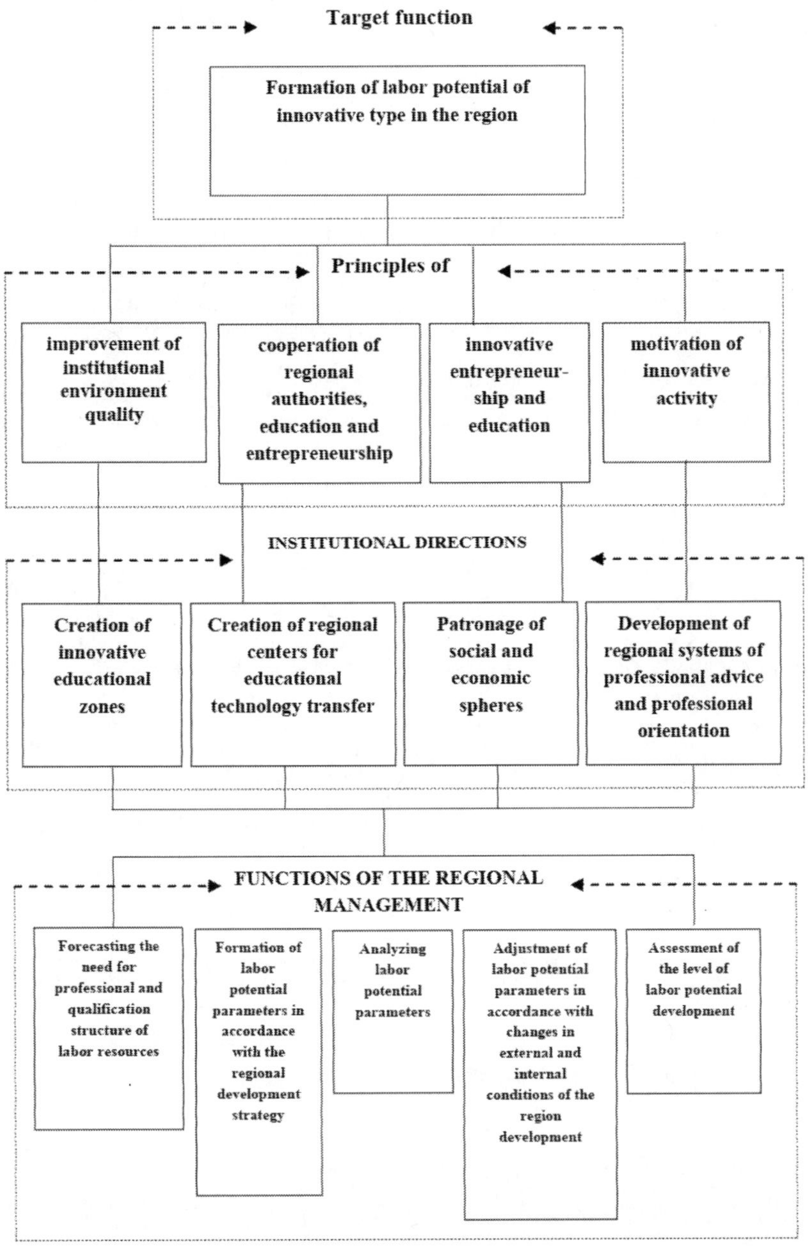

Fig. 2. Organizational and economic conditions to form and develop labor potential of innovative type in the region
Source: Authors.

Digital Platform of Industrial Cooperation – Innovative Direction of Regional Industry Development

O. Pudovkina[1,*], and *S.* Sharokhina[1]
[*] Corresponding author: olechkasgeu@mail.ru.

[1] Syzran Branch of Samara State University of Economics, Syzran, Russia

Abstract: The authors consider the role of the digital economy in Russia and some related processes. The aim of the study was to develop recommendations for the formation of a digital platform for the industrial cooperation. It is obvious that Russia is not a leader by the index of business digitalization (the leading countries are Finland, South Korea, and Netherlands). The authors make a conclusion about the feasibility of using a digital platform for industrial cooperation, which allows linking resources and consumers of products and services in a single information space. They also consider the creation of a regional subcontracting centre as the organizational model focused on the cooperative relations formation for production chains of large, small, and medium business, as well as on the stimulation of innovations through technology transfer, promotion opportunities for products of industrial enterprises on regional markets.

Keywords: digitalization, subcontracting, industrial enterprises, cooperation, digital technologies.

1 Introduction

The "digital economy" is the subject of endless debate and discussions. Today, we are witnessing fundamental changes in the society and economies due to digitalization. The transformation of national economy sectors should be the basis, because this transformation is one of the factors that is the key to the success of the Russian economy: its highly innovative and future-oriented industrial potential.

Industrial production and related services are at the heart of the national economy. They represent more than a half of its economic indicators. 2019 is declared as a year of regional development and digitalization of the country. Nowadays, we are forming a new industrial structure based on various digital technologies. Digitalization is not a new trend, but a new way of life.

The development of digital technologies plays a key role in the development of the modern society. In recent years, all the languages of the world were penetrated by such terms as 'smart city', 'smart CRI' and, in this context, Russia is no exception. Digitalization has become the main topic of news, and state programs 'Digital economy', 'Digital energy', 'Digital agriculture' and others have become a fundamental development trend not only of the state, but also of all major industrial holdings.

This course on the industry reorientation to perspective technologies is caused by the need for developing domestic technologies, reducing the "brain drain" and aging of qualified personnel, and also stimulation of digital development of industrial sector in the Russian Federation [1].

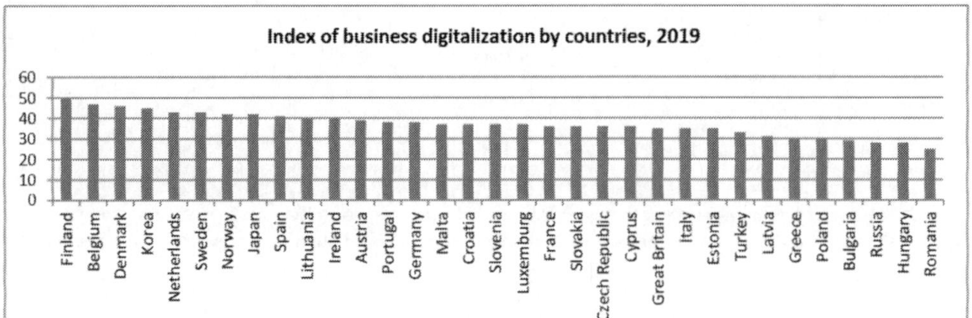

Fig. 1. Index of business digitalization by countries (Source: Authors based on Rosstat data [2])

Figure 1 shows the business digitalization index by countries. As we can see, Russia is in the group of developing countries implementing technologies for economic relations. Leading positions are occupied by countries such as Finland, South Korea, and Netherlands. Taking into account development trends of the digital future, which was emphasized by the Russian President who said that "digitalization is the general path of economic development", Russia has an exceptional opportunity to realize its potential during the digital revolution and take its rightful place among the leaders, effectively implementing the basic principles and goals of the digital economy.

The digital transformation trends from 1998 to 2018 are presented in figure 2. 1998 – the first stage of automation (30% of the total number of business processes of the company). The further process was marked by the transition from the automation of individual processes to the digital components of Industry 4.0. The process of forming Industry 4.0 in Russia has its own specifics [3]. In some studies, it is noted that Industry 4.0 is the "4th industrial revolution", which is characterized by advanced digitalization and integration of industrial production and logistics processes, the use of the Internet and "smart" objects (machines and products) [4, 5].

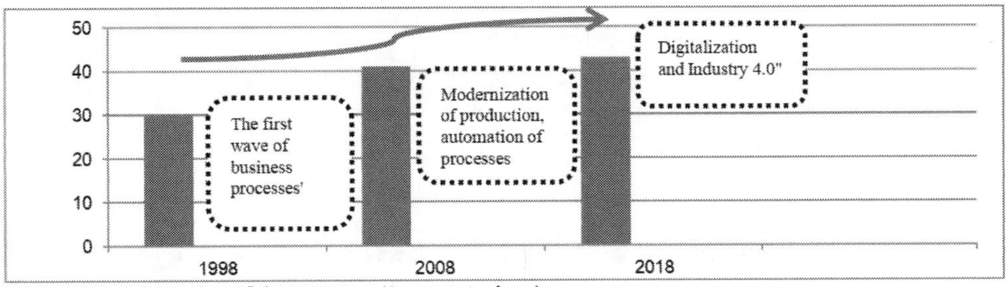

Fig. 2. Digital transformation of the economy (Source: Authors)

2. Research Methods

To achieve research tasks the authors have used a complex of scientific methods. The main research method was content analysis. This method allowed identifying the most functional information resources necessary for this study.

3 Literature review

Analysis of the scientific works revealed the fact of significant interest to the topic "industry 4.0, digital economy", its various aspects. At the same time, most scientists and practitioners agree that the digital economy is a complex concept that is based on the convergence of technologies, but in real, there are no scientific works are in both domestic and international databases about it [6]. As a result, the aim of the study is to analyze the main trends in the digital economy in Russia and develop recommendations for the formation of a digital platform for the industrial cooperation. In addition, the authors have formed a list of key stakeholders for the implementation of the digital transformation projects on the regional level that have the potential for integration and cooperation at various interaction levels to exchange large data amounts to make reasonable management decisions.

The history of industrial automation begins in the 80-s of the XX century, when the first programmable controllers for maintaining various technological processes appeared. Over time, they began forming an automated information systems that have large input / output modules, which allows you to receive information about the process in real time.

With the development of digital technologies, automated enterprise management systems (AEMS) were connected to the so-called MS systems that allowed collecting information on the optimization of the quality control process. In fact, MS system is software allowing you synchronize and manage the quality within any production. Despite the fact that MS systems are shop floor systems, they can be integrated with each other within the whole enterprise. MS system tasks include process equipment management, load tracking, personnel management and others.

The next level of software is ERP. It is a system that manages several production shops or technological platforms in order to manage production assets (such as personnel, finance, industrial resources, raw materials, etc.). Today, most enterprises in the Russian Federation use the level of MS and ERP systems to manage their production. Of course, modern digital technologies have an impact on the structure of the production management system, from the level of control systems and to the level of ERP system, making it more perfect, optimal and convenient. These technologies include cloud computing, artificial intelligence, machine learning, 3-D printing and others.

The merger of industrial automation and information technologies has led to the emergence of a new ecosystem of industrial Internet broadcasting. Participants of this system are engaged in the development of new approaches to the production management, testing and commercialization of these solutions, they produce many new innovative products and implement them in various industries. For example, an innovation that originated within a University and was picked up by talented technology entrepreneurs (the pattern recognition technology in the assembly of units can be tested on the engineering company and subsequently enter the list of products – from digital design to assembly, and the customer will be able to use it as a complete solution). The ecosystem is based on a large amount of data that is generated every second in industrial enterprises. Data is becoming a valuable asset and 'fuel' for all digital innovations. Their competent use allows receiving new products and services, making correct administrative decisions. The assessment of the benefits of using IT in the industry is shown in figure 3.

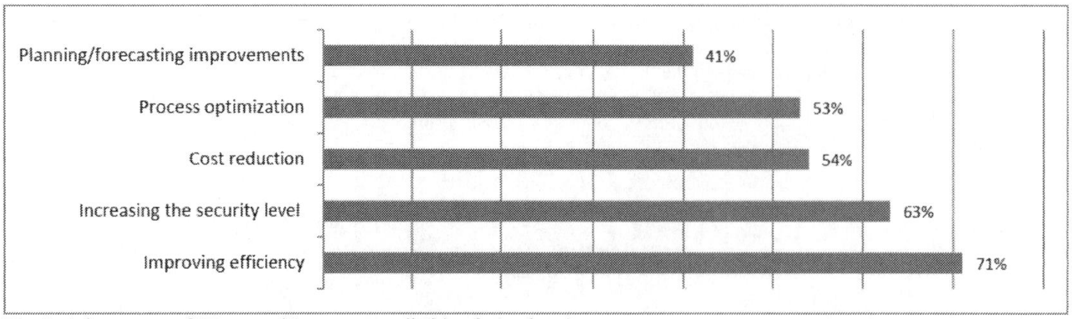

Fig. 3. Advantages of IT usage (Source: compiled by the authors)

As a result, in our country there is a digital transformation of industry, which means the process of transition of the industrial sector from one technological mode to another through digital technologies in order to improve the efficiency and competitiveness of enterprises. Why are digital technologies attractive for the industry? They allow increasing the production efficiency. Modern digital technologies enable not only to collect information from a large number of sensors and devices, but also accumulate, structure, analyze these data by artificial intelligence methods, make conclusions, offer practical steps to optimize industrial production with a ready financial calculations. Thus, we get an assessment of production efficiency not only for the technical staff of the organization, but also for financiers and management. In general, the concept of Internet of things is a concept that allows you to collect data from different enterprise systems sensors into IT systems and more effectively manage the production.

An important characteristic of the regional economic development is the innovative activity of enterprises. In December 2014, the Russian government adopted Federal law No. 488-FZ of 31 December 2014 "On industrial policy in the Russian Federation" [7]. According to this law, one of the main goals is the formation of high-tech competitive industry, ensuring the transition of the state from export of raw materials to a new innovative type of the development. The law defined the industrial information and digital system as one of the main tools for the implementation of the industrial policy.

The above circumstances are the basis for the recommendations development to form a digital platform for industrial subcontracting and technology transfer within the region, which will strengthen industrial capacity and implement national projects in the industrial field. The launch of a digital project will solve the cooperation problem. Those, who have access to the platform of digital products and services, will get undoubted advantages. The digital platform is an innovation that changes the existing market structure and creates new markets [8].

4 Results

The essence of the proposed concept is to create a uniform digital platform for subcontracting industrial enterprises in the region. A fundamental prerequisite for determining effective prospects of the digitalization projects realization in the context of cooperation is that all industrial enterprises are built and working according to common standards and use common standards for reporting information about the life cycle of products (standards SPDP (System of Product Development and Production), USDD (Unified System of Design Documentation), USTD (Unified System of Technological Documentation), USTPP (Unified System of Technological Preparation of Production, etc.). Therefore, the key conclusion in the context of the definition of the digitalization strategy is a principal possibility of development of a standard, configured under specific features of the economic activity, program of digital cooperation of an industrial complex. The platform is a virtual territory where business connections are formed between different participants who build their business and innovations in this virtual information space.

Subcontracting has proved its effectiveness in the international practice, where the main way of organizing production is to rely on the cooperation of small, medium and large businesses. The use of the subcontracting mechanism provides an opportunity for a parent company (contractor) to avoid unproductive costs for the maintenance of underutilized productions and to concentrate efforts on the most important tasks - modernization, updating of a model range of products, technological re-equipment. Subcontractors (usually small and medium enterprises) achieve a high level of equipment utilization and high productivity performing work on subcontracts. The use of the mechanism of subcontracting allows to optimizing the production process and significantly increasing competitiveness, both at the enterprise and regional levels.

The organizational model of the regional digital economy platform includes as a basis a technological platform where the main managing tools can be implemented to provide functioning of the economic digital space. This model provides interrelation of needs and opportunities of suppliers and consumers of resources, products and services on the basis of innovative technologies and digital economy principles. The fundamental tasks of the platform of industrial cooperation and subcontracting are optimization of production processes by placing orders by industrial companies for the development, production and maintenance of industrial products, as well as implementation of technological processes at other industrial enterprises, which makes it possible to build a more effective organizational structure of the production and optimally load production facilities.

The database of the digital platform of industrial cooperation contains information about industrial enterprises and their products. All this information will be available at the centre for subcontracting and industrial cooperation. Through the center, orders will be placed by industrial giants for the development, production and service of products. The regional subcontracting center will coordinate the work of information management systems of its member enterprises, based on the strict centralization of management functions, caused by the need to control the most important elements of the unified system at the stage of its formation and early development.

In the conditions of a volatile economy, enterprises face some common problems, to overcome which the state provides support. For example, when there is a shortage of working capital. In order to help a company to find these funds and, for example, apply for a subsidy for the production development or for the development of new technologies, we propose to create a navigator of support measures. These are three simple questions that the company needs to answer in order to form a suitable list of services (the choice of industry, enterprise size and a problem type). On the basis of these data, a huge list of support measures will be formed on the digital platform, where there is information about the contact person, regulations on the provision of support measures, and the necessary documents. In the case of enterprise authorization, the enterprise can subscribe to this application. For each measure of the state support, it will be possible to obtain a list of all necessary documents. The electronic signature confirms the legal significance of the application. It should be noted that the successful implementation of this idea requires a fairly equal contribution from representatives of both the public and private sectors of the economy [9].

Further, after receiving subsidies, the enterprise is obliged to report on the implementation of projects supported by either the regional industrial development fund or the federal one. On the basis of the digital platform, here will be enterprises with import substitution projects providing information on the monthly and quarterly basis. These data are the basis of reports that will be considered by the government, all the representatives of the federal and regional authorities will have access to it. It is possible for regional ministries of the industrial development and federal sectoral departments to introduce import substitution plans here.

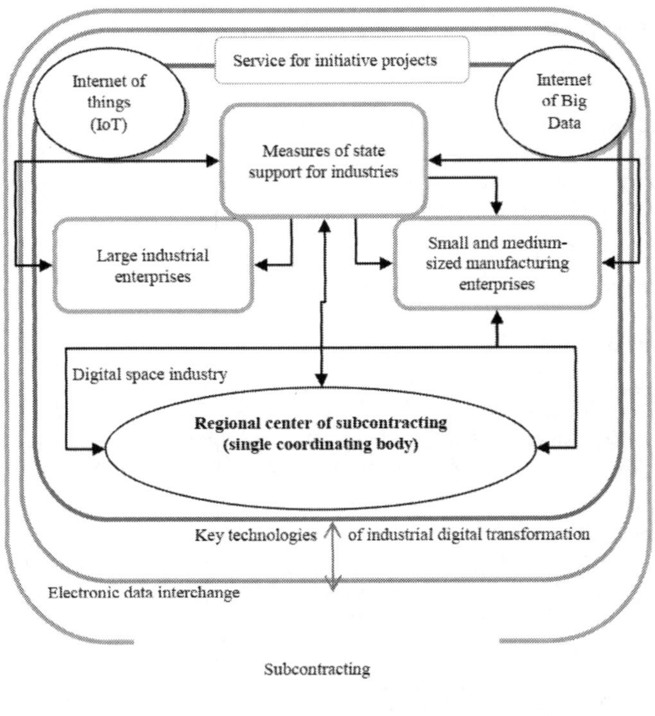

Fig. 4. Organizational model of the digital platform of the industrial cooperation

As an option for the possible structure of the organizational model of this platform, we can consider the set of ecosystems presented in figure 4 with the allocation of a regional center for subcontracting. The service of initiative projects is necessary in the case if an enterprise (having a project or an idea) needs to attract investors and a development institution for financing. The logic of the project is so that the company fills in the application form and this form is accessed by representatives of different development institutions. For example, the industrial development funds, these funds are interested in the development and investment issues in a particular sector of the economy. This service allows combining businesses with investors.

As a platform, the industrial information system should combine a number of important thematic blocks related to the monitoring of import substitution, monitoring of the financial and economic condition of the backbone enterprises. The digital platform will create conditions for effective interaction, reducing the time spent on investment projects. Digital industrial cooperation is able to integrate the production processes of many enterprises, coordinated actions of which will make the competitiveness of their business much higher [10].

When creating a platform, it is necessary to lay down a number of principles:

- openness of data and openness of services in order to allow any participant of this platform to publish data on measures of state support both at the regional level and at the level of industrial development institutes;

- single entry – once collected data about the company, they can be reused, providing a single view for all participants.

In order to form a digital platform for industrial cooperation and subcontracting, members implement the activities presented in figure 5.

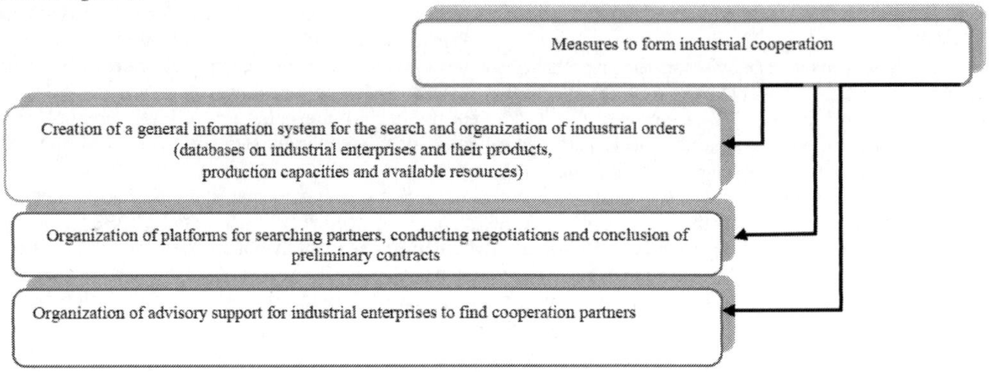

Fig. 5. Measures for the formation of the industrial cooperation (Source: Authors.)

5 Conclusion

As a result of the analysis the following conclusions can be made:

1. The processes of digital transformation of economic relations, which are the essence of the 4th technological revolution (Industry 4.0), are actively developing in the world nowadays. Russia is not a leader by the index of the business digitalization.

2. The platform of digital products and services is the most important element of the regional digital economy, where different business relations are formed between participants (business entities) and innovations in this virtual information space, coordinated actions of participants will make the competitiveness of their products much higher.

3. The proposed organizational model of the digital platform for the regional industrial cooperation is universal and can be applied to most industries in different regions of the country. It is designed for the effective digital interaction and connects businesses with authorities and with each other. Due to this fact, subjects of the Russian industry have an opportunity to receive the state support, find producers and suppliers, take part in purchases. The results of the introduction of a unified interaction system should lead to an increase in the competitiveness of domestic enterprises, activation of sales, growth of research and development practices and their implementation in the real business practice.

References

1. D.S. Lologi, T.H. Mirzaev, The method of technology's transfer in services of industrial cooperation on the basis of the state digital platform. *Innovation Economy*, **3**(16) (2018). URL: http://37.18.92.27/web-local/fak/rj/index.php?id=25&mod=rubric&mm=36. Accessed: 30.09.2019. [in Rus.].
2. Rosstat, *Digital Economy* (2019). URL: https://issek.hse.ru/data/2019/02/27/1193920132/NTI_N_121_27022019.pdf. Accessed: 1.10.2019. [in Rus.].
3. T.G. Stroiteleva, E.Y. Kalinicheva, G.G. Vukovich, V.S. Osipov, Peculiarities and problems of formation of Industry 4.0 in modern Russia. In Popkova E., Ragulina Y., Bogoviz A. (Eds.), *Industry 4.0: Industrial Revolution of the 21st Century. Studies in Systems, Decision and Control*, **169** (pp. 145-153). Cham: Springer (2019).
4. J. Schlick, P. Stephan, D. Zühlke, Produktion 2020: Auf dem Weg zur 4. industriellen Revolution. *IM – Fachzeitschrift für Information Management und Consulting*, **27**, 26-33 (2012).
5. E. Bucherer, U. Eisert, G. Gassmann, Towards systematic business model innovation: Lessons from product innovation management. *Creativity and Innovation Management*, **21**(2), 183-198 (2012).
6. E.B. Babayan, R.R. Timergaleeva, I. Yu. Grishin, Distributed infrastructure of the industrial region digital economy ecosystem. *International Journal of Open Information Technologies*, **7**(7) (2019). [in Rus.].

7. The State Duma of the Russian Federation, Federal law No. 488-FZ of 31 December 2014 "On industrial policy in the Russian Federation" (2014). URL: http://base.garant.ru/70833138/. Accessed: 1.10.2019. [in Rus.].
8. M.N. Polkovskaya, N.I. Fedurina, On the development of digital platforms in Russia. In *Proceedings of All-Russian scientific-practical conference with international participation "Formalization as the basis of the digital economy»* (pp. 9-15). Irkutsk: Publishing house of Irkutsk SAU (2018). [in Rus.].
9. M. Kassen, Adopting and managing open data. *Aslib Journal of Information Management*, **70**(5), 518-537 (2018). DOI: 10.1108/AJIM-11-2017-0250.
10. E.S. Kopkova, E.K. Grechkin, Digital industrial cooperation. Development of digital industrial platforms. *Synergy of Sciences*, **31**, 27-34 (2019). [in Rus.].

Modernization of Managing a "Smart City" on the Basis of Digital Platforms

E.V. Pogorelova[1*], *T.B.* Efimova[1], and *V.A.* Haitbaev[2]
*Corresponding author: jour.ru@gmail.com.

[1] Samara State University of Economics, Samara, Russia
[2]Samara State Transport University, Samara, Russia

Abstract. The main challenges facing modern cities initiate review of approaches to management of city development. Combining of information systems and "smart" equipment used in various spheres of activity is analyzed. It causes digitization of social and economic processes which are not the optimal ones, and management mistakes are transferred to the sphere of digital transformation. In this research it is proposed to consider the development of digital platforms of a "smart city" according to systematic transformation of social and economic processes into the view optimal for management whilst taking into account psychological and cognitive characteristics of city dwellers. This issue may be solved by creating digital platforms of a new class which are based on neural management of knowledge and artificial intellect.

Keywords: artificial intellect, knowledge neural management, smart city.

1 Introduction

Stable development of a "smart city" must be oriented towards achievement of high living quality level whilst a complex of various indicators of social, ecological and economic character possesses positive dynamics. At present managing a "smart city" demands not only application of new theoretical approaches, but also development of the corresponding methodology and techniques. The suggested research is devoted to development of theoretic and methodological bases and practical recommendations in creation and development of a new class of digital platforms built on the basis of the concept of neural management of knowledge and artificial intellect and increasing efficiency of managing a "smart city".

2 Problem Statement

The main challenges which modern cities encounter (migration growth, excessive density, transport problems, growing ecological pressure, changing of demands of dwellers and business to the quality of the city environment and provided services) are the reason of revision of approaches to management of the city development which is carried out on the level of a "smart city", as a rule. Therefore combining information systems and "smart" equipment used in the housing and utilities sector, health care, education, security and other spheres of activity is considered. This often leads to digitization of the social and economic process which is not the optimal one, as a rule, i.e. management mistakes are transferred to the sphere of digital transformation. Therefore using another approach which could bring digitization of social and economic processes to the end user – population - is necessary here. The centre of strategic developments issued the report "A state as a platform. (Cyber)state for digital economics" in May 2018. It states that a citizen that will interact with the digital ecological system and obtain digital services from it in accordance with his demands is located in the focus of development of the state being a platform [1, p 1].

For this very reason in this research it is suggested to consider the development of digital platforms of a "smart city", as digital platforms serve as catalysers of economy development [2, p.10].

The specific feature of our approach consists in the fact that digital platforms of a "smart city" are created on the basis of systematic transformation of social and economic processes into the view optimal for management, with due regard to psychological and cognitive characteristics of the city dwellers. This problem may be solved by means of creation of new class digital platforms based on neural management of knowledge and artificial intellect. Therefore the main purpose of the research is development of theoretic and methodological fundamentals and practical recommendations for creation and development of the new class of digital platforms built on the basis of the concept of neural management of knowledge and artificial intellect increasing the efficiency of management of a "smart city".

3 Research Questions

For the purpose of solving the set goal it is necessary to solve the following items: find out and systematize global trends of creation and development of digital platforms of a "smart city" based on the research of international experience, digital products and services markets; investigate areas of "Neural net" under the conditions of digital economics of a "smart city"; form classification of digital platforms used within the scope of a "smart city" and find out locations and characteristics of use of the new type of digital platforms functioning in accordance with "Neural net" principles; develop the concept of neural management of knowledge and form instrumentation of the artificial intellect realizing its basic principles; form the model of the digital platform of a "smart city" on the basis of the concept of neural management of knowledge and artificial intellect, validate the methodology of the life cycle; develop mechanisms of management of digital platforms development according to the spiral (from a separate digital platform to a bush of platforms, from a bush of digital platforms to the digital ecological system, and from the digital ecological system to new kinds of digital platforms); develop the mechanism of forming the pool of digital platforms of a "smart city" built on the basis of the concept of neural management of knowledge and artificial intellect taking into account demands of all spheres of the city activity; create models of designing digital platforms of a "smart city" on the basis of the concept of neural management of knowledge and artificial intellect; investigate and form instrumentation of designing digital platforms of a "smart city" on the basis of the concept of neural management of knowledge and artificial intellect.

4 Purpose of the Study

For the purpose of solving the set goals global trends of creation and development of digital platforms of a "smart city" based on the research of international experience and markets of digital products and services were determined. Models and algorithms of channels of neural communications in the fields of education, health care, housing and utilities sector, transport are proposed. A concept of neural management of knowledge including principles, priorities, orienting points, criteria, mechanisms, instruments of realization in the fields of education, health care, housing and utilities sector, transport is suggested.

This concept includes models of neural transfer of knowledge, models of neural generation of new knowledge and (or) improvement of available knowledge according to the schemes "man-man", "man-collective", "man-digital system", principally new complex crowd sourcing models of generation of new knowledge synthesizing the schemes "man-man", "man-collective", "man- according to the schemes "man-man", "man-collective", "man-digital system", models of search of inanimate sources by a subject, models of search of animate sources of knowledge by a subject, models of finding knowledge in animate sources of knowledge on the basis of neural preferences, models of finding knowledge from network sources, knowledge and data bases on the basis of visual and psycho-emotional characteristics of subjects.

The digital system of neural management of knowledge is a complex system functioning on the basis of self-organization and evolution approach. Self-organization is carried out by means of coordinated two-sided interaction of digital platform users on the basis of incoming neural signals. The evolution approach consists in gradual self-training of a digital platform on the basis of incoming neural signals and support of decision-making on the basis of the usefulness function.

The digital platform of neural management of knowledge is convenient for the user from the cognitive point of view, it also decreases transaction costs. Structures and properties of digital platforms containing channels of neural communications in the fields of education, health care, the housing and utilities sector and transport are summarized [3, p.55]. In connection with the fact that non-verbal meta communication is an essential addition to verbal communication, models of digital platforms of a "smart city" based on neural management of knowledge for the fields of education, health care, the housing and utilities sector and transport are proposed [3, p.55].

The life cycle of a digital platform with channels of knowledge neural communications and neural management, as well that of a bush of digital platforms with channels of knowledge neural communications and neural management and a digital ecological system with channels of knowledge neural communications and neural management are shown. An algorithm of planning the pool of digital platforms of a "smart city" built on the basis of the concept of knowledge and artificial intellect management is developed in accordance with demands of all spheres of the city activity. Notations (models) describing designing of digital platforms using knowledge and artificial intellect neural management are developed. Instrumentation of designing digital platforms of a "smart city" on the basis of the concept of neural management of knowledge and artificial intellect is proposed.

5 Research Methods

In this research a complex of economic methods and methods concerning artificial intellect and data analysis was considered. We investigated digital platforms, the use of which decreases transaction costs, connection with voluntary mutually profitable cooperation and transaction costs is described as Coase theorem. In accordance with the latter "it is always possible to change the initial legal rights differentiation with the help of market transactions. It goes without saying, if such transactions are concluded without excessive costs, this rights differentiation will always take place, in case it opens the way to the growth of production value" [4, p. 100].

Methods of bilateral markets are also considered, because any digital platform of knowledge neural management functions under the conditions of the bilateral market, and the platform is capable of influencing the volume of transactions by increasing the price on one side of the market and reducing it on the other one- in other words, the pattern of prices is of great importance in such market [5, p. 385].

Approaches of organizational transformation taking place when developing and introducing digital platforms of knowledge neural management were used. The essence of organizational transformation consists in closer involvement of the electronic site into the organization of interaction of government customers and suppliers. This will increase the cost of its services for both parties, but we believe that costs growth will be compensated due to the improvement of interaction quality [6, p.57].

Out of fields of the artificial intellect and data analysis, possibilities of application of various neural networks – multilayer perceptrons, recurrent neural networks, radial basic functions – were considered. Characteristics of direct methods based on direct management and indirect methods when the neural network is used for performing auxiliary management functions are investigated [7, p.79]. One of the popular approaches to the management systems synthesis is the predictive control named in foreign literature as «Model-Based Predictive Control» (MPC) [8, p. 110].

Artificial Neural Networks are artificial and simplified models of the neurons that exist in the human brain. They can be used as a black box approach to create models of systems profiting of the facility to model non linear (as well as linear) systems [9, p.251]. Artificial Neural Network (ANN) is a set of artificial neurons which are mutually connected similar to the biological neural network [10, p. 20]. Artificial neural networks are a flexible instrument for solving a lot of problems including non-linear regression, supervised learning and pattern recognition, unsupervised learning, associative memory, optimization tasks etc. [11, p.186]. Most neural network researchers would agree that the question of whether batch or on-line training is faster and/or 'better' has been settled [12, p.1429].

According to its composition and designation each system is functionally restricted, as it solves certain tasks and requirements [13, p.62]. Many methods of inductive learning were developed as an alternative to classical statistic approaches and are closely connected with the methods of information extraction and methods of intellectual data analysis (data mining) [14, p.66].

6 Findings

Models and algorithms of neural communications in the fields of education, health care, the housing and utilities sector and transport are investigated. The concept of knowledge neural management is formulated. Digital platforms are treated as the main digital object of a "smart city" due to the fact that one of the advantages of the platform business model is provision of direct cooperation and exchange between subjects and decrease of their transactions and other costs [15, p. 22]. Generalized structures of digital platforms including neural communications channels in the fields of education, health care, the housing and utilities sector and transport are developed. These digital platforms are defined as digital platforms of knowledge neural management. Their life cycle and algorithms of development management are described. New models describing designing of digital platforms using artificial intellect are created.

7 Conclusion

Thus, in this research global trends of creation and development of digital platforms of a "smart city" based on thorough analysis of world trends are analyzed. Variants of new solutions application based on the principles "Neural net" are proposed. Of special importance is the concept of knowledge neural management. These developments may be useful in the modern environment of transformation of RF cities into "smart" ones; they will facilitate realization of the program "Digital economics of RF" and will form the basis of further research in this aspect.

References

1. M. Petrov, V. Burov, M. Shklyaruk, A. Sharov, *State as a platform. (Cyber)state for digital economics* (Center for Strategic Research, Moscow, 2018). [in Rus.].
2. V.Yu. Anisimova, L.N. Semerkova, Development of digital platforms and information technologies in the Russian Federation. *Journal of Samara University. Economics and Management*, **9**(4), 7-11 (2018). [in Rus.].
3. D.V. Popov, Consciousness as interface in communication structure. *Manuscript.* **12**(2), 54-60 (2019). [in Rus.].
4. V.V. Dementiev, Innovation: Between the coase theorem and the Hobbes theorem. *Journal of Institutional Studies*, **11**(1), 95-114 (2019). [in Rus].
5. I.A. Strelets, Features of bilateral markets network benefits. *Economic Relationships*, **9**(1), 383-392 (2019). [in Rus.].
6. R.A. Knyaznedelin, I.D. Bekmurzaev, V.A. Titov, Digital platforms as a tool for efficiency enhancement in the system of state procurement in Russia. *Journal of Tver State University. Economics and Management,* **2**(46), 53-61 (2019). [in Rus.].

7. A. Chernodub, D. Dzuba, Overview of neurofeedback techniques. *Programming Problem*, **2,** 79-94 (2011). [in Rus.].
8. K.V. Zmeu, N.A. Markov, B.S. Notkin, I.A. Shipitko, Model-free predictive inverse control with recycled reference step response. *Computer Science and Control Systems*, **3**(21), 109-117 (2009). [in Rus.].
9. J. Vieira, F. Dias, A. Mota, Comparison between artificial neural networks and neurofuzzy systems in modelling and control: A case study. Paper presented at the *IFAC Proceedings Volumes (IFAC-PapersOnline)*, **36**(12), 249-255 (2003). DOI: 10.1016/S1474-6670(17)32543-0.
10. A.S. Sorokin, V.Yu. Bondarev, E.L Krotova, Artificial neural network as a means and method of statistical data processing. *Journal of the Urals Federal District. Security in the Information Sphere,* **2**(20), 19-22 (2016). [in Rus.].
11. V.G. Tsaregorodtsev, Parallel implementation of back-propagation neural network software on SMP computers. In V. Malyshkin (Ed.), *Parallel Computing Technologies. PaCT 2005. Lecture Notes in Computer Science*, **3606** (pp.186-192). Berlin, Heidelberg: Springer (2005). DOI: 10.1007/11535294_16.
12. D.R. Wilson, T.R. Martinez, The general inefficiency of batch training for gradient descent learning. *Neural Networks,* **16**(10), 1429-1451 (2003).
13. S. Tarabrins, Review on the research topic "Modeling the system of knowledge management on the basis of neural networks". *Young Scientist,* **22**(208), 58-62 (2018).
14. D.M. Shotylo, V.E. Krainova, A.V. Skurydin, Trends in the development of artificial neural networks in digital economy. *Econominfo*, **15**(4), 65-69 (2018).
15. I.Z. Gelishanov, T.N. Yudina, A.V. Babkin, Digital platforms in economics: Essence, models, development trends. *Scientific and Technical Bulletin of St.Petersburgh State Polytechnical University*, **11**(6), 22-36 (2018). [in Rus.].

Forming Legal Definitions of Modern Information Technologies under Economic Digitalization

K. K. Taran[1]*
*Corresponding author: taran.kira@yndex.ru

[1] Moscow State Institute of International Relations (University) of the Ministry of Foreign Affairs, Moscow, Russia

Abstract. During the intensive development of information technologies (IT), the need for their regulation gradually arises. For forming basic regulation of IT-sphere a basis is needed, which determines the subject of regulation. It's a complicated task, but its solution will help to determine legal relations stemming from IT-usage. The existing technical definitions are accurate, since they are technical in nature. In some situations there are difficulties in their legal application, it complicates the legal analysis of a specific situation. The legal definition of a particular technology will assist legal professionals and theorists, as it will clarify the essence of relations, help to avoid misconception, support Internet law developing. This study discusses a short list of technologies used in practice in legal, medical and other fields. The author provides an overview of legal definitions at various levels: international, regional and national. The author examines the formation of definitions in Russia and the USA. A comparative analysis of terminology allows us to highlight the laws and features inherent in these countries. The author discusses the definitions of well-known, basic concepts such as the Internet, the user and the definition of technology that is currently widely used - a cloud storage of information.

Keywords: legal definition, information technology, Internet law, understanding, legal relations on the Internet

1 Introduction

In the modern world, technology is developing very rapidly. The number of technology users among such entities as individuals, legal entities, government bodies is growing, and with this the use of IT is spreading in various fields: education, medicine, law, the media (film industry, news services, etc.), trade etc. When technologies begin to be applied as actively and even rather globally, as the Internet itself, there is a need not only to regulate new legal relations, but also to create a conceptual framework that would be useful and introduce clarity to lawyers, law enforcers and the subjects that use technology. In this study, the author minimizes the use of technical definitions. The study discusses the most used, according to the author, technology.

2 Problem Statement

The problem of this study is how to formulate the most optimal legal definition of technology, how to be guided when creating this definition, whether to use the experience of foreign countries, whether it is necessary to specify the existing definitions (for example, in the situation of a consumer and a user), etc. The definition should be formulated in such a way that it is the basis for regulation, so that it is understandable for people working in the legal sphere. Therefore this problem is of great importance for the development of legal thought. Moreover, it is important how the definition will be voluminous in content. The less it will be voluminous, the more freedom will be given to the development of legal regulation.

3 Research Questions

The author considers the following issues:How were different definitions formed? What conditions influenced them? What are differences in definitions of specific technology by different actors?

4 Purpose of the Study

The author sets the task of studying legislation at various levels (international and national, using the example of Russia and foreign countries), judicial practice and doctrine. Based on the material studied, the author presents the most optimal definitions and suggests a possible version of a term.

The purpose of this study is to compare the formation of definitions in Russia, the USA and the EU, to highlight their features and patterns. The purpose of this study is important both for the theory of law and for law enforcement. It is not easy at all to formulate the exact definition after the technology has appeared. The technology needs time to "take root" in society, and to become widespread in practice. The formation of the conceptual definition and its development will create the basis for new definitions of new technologies.

5 Research Methods

When conducting the study, the author used the following general scientific and special methods. The general scientific methods are: analysis - the study of key components of the definition of a particular technology; inductive and deductive methods; modeling methods, etc. Special methods include: comparative legal methods for comparing approaches to formulating definitions in Russia, the USA, and the EU; logical and legal methods, methods of systematic interpretation of legal norms; synergistic method; systematization; and others. It should be noted that the topic of formulating legal definitions affects not only jurisprudence, but also other sciences, such as sociology and cultural studies.

4 Results

The author considered the most widely used terms, which, on the one hand, are known to everyone, and on the other hand, not everyone can explain what they represent from a legal point of view. The Internet. The term Internet is very common and has become very firmly established, but there is no definition of the Internet. In international documents, for example, in the UN documents in report A / HRC / 17/27 of May 16, 2011, which recognizes the human right to access the Internet, the Internet is considered as a means of communication, but the definition of the Internet is missing. We can observe a similar situation in Resolution A / RES / 70/125, adopted by the General Assembly on December 16, 2015, which refers to the management of the Internet, its development and dissemination, but its definition is not given. Most likely, this is due to the fact that before giving a definition of this technology, the UN enables states to formulate their own definition and, on the basis of these definitions, they will draw up an optimal unified, legal definition of the term Internet that takes into account all the features.

The US law represents the definition of the Internet. For example, in U.S. Code 47 Section 230 of the Decency Act in the Media (Eng. Communications Decency Act 47 U.S.C. Section 230), the Internet is seen as a service that provides access to educational and information resources and it is implemented with minimal government regulation.

The Regulation 2015/2120 of the European Parliament and of the Council of the European Union of November, 25, 2015 on the establishment of measures regarding open Internet access and the amendment of Directive 2002/22 / EC on universal services and user rights in relation to electronic communications networks and services and the Regulation (EU) 531/2012 on the roaming of public mobile networks within the Union also do not contain a definition of the Internet; the Internet is more likely defined as a means of communication to which end-user access should be provided.

In Russia, there are some definitions of the Internet. In the Federal Act of July 27, 2006 N 149-FZ "On Information Technologies and Protection of Information", the Internet is defined as an information and telecommunication network.

Taking into account how the Internet is widespread now and taking into account how many different services are provided through it, you can define it as a social phenomenon characterized by the universal global availability of bidirectional digital communication services, both between people and between their belongings. The Internet is by far the first and only global public service.

The second term that also needs to be considered is the user. As mentioned in earlier documents, the human right to access the Internet is recognized at the international level. The international community calls for the spread of the Internet as a means of expressing freedom of a person and citizen. But still there is the question - who the user is. Often there are terms Internet user, end user, service user, user agreement. Most international documents do not qualify users. Moreover, the term user has not been fixed in the legislation of the Russian Federation and many foreign countries yet. There are difficulties associated with who falls into the category of user, only individuals or legal and state structures will also be users. The answer is rather positive, since both legal entities and government agencies use information technology.

On the other hand, if we talk about users who are individuals and consumers, then what the ratio between the user and the consumer is. According to the Law of the Russian Federation of February 7, 1992 No. 2300-I "On Protection of Consumer Rights", a consumer is a citizen who has the intention to order or purchase, or who orders, purchases or uses goods (works, services) exclusively for personal, family, household and other needs not related to business activities. That is, the user will be considered as a consumer when he is exclusively an individual; legal relations on the use of Internet technologies will be governed by a service agreement, the use of which should not be related to business. It seems that the concept of the consumer will intersect with the concept of the user in the context when the user acts as the consumer. In this regard, numerous questions will arise, what rights the user has, how the protection of consumer rights and the protection of user rights will relate, etc.

Gradually, documents appear that develop the regulation of user rights in general. For example, Recommendation of Ministerial Committee of Council of Europe CM / REC (2014) 6 for States-members about Guidance of Human Rights for Internet Users states that human rights that are universal and indivisible, as well as related standards, take precedence over general provisions and conditions that are imposed on Internet users by any private sector entities. The document recommends the provision of Internet services without discrimination, at a reasonable price, safely, reliably and without interruption. At the same time, users can receive assistance in protecting their rights and use legal remedies. The document is aimed at developing regulation of user rights in both online and offline space.

In the European Union, the regulation of user rights is addressed in Regulation of the European Parliament and the Council of the European Union 2015/2120 of November 25, 2015 on the establishment of measures regarding open

Internet access and the amendment of Directive 2002/22 / EC on universal services and user rights with respect to networks electronic communications and services and Regulation (EU) 531/2012 on roaming of public mobile networks within the Union. These Regulations establish general rules to ensure an equal and non-discriminatory approach to traffic when providing services for the Internet access and the rights of end users associated with it. The end user refers to the actual user of the service. The Regulations contain general provisions aimed at freedom of access and freedom to use services on the Internet, to disseminate information, the right to defense, including judicial protection. The documents refer to monitoring for the quality of provided services by authorized state bodies. Moreover, as part of the monitoring and enforcement functions, national regulatory authorities and other competent authorities should intervene in contractual and commercial practices if they undermine the nature of the rights of end users.

It is interesting that in spite of the fact that the development of user rights is very active now, opinions are expressed that the user has very few opportunities to protect his rights [15].

Based on the foregoing, it is possible for a user to determine the subject of law who uses the Internet or services provided on the Internet, concluding relevant agreements. Most likely, in the future the states will not only specify the existing acts taking into account such a subject as the user, but will also adopt special acts in which the rights and obligations of the user will be determined.

The above concepts of the Internet and the user are basic for regulation and for the formation of Internet law. Due to the increasing number of technologies and their complication, basic concepts are intertwined with more complex ones, as new applied technologies appear, for example, such as cloud storage. Formation of the legal definition of such a service should be based on basic concepts and should include more complex legal definitions (for example, understanding the user agreement when using cloud storage is based on the legal understanding of the term user).

Cloud storage of information is a server on which user information is stored, and to which the user can access anywhere in the world. The Russian legislation and the legislation of most foreign countries have not developed a legal definition of cloud storage. Cloud storage is an integral part of the cloud service, which includes a wide range of functions, including information storage (the cloud service - came from cloud computing), which includes the storage service).

One of the most relevant definitions of cloud computing was developed by the National Institute of Standards and Technologies (NIST): cloud computing is a model of providing a convenient network access in the "on demand" mode to a collectively used set of configurable computing resources (for example, networks, servers, data storages, etc.) that can be quickly activated and released with minimal interaction with the service provider and minimal own control management efforts (Mell P., 2019). However, this definition is not fixed in the legislation, but it is used in the scientific literature of the USA and other states. This definition can be called generally accepted, but it does not reflect the legal aspects of this technology, as it was given by the technological organization and was not originally planned as legal.

Recently, a definition of cloud computing has been developed in Russia. Cloud computing is an information technology model for providing ubiquitous and convenient access via the Internet to a common set of configurable computing resources (cloud), data storage devices, applications and services that can be quickly provided and freed from the load with minimal operating costs or practically without the participation of the provider (Decree of the President of the RF "On Information Society Development Strategy in the Russian Federation for 2017-2030" of 09.05.2017 N 203). This definition is new, not enshrined in law yet. But this is a significant step in the development of legal regulation of the cloud information storage technology.

The formation of the definition of cloud storage can be traced in judicial practice. The decision of the Moscow City Court of January 21, 2019 in case No. 3-0011 / 2019 states that the Cloud Mail.Ru service is cloud storage that allows registered users (individuals) to download and store files in their personal cloud storage for personal use, access such files in the "cloud" at any time when connected to the Internet, and also provide access to such files to third parties by linking and providing such a link to a third party.

In the ruling of the Thirteenth Arbitration Court of Appeal of April 2, 2019 N 13AP-3256/19, cloud storage of information was defined as a virtual data room (cloud storage), which will be downloaded electronically. And the data for entering cloud storage was designated as login details.

Interestingly, courts do not use cloud storage classifications that match their functionality. For example, in the decision of the Eleventh Arbitration Court of Appeal of February 8, 2017 N 11AP-18802/16, public cloud storages are called cloud Internet storages, in the definition of the Moscow City Court of July 20, 2017 N 4g-7032/17 it was called external "cloud" file storage. The decisions of the courts of the Russian Federation often contain formulations of personal cloud storage. The writing of cloud storages is also unsteady. Sometimes the word cloud is used in quotation marks when writing, although spelling without quotes is more common.

The name cloud service (English) is also ambiguous. Since the use of the term service is from English, the service already implies the use of a service contract design. But in many law and order, the agreement on the use of cloud storage of information has a different legal nature (discussed later in the work). In this regard, many authors do not use translation, but use the term cloud service, which is not exact, but at the same time it is not misleading about the nature of the contract.

Cloud storage can be defined as a cross-border Internet service that provides access (including remote) to the server on which user information is stored.

7 Conclusion

This study examined important definitions of information technology, but not all of them, since there are many more technologies now. It can be concluded that such general definitions or their absence was a thoughtful step. Taking into account the practice of applying technology and its impact on social relations, we can draw up the most accurate legal definition of both technologies and legal relations. Most likely, definitions given by lawyers in the scientific literature and definitions that are used in judicial practice will serve as the basis for the formation of the conceptual definition in legislative acts.

References

1. Act of the RF of 7.02.1992 № 2300-I "About Protection Consumers' Rights". URL: http://base.garant.ru/10106035/. Accessed: 22.09.2019. (2019).
2. Decision of the Moscow City Court of 21.01.2019 in case N 3-0011/2019. URL: https://www.mos-gorsud.ru/mgs/services/cases/first-civil/details/996abf18-f457-4f92-bad4-53193dab23ef. Accessed: 22.09.2019. (2019).
3. Decree of the President of the RF "About Information Society Development Strategy in the Russian Federation for 2017-2030" of 09.05.2017 N 203. URL: https://base.garant.ru/71670570/. Accessed: 22.09.2019. (2019).
4. Federal Act of 27.07.2006 г. N 149-FZ "On Information, Information Technologies and Protection of Information". URL: http://ivo.garant.ru/#/document/12148555/paragraph/1844/doclist/0/selflink/0/highlight/%D0%98%D0%BD%D1%82%D0%B5%D1%80%D0%BD%D0%B5%D1%82:0. Accessed: 22.09.2019. (2019).
5. I.D. Kuzmina, & I.S. Bogdanova, Some problems of digital environment influence on the content of civil law object and subject categories. *Journal of Siberian Federal University "Humanities & Social Sciences"*, **3**(12), 393-409 (2019).
6. P. Mell, & T. Grance, The NIST definition of cloud computing: Recommendations of the National Institute of Standards and Technology. Special Publication 800-145. URL: http://nvlpubs.nist.gov/nistpubs/Legacy/SP/nistspecialpublication800-145.pdf. Accessed: 22.09.2019. (2019).
7. Recommendation of Ministerial Committee of Council of Europe CM/REC (2014)6 for States-members about Guidance of human rights for Internet users (adopted by Ministerial Committee of Council of Europe on 16.04.2014 on 1197th meeting of Deputy-Ministers). URL: https://www.refworld.org.ru/pdfid/55cb537c4.pdf. Accessed: 22.09.2019. (2019).
8. Regulation (EU) 2015/2120 of the European Parliament and of the Council of November 25, 2015. URL: https://eur-lex.europa.eu/legal-content/EN/TXT/PDF/?uri=CELEX:32015R2120&from=EN. Accessed: 20.09.2019. (2019).
9. Report of the Special Rapporteur on the Promotion and Protection of the Right to Freedom of Opinion and Expression, Frank La Rue. A/HRC/17/27. URL: https://www2.ohchr.org/english/bodies/hrcouncil/docs/17session/A.HRC.17.27_en.pdf. Accessed: 20.09.2019. (2019).
10. Resolution adopted by the General Assembly on December 16, 2015. A/RES/70/125. URL: https://undocs.org/en/A/RES/70/125. Accessed: 20.09.2019. (2019).
11. The decision of the Eleventh Arbitration Court of Appeal of 8.02.2017 N 11AP-18802/16. URL: http://base.garant.ru/61761743/. Accessed: 22.09.2019. (2019).
12. The determination of the Moscow City Court on 20.07.2017 N 4g-7032/17. URL: http://ivo.garant.ru/#/basesearch/%D0%9E%D0%BF%D1%80%D0%B5%D0%B4%D0%B5%D0%BB%D0%B5%D0%BD%D0%B8%D0%B5%20%D0%9C%D0%BE%D1%81%D0%BA%D0%BE%D0%B2%D1%81%D0%BA%D0%BE%D0%B3%D0%BE%20%D0%B3%D0%BE%D1%80%D0%BE%D0%B4%D1%81%D0%BA%D0%BE%D0%B3%D0%BE%20%D1%81%D1%83%D0%B4%D0%B0%20%D0%BE%D1%82%2020%20%D0%B8%D1%8E%D0%BB%D1%8F%202017%20%D0%B3.%20N%204%D0%B3-7032|17:0. Accessed: 22.09.2019. (2019).
13. The ruling of the Thirteenth Arbitration Court of Appeal of 2.04. 2019 N 13AP-3256/19. URL: http://ivo.garant.ru/#/basesearch/%D0%9F%D0%BE%D1%81%D1%82%D0%B0%D0%BD%D0%BE%D0%B2%D0%BB%D0%B5%D0%BD%D0%B8%D0%B5%20%D0%A2%D1%80%D0%B8%D0%BD%D0%B0%D0%B4%D1%86%D0%B0%D1%82%D0%BE%D0%B3%D0%BE%20%D0%B0%D1%80%D0%B1%D0%B8%D1%82%D1%80%D0%B0%D0%B6%D0%BD%D0%BE%D0%B3%D0%BE%20%D0%B0%D0%BF%D0%B5%D0%BB%D0%BB%D1%8F%D1%86%D0%B8%D0%BE%D0%BD%D0%BD%D0%BE%D0%B3%D0%BE%20%D1%81%D1%83%D0%B4%D0%B0%20%D0%BE%D1%82%202%20%D0%B0%D0%BF%D1%80%D0%B5%D0%BB%D1%8F%202019%20%D0%B3.%20N%2013%D0%90%D0%9F-3256|19/all:1. Accessed: 22.09.2019. (2019).
14. U.S. Code 47 Section 230 (Communications Decency Act). URL: https://www.law.cornell.edu/uscode/text/47/230. Accessed: 20.09.2019. (2019).
15. M. Zalnieriute, & S. Milan, (2019). Internet architecture and human rights: Beyond the human rights gap. DOI: 10.1002/poi3.200. URL: https://onlinelibrary.wiley.com/doi/full/10.1002/poi3.200. Accessed: 22.09.2019. (2019).

Directions of Priority Industries Development as a Basis of Economy Modernization

O.A. Bulavko[1], *L.R.* Tuktarova[1], and *E.Yu.* Nikulina[1,*]
*Corresponding author: katerina_nikulina@list.ru.

[1]Samara State University of Economics, Samara, Russia

Abstract. Within the framework of this study the priority areas of the development of industries are considered as transformation technologies integrally distributed with the aim of effective implementation of the new industrial policy. The analysis of the identified problems is carried out and the consequences of the introduction of effective solutions for the real sector of the economy are examined. On the basis of the assessment the authors substantiate an approach to the study of the potential advantages of the directions of development of industrial policy in accordance with social needs and social values within the framework of a changing institutional paradigm stipulated by the transition to a new economic policy.

Keywords: modernization, high technology production, investment resources, digital economy, industrial policy, economic growth.

1 Introduction

Currently, the economy of many countries is undergoing a transformation associated with the new industrial revolution «Industry 4.0». The concepts «transformation» and «modernization» are closely connected. Modernization is an updating of an object, bringing it in line with new requirements and standards, technical conditions and quality indicators. According to S.N. Gavrov the concept of modernization is, as a rule, viewed in several meanings [1]. We are mostly interested in the following: catching up modernization, practiced by the countries, which try to catch up with the most developed countries; modernization as a permanent process, realized through reforms and innovations and modernization as a process of social system reconstruction aiming at development acceleration. In our opinion, the issues of modernization of the economy and society as a whole are much wider, as well as the range of meanings of modernization. A retrospective analysis of the Russian economy makes it possible to single out mobilization modernization as an independent type (of the industrial economy, the territorial-production complex, the corresponding regional varieties, etc.), i.e. modernization, the main characteristic of which (while maintaining the goals) are short implementation periods, maybe with slightly lower economic efficiency, dictated by the change in external economic and political conditions. These conditions include the need to provide military (modernization of the USSR industry in the 30-s of the last century), food (food program of the 80-s), economic security, etc.

2 Problem Statement

At present many Russian and foreign authors have research works on the processes of modernization, the solution of technological problems proposed for the implementation of effective industrial policy. In this regard issues related to the new economic policy are being investigated. Many authors study the concept. Currently, there are practically no studies on the regulation of the new economic policy, identification of strategic directions of the development. There are some studies of the management sphere, however, they cover only certain areas or are limited by public regulation. At the same time, technologies are developing rapidly, and the level of the development of the society, as well as the rate of the economic growth, directly depend on the nature and directions of the industrial policy development. Therefore, planning and identifying of the priority areas of the development in the field of industrial and investment policy are the key tasks of the state. The article is devoted to these aspects and actual problems.

Andreeva& Shevchik [2]; Baranova, Shevtsova, N. A., Dmitrieva & E.G. [3] and also Gabrielova [4] discussed the peculiarities in the trends and mechanisms of the industrial policy in Russia. These questions are also discussed in the works of C.N. Pitelis, P. Kelmendi [5] and O.N. Soboleva, T.A. Burtseva, A.A. Nabokih & A.V.Shpengler [6]. The role of the national services sector in the building of the country's external competitiveness were assessed; the instruments and directions of industrial relations modernization are offered by E.V. Andreeva [7] and Wu Yiyun, Zhu Xiwei, Nicolaas Groenewold [8]. D.Kiselev, Y.Popov, O.Kanygina [9] and Salfetnikov D.A., Salfetnikov A. et al. [10], consider the reasons of inefficient modernization of industrial enterprises. Industrial complex are discussed in the works of N. Rogalina [11] and T. Kolmykova, A. Telizenko, V. Lukianykhin [12].

3 Research Questions

According to the authors, the problem of industrial modernization in the broad sense of the word arose in the first years of the existence of the independent Russia, as a problem of old-fashioned structure of industry inherited from the USSR. Moreover, while in the USSR, up to the beginning of perestroika, a number of manufacturing industries were on the rise, the decline in industrial production in Russia further aggravated the structural gap between the existing industry and the required industry. A number of branches of industrial production, in particular, electronic industry, machine tool building, certain types of instrumentation production disappeared; the volume of production in the military-industrial complex significantly decreased; and the remaining enterprises were surviving due to export and conversion, that, of course, can be attributed to the lack of competitiveness of the industry in conditions of Russia's entry into the world market. The authors revealed the problems of identifying priority industries as the basis for modernization and transition to a new technological structure contributing to the effective development of the industrial production, and, in general, the industrial policy.

4 Purpose of the Study

The motivation for choosing modernization mechanisms of the formation of a new technological structure in industrial and investment policy as the topic of the research was the multifaceted nature of the topic, covering a wide range of issues related to the transformation of the system of economic relations; the formation of coalitions as a separate mechanism for accelerating modernization processes; the transformation of risk into a specific type of resource for the development of innovative systems; the use of human capital in the interests of investment development, elements and methods of control of the information and innovative economy. Innovative and informational economic growth is characterized by making investments in information and communication technologies (ICT), which gives a multiplicative effect to the development of industrial enterprises. Industrial policy should be aimed at the implementation of interrelated elements between management bodies and business entities regarding investment and innovation activities as well as competitiveness of industrial production sectors.

The main suggestions for this study are issues related to the possibility of increasing investment activity, since investments are the key link, and investment policy is the most important component of industrial policy, including industry. The main proposals for this study are issues related to the possibility of increasing investment activity, since investments are the key link, and investment policy is the most important component of industrial policy, including industry. It is the volume of investments that determines sustainable and dynamic development of industrial production sectors and the increase in their competitiveness in the world markets. Industrial production is characterized by a relative decrease in gross output of the manufacturing industry, while maintaining stability and growth in the sectors of the fuel and energy and raw materials complex. In Russia, the development of industries of the industrial production is associated with the transformation of financial resources into investment instruments of a long-term nature, which ensures, in practice, the connection between the volume (and efficiency) of the industrial production and the level of investment. The second suggestion of the authors is the revealing of the problems associated with the increase of competitiveness, the ability to innovate and attract direct investments to the real sector of the economy. Particularly promising directions of development are technological changes that strengthen the role of innovations and investments in the economic development of the industrial production.

5 Research Methods

The methods of system and factor analysis, empirical, diagnostic, retrospective, predictive, stochastic and others methods are applied. The goal of the research is to analyze the identified problems and consider the consequences of implementing effective solutions for the sectors of the real economy. The scope of the research is the highlighting priority industries as the basis for modernization. The information base of the research are the articles of Russian and foreign authors, highlighting the main problems of modern industrial policy.

6 Findings

An important stage in the development of industries in the real sector of the economy is the possibility of forming a new technological base of economic systems based on the use of the latest achievements in computer science, nanotechnology and bio-technology. The research potential and high-tech industries create conditions for the use of competitive advantages associated with the distribution of new technologies and the modernization of the traditional sectors of the economy, also on the account of the deployment of globally oriented specialized industries.

The very concept of modernization of a particular facility has not been unambiguously defined. For example, Professor V.A. Tsvetkov [13] believes that it is possible to use various interpretations depending on the object of modernization, and, probably, he is right, and the formulation proposed by him is good. Naturally, in view of a significant number of modernization facilities (for example, vertically: a machine tool, a site, a workshop, an enterprise, a territorial-production complex, a region's economy, an economy and a society of a state), a single model of modernization is not possible either vertically or horizontally. As it was already mentioned, Russia needs a catching up modernization now. And the optimal start time of it has already been missed. For more than ten years, there has been a talk about the need for this modernization, and only now there are some perspectives of it. V.V. Ivanter [14] defined the tasks of the modernization of the Russian economy as the diversification of domestic production, transformed in accordance with changing external conditions on the basis of innovative orientation. Accordingly, in order to create and activate new factors of economic growth in the framework of modernization processes that meet the challenges of the long-term period, it is necessary to consider the development of the main industries from the perspective of investment and innovation orientation, i.e. to mark the main directions and instruments of industrial policy at the state and regional levels of government. It is industrial policy that contributes to the formation of a new type of sectoral structure of industry; at the stage of economic growth contributes to its development, and at the stage of stagnation or recession does not allow the realization of an avalanche-like fall in the real sector of the economy. From our point of view, industrial policy should provide a combination of all its components and interact with other types of policies - macroeconomic, foreign economic, investment, innovative, regional, environmental and social policies. At present Russian industrial sectors (although formally there are no sectors after replacing All-Russian Classifier of Economy Branches with All-Russian Classifier of Types of Economic Activity) can be divided into the following categories:

basic (main) industries that provide the export of goods and services outside the region (mechanical engineering in the broad sense, as an integrated industry; metallurgy, oil refineries, wood processing enterprises, food industry enterprises, etc.);

industries of regional importance serving regional needs (production and distribution of water, gas, electricity, heat; industrial services);

industries in which the region does not specialize in the given period (import-substituting industries, sectors of possible future development of the region), but due to changes in the development of scientific and technological progress, it is quite possible to develop them, if an appropriate resource base is provided.

An important strategic task of industrial policy implementing is to overcome the structural crisis on the basis of identifying and supporting priority industries of the region (state), taking into account its existing scientific, technical, industrial and geopolitical potential.

The structural crisis of the industrial complex of a state or any region is caused by the discrepancy of its basic structural elements (the structure of commodity products, the organizational, technical and technological structure and the structure of the distribution network) with the requirements of effective industrial activity. At the same time, one should take into account the fact that the structure of the region's economy and its core, the industrial complex, formed under pre-reform conditions, is a carrier of certain qualities that enable the region to realize one or another set of functions and at the same time act as factors supporting inertial development. As we have already said, in the current economic environment, the structure of the region's economy and the regional industrial complex often contradict to the goals of its long-term comprehensive socio-economic development and with its new strategic choice. The resolution of this contradiction is the formation of a new structure of the economy by the region which makes it possible to solve a wide range of problems of social and economic development effectively. As you know, a very significant place in the regional economy system is played by the assessment of the prospects of the industries of the existing specialization of the region, as well as the justification of the direction of its effective specialization. Accordingly, a number of indicators are used for this, in particular, the localization (concentration) coefficient of the industry on the territory of the region, calculated on the basis of comparing of the sectoral structure of the region's economy with the similar structure of the country's economy as a whole, and representing of the ratio of the sector share in the region's economy to the share of this industry in the economy of the country. From such a brief review it follows that the renewal of fixed assets of the industry as a factor in increasing competitiveness requires a selective investment policy of the state.

On the basis of the above-mentioned we can distinguish the main problems of the industrial policy of Russia:

- the lack of a single concept and mechanism for the implementation of industrial policy built on common goals and objectives; and first of all, the absence of the Law on Industrial Policy;
- insufficiency of the regulatory framework regulating the mechanism of the industrial policy; almost in every subject of the federation in one form or another there is a regional law on industrial policy, and these legislative acts are in some cases similar as twins, and the specifics of the region's potential, the level of innovative development and the existing resource base are not taken into account;
- raw material export orientation of the development of the Russian industry; avoiding the raw material orientation of the development of Russia and, accordingly, changes in the structure of export and import;
- uneven development of the industry in the regional context;

the undeveloped transport infrastructure of most regions, as well as interregional transport links, which obstruct the development of industrial production, because of the impossibility of entering foreign markets, as well as of accessing the resources;

- low level of investment in fixed assets of industrial enterprises;
- territorial and sectoral unevenness of state support for enterprises.

The solution to this problem should be closely related to the possibility of stimulating economic growth on the basis of modernization with the use of available resource-efficient technologies [15] . The result of the socio-economic revolution taking place in the modern world, according to some authors [16], is the construction of a post-industrial society, in which information technologies, computerized systems, high tech production and innovative technologies play an important role.

7 Conclusion

Investment is a key part of industrial policy, investment policy is an important component of industrial policy, including sectoral. The most promising areas of development are technological changes that reinforce the role of innovation and investment in the economic development of industrial production. In conclusion we can say that the diversification of the Russian economy on the basis of modernization of its technological, institutional, personnel and management base is an essential condition for the implementation of national projects and an effective factor of integration into the global economy.

References

1. S.N. Gavrov, *Modernization of Russia: Post-imperial transit* (MSDTU, Moscow, 2010). [in Rus.].
2. O. Andreeva, E. Shevchik, Organizational and financial mechanism of new Russian industrial policy: Systemic approach. *Journal of Applied Economic Sciences*, **11**(3), 351-363 (2016).
3. N.M. Baranova, N.A. Shevtsova, E.G. Dmitrieva, Industrial policy as a tool of restructuring the Russian economy. In: V. Mantulenko (Ed.), *Proceedings of GCPMED 2018 – International Scientific Conference "Global Challenges and Prospects of the Modern Economic Development". The European Proceedings of Social & Behavioural Sciences EpSBS,* **LVII** (pp. 1877-1887). London: Future Academy (2019). DOI: 10.15405/epsbs.2019.03.191.
4. H. Gabrielova, New trends in European industrial policy. *Ekonomicky Casopis*, **58**(9), 888-908 (2010).
5. C.N. Pitelis, P. Kelmendi, European industrial policy: Perspectives, trends and a sustainability-focused new framework. In M.N. Jovanović (Ed.), *International Handbook on the Economics of Integration*, **II**, chapter 2 (pp. 21-44) (Edward Elgar Publishing, Cheltenham, 2011).
6. O.N. Soboleva, T.A. Burtseva, A.A. Nabokih, A.V. Shpengler, Factors affecting innovations development in priority industries of a region. *International Review of Management and Marketing*, **6**(5), 91-95 (2016).
7. E.V. Andreeva, Service industries development as a priority for national post-industrial modernization. *Actual Problems of Economics*, **175**(1), 44-51 (2016).
8. Y.Wu, X. Zhu, N. Groenewold, The determinants and effectiveness of industrial policy in China: A study based on five-year plans. *China Economic Review*, **53**, 225-242 (2019). DOI: 10.1016/j.chieco.2018.09.010.
9. D. Kiselev, Y. Popov, O. Kanygina, Problems of ineffective modernization at industrial enterprises and ways of their solution. In S. Bratan, S. Gorbatyuk, S. Leonov and S. Roshchupkin (Eds.), *Proceedings of the ICMTMTE 2017 - International Conference on Modern Trends in Manufacturing Technologies and Equipment. MATEC Web of Conferences,* **129.** Les Ulis: EDP Sciences (2017). DOI: 10.1051/matecconf/201712901029.
10. D.A. Salfetnikov, A.A. Salfetnikov, E.M. Kharitonov, A.A. Popov, S.V. Khoruzhaya, Stages and problems of Russian modernization and features of soviet industrial construction in the 1920s and 1930s. *Terra Sebus*, **9**, 337-353 (2017).
11. N. Rogalina, Industrialization as a national economy problem (On the issue of the initial level of soviet industrial modernization). *Voprosy Ekonomiki*, **2007**(2), 127-140 (2007). DOI: 10.32609/0042-8736-2007-2-127-140.
12. T. Kolmykova, A. Telizenko, V. Lukianykhin, Problems of modernization and development priorities for industrial complex. *Problems and Perspectives in Management*, **11**(4), 28-34 (2013).
13. V.A. Tsvetkov, To the problem of modernization of the Russian economy. *Economy of Metropolises and Regions*, **2**(38), 4-14 (2011). [in Rus.].
14. V.V. Ivanter, The necessity of modernization and the speed of the economic growth. *Innovations*, **8**(154), 6-7 (2011). [in Rus.].
15. O.A. Bulavko, *Industrial and investment policy in the post-crisis modernization of Russian industry* (St. Petersburg State University, Saint-Petersburg, 2013). [in Rus.].
16. E.V. Ioda, O.A. Bulavko, G.A. Khmeleva, Yu.V. Ioda, Modernization mechanisms of new technological state of things formation. Monograph (Samara Academy of State and Municipal Management, Samara, 2013). [in Rus.].

Individualization of Training Using the Resources of International Digital Platforms

A.I. Popov[1], and *R.V.* Fedorenko[2,*]
*Corresponding author: fedorenko083@yandex.ru.

[1]Tambov State Technical University, Tambov, Russia
[2] Samara State University of Economics, Samara, Russia

Abstract. The article is devoted to the definition of psychological and pedagogical conditions and design methods of formation of a creatively thinking specialist using international digital platforms. The following universal abilities making up a specialist's competitiveness are defined: creativity, digital literacy, and psychological stability. The importance of tolerance for international innovation is shown. The state and directions of development of education digitalization are analyzed. Psychological and pedagogical conditions are revealed, and provisions of the concept of training individualization are defined. A set of measures are considered to facilitate the passage of students personal educational trajectory. The experience of organizing individual creative development in the digital space of the Union state is described. The introduction of the developed theoretical and methodological conceptual provisions of training individualization will provide conditions for improving the quality of vocational education and training for the implementation of innovative programs in the framework of international cooperation.

Keywords: education digitalization of education, professional creativity, international cooperation, distances learning.

1 Introduction

Satisfying the growing needs of individuals and society in competitive education presupposes a dynamic renewal of the education system. Vocational education should not only prepare for the performance of standard labor functions, but also develop the creativity of the individual, motivate her to further self-improvement. Improving the quality of education can be achieved with the active use of the potential of digitalization. The use of digital educational platforms and technologies is caused by the need for flexible updating of the content of training in accordance with the achievements of scientific and technological progress and taking into account the trends of the emerging innovative economy [1]. The digital transformation of a company can be successful only if there are people in the company who can competently carry it out [2]. High competition in the labor market and the need for lifelong education throughout life increase the relevance of the formation of a personal educational path and adaptive management of the process of creative development of the individual.

2 Problem Statement

Modern globalization, growth and competitiveness of the national economy, geopolitical and geo-economic factors determine the level of intellectualization of the main factors of production [3]. Globalization of all economic processes and the establishment of the sixth technological order in industry determine new requirements for a competitive specialist: readiness for creative non-standard activities in a competitive environment; the ability to take into account the specifics of economic and legal support of professional activities in the implementation of international innovative projects [4]. Economic reasons that encourage a significant number of people to get a new education in parallel with work, and the need to continuously improve their skills increase the relevance of using the resources of digital educational platforms, especially in online learning [5]. The content of education and the technologies used should take into account both the requirements of specific employers [6] and the individual characteristics of the knowledge of a particular student.

The problem of developing a methodology for individualizing training in vocational education with the optimal use of the resources of international digital platforms requires special attention and a separate scientific research. A deeper study of this problem from new theoretical and methodological positions should be carried out taking into account the contradictions identified on the basis of research and practical experience in higher education:

– the need of the individual for creative self-realization and taking into account individual characteristics, and the limited possibility of the educational environment of one university to ensure the corresponding realization of creative potential and personal educational requests;

– a potentially international sphere of activity in the context of the globalization of the economy and the focus of the educational process at the university on the solution of specific narrowly targeted problems.

3 Research Questions

In the process of psychological and pedagogical research, the main attention was paid to determining the conceptual provisions of the methodology and developing a model for personified learning in the context of digitalization of education. Of interest is the problem of combining competence and activity approaches in distance learning on international educational platforms and overcoming the psychological deformation of students when working in virtual space [7, 8]. In the context of the development of the personnel component of innovation in work, the mechanisms of creativity development when working in digital space were investigated.

4 Purpose of the Study

The aim of the study is to determine the psychological and pedagogical conditions and design a methodology for the formation of a creatively thinking specialist when using international digital platforms. The developed theoretical and methodological conceptual provisions for the individualization of education should provide conditions for improving the quality of vocational education and training for implementation of innovative programs in the framework of international cooperation. The provisions and conclusions of the study can become part of the theoretical basis for improving the professional training of an innovative specialist for creative activities in the digital economy.

5 Research Methods

The main methodological approaches used in the design of technologies for individualizing instruction through the use of international educational platforms are competency-based and personality-active. The competency-based approach was implemented through the presentation of the goals and objectives of training in the form of a list of creative competencies of a specialist, determining the ability to realize oneself in the field of international cooperation. The process of creative professional development of a student is considered as personality-related. The development of a personal educational trajectory is the result of the student's educational activity in the digital educational space. To solve the tasks, research methods were used: analysis and synthesis; modeling of pedagogical phenomena; direct and indirect observation of the process of cognitive activity of students in the digital educational space; study and generalization of the experience of digitalization of education.

6. Findings

6.1 Competitive Model

The development of the digital economy requires appropriate training of specialists who are able to think creatively and maximize their intellectual and creative potential in the implementation of professional activities. At the same time, they must work effectively in the digital space, introducing advanced digital technologies in the production process [9].

The study of the dynamics of the formation of an innovative economy and globalization processes, an analysis of development trends and structural transformation of the labor market made it possible to determine the universal abilities that determine the specialist's competitiveness:

- the creative nature of thinking and the willingness to make innovative decisions in professional activities;
- digital literacy and knowledge of the tools of the digital economy;
- readiness for self-education and self-development, obtaining a new profession and qualification in the digital educational space;
- willingness to work in the face of restrictions and psychological stress.

For effective professional activity in the context of a globalized economy, a specialist needs to have economic and legal competence both at the national level and at the level of countries - potential partners. An important quality of a competitive specialist will be his genuine tolerance for cultural values and psychological characteristics of representatives of a multinational and multiconfessional modern society, included in innovative activities at the supranational level.

The ego-identity formed in childhood and adolescence (low self-sufficiency, lack of willingness to take responsibility, poor creative initiative) and group identity, as well as hypertrophied involvement of some representatives from real processes in virtual space are the constraints on the manifestation of the abilities required by the economy youth.

6.2 Problems and directions of development of digitalization of education

The digitalization of education is constrained by objective and subjective reasons. Let us single out a number of problematic issues that impede the optimal use of the possibilities of digitalization of education.

1. Narrow specialization is part of the educational content. The ability to use content and technology.

2. Advantages of reproductive-oriented development of the educational resource, insufficient creative orientation of the student in the learning process.

3. The lack of sophistication of the legal framework for e-learning and the use of distance learning technologies, legal difficulties in the formation of a personal educational trajectory.

The most popular areas in the field of digital education technologies are:

- the optimal combination of traditional learning (with the direct interaction of the student and teacher) and digital learning;
- creating a system of adaptive management of personal educational trajectory based on digital traces of students, their needs and personal qualities;
- use of resources of international educational platforms for phased acquaintance with national cultural and mechanisms of economic and legal regulation of activities of interest for the development of international cooperation.

6.3 The concept of individualization of learning when using the resources of international educational platforms

Digital professional education in modern socio-economic and political conditions should be integrated with traditional and innovative teaching technologies, and based on the psychological and pedagogical characteristics of specific groups of students and the mastered subject area, ensure the rational use of the intellectual and creative potential of students. The key is the understanding of the student as a creative person, the orientation on the development of his integrity and the diversity of individual manifestations, the outstripping development of his creative activity. The main goal of the digital educational process is to increase the level of intellectual activity of the student (to heuristic and creative levels). This will allow the specialist to most fruitfully realize their abilities in professional activities and in the process of creative self-development.

In the process of research, psychological and pedagogical conditions were identified that determine the effectiveness of the formation of readiness for innovative activities in the framework of international cooperation:

- the content of education presented on the used international educational platforms should be aimed at developing creativity and the manifestation of leadership qualities in conditions of psychological stress;
- the training tools are problem situations that reflect the subject and social contexts of activity and are of an interdisciplinary nature;
- educational activities should have an element of competitiveness, aimed at the formation of psychological stability;
- a system for monitoring the digital footprint of students in the context of designing a personal development path is needed;
- the creation of a favorable psychological background of cognitive activity in the digital space is ensured by the alternation of periods of intense intellectual activity and stages of reflection and relaxation.
- a correction of the teacher's thinking is necessary, suggesting a change in his role in the digital educational process from a leader to an equal participant, coordinating the process of creative self-development of students and developing along with them.

The solution to the problem of individualization of learning in the context of digitalization is carried out through a set of measures, including:

- the organization of advanced training in accordance with the educational needs and capabilities of students, the adjustment of the content of training taking into account economic development trends;
- strengthening the spiritual and moral development of students through the expansion of digital elective courses focused on universal values and the development of tolerance; support of informal education in the electronic environment;
- ensuring the availability of technical education for persons with disabilities;
- expansion of economic and legal training of technical specialists;
- accompaniment of the creative development of gifted students through their access to the bases of creative tasks and the organization of interactive work of informal groups in a remote format;
- organization of project training in the framework of interaction with industrial enterprises and research organizations in the digital space; the creation of virtual study groups from students of various forms and courses, the involvement of leading practitioners as experts.

6.4 Practical implementation

The globalization of the economy, affecting all aspects of the ongoing economic processes, involves reaching a new level in the field of economic and legal support for the work of joint ventures, support for business entities to enter international markets. In modern political conditions, one of the promising areas for using international educational platforms is the training of specialists for the Union State of the Russian Federation and the Republic of Belarus. The main attention is paid to the formation of economic, legal and digital competencies of specialists. The design of the content of the educational platforms of the universities of the two countries, ensuring the implementation of the components of the personal educational path of specialists in the innovation sphere, involves the following actions.

1. The formation of a set of problem situations in the organization of joint economic and scientific activities.

2. Creation of digital platforms for the implementation of joint projects in priority areas of the economy. Ensuring both a regime of intensive generation of ideas and development of a program for the implementation of the project, as well as stages of reflection and relaxation.

3. Integration into a single database of all regulatory documents governing the implementation of joint projects or interaction with partner organizations within the Union State. It is necessary to bring the structure of this information as close as possible to use by enterprise managers. For this, a comparative analysis of the documents regulating the same business process in each of the countries of the Union State is carried out, recommendations are prepared for resolving possible contradictions in the legal framework.

4. Development of an information system that allows you to find a potential partner or buyer of products, as well as organizations performing stages of applied research to create new equipment or technologies.

As part of the interaction between universities of the Russian Federation and the Republic of Belarus, approaches have been taken to create an open digital educational environment in the form of a set of international educational platforms.

The organization of personalized training when using the resources of digital educational platforms allowed at a higher level to form competencies that determine the innovative readiness of a specialist:

- knowledge of information and communication technologies;
- free orientation in legal relations and the regulatory framework of the economic processes of the Union State countries; understanding of the need to take into account the characteristics of the economic policies of partners and socio-political processes;
- knowledge of the specifics of the mentality of potential economic partners and the ability to establish trustful and mutually beneficial relations, sociability and tolerance;
- possession of the skills to carry out joint activities at the level of educational and student research, project activities.

7 Conclusion

The competitiveness of a specialist and readiness for international cooperation determine his creativity, digital literacy and psychological stability. To prepare specialists for the implementation of international innovative programs, it is necessary to create a system for managing personal educational trajectories and optimal use of the resources of international educational platforms. Learning in the digital space involves the use of problem situations and competitive activities to enhance thinking. The effectiveness of individual learning is enhanced by the alternation of intense cognitive activity and the stages of reflection and relaxation. Using the resources of international educational platforms ensures the passage of a personal educational trajectory and the formation of readiness for international cooperation in the era of globalization.

References

1. S.D. Starygina, N.K. Nuriev, E.A. Pechenyj, A.A.M. Obadi, Designing smart education systems with digital technology. *Educational Technologies and Society,* **22**(3), 54-66 (2019). [in Rus.].
2. T.O. Tolstykh, E.V. Shkarupeta, I.A. Purgaeva, R.V. Fedorenko, Transformation of positions, competences and skills in the digital economy industry. In: V. Mantulenko (Ed.), *Proceedings of GCPMED 2018 - International Scientific Conference "Global Challenges and Prospects of the Modern Economic Development", The European Proceedings of Social & Behavioural Sciences EpSBS,* **LVII** (pp. 953-959). London: Future Academy (2019). DOI: 10.15405/epsbs.2019.03.94.
3. A.A. Polidi, Z.O. Goukasyan, I.A. Maslova, R.V. Fedorenko, Some aspects of the quality of corporate governance in digital economy. In: V. Mantulenko (Ed.), *Proceedings of the 17th International Scientific Conference "Problems of Enterprise Development: Theory and Practice" 2018. SHS Web of Conferences,* **62,** 04002 (2019). DOI: 10.1051/shsconf/20196204002.
4. N.V. Molotkova, A.I. Popov, Organization of training of engineers for innovation. *Alma mater: Bulletin of High School,* **4,** 9-14 (2019). DOI:10.20339/AM.04-19.009. [in Rus.].
5. S. Marathe, Digitalization in education sector. *International Journal of Trend in Scientific Research and Development* (IJTSRD), Special Issue: International Conference on Digital Economy and its Impact on Business and Industry, 51-56 (2018). DOI: 10.31142/ijtsrd18670.
6. M. Douse, P. Uys, TVET teaching in the time of digitization. In S. McGrath, M. Mulder, J. Papier, R. Suart (Eds.), *Handbook of Vocational Education and Training* (pp. 1-16). Cham: Springer (2019). DOI: 10.1007/978-3-319-49789-1_75-1.
7. J. Böhner, M. Scholz, J. Franke, A. Sauer, Integrating digitization technologies into resource efficiency driven industrial learning environments. *Procedia Manufacturing,* **23,** 39-44 (2018). DOI: 10.1016/j.promfg.2018.03.158.
8. F. Li, J. Yang, J. Wang, Sh. Li, L. Zheng, Integration of digitization trends in learning factories. *Procedia Manufacturing,* **31,** 343-348 (2019). DOI: 10.1016/j.promfg.2019.03.054.
9. A.I. Popov, A.D. Obukhov, Conceptual approaches to automated management of professional education. *Transactions of the Tambov State Technical University,* **24**(2), 246-257 (2018). DOI: 10.17277/vestnik.2018.02.pp.246-257. [in Rus.].

Innovative Approach to Improving the Efficiency of Labor Potential: Talent Management

I.N. Makhmudova[1,*], *L.L.* Kifa[2], and *S.E.* Vasilieva [3]
*Corresponding author: Mahmudova.I@yandex.ru.

[1]EDP Samara State University of Economics, Samara, Russia
[2]Togliatty National Research University, Russia
[3]Togliatty National Research University, Russia

Abstract. The paradigm of the personnel management concept has been changed based on the introduction of digital technologies in the labor process. The article reveals the impact of new information technologies on the formation and usage of laborpotential in the renewed organizations. The new approach is not knowledge management, but talent management. The socio-psychological analysis of the structural elements of labor potential is also used. Terminological definiteness of such concepts as loyalty and reliability of employeesis carried out according to their participationactivity in the labor process. Distinctions of highly professional (HiPro) and high-potential (HiPo) personnel are given. High level of motivation among high-potential employees is emphasized as a condition that increases the level of labor potential efficiency. The role of employees' involvement in the formation of the personnel reserve is defined. Modern tools, based on digital platforms, developing labor potential were used. The mechanism of HR-technologies is revealed, in which an employee with greater potential is unable to reveal his labor potential, and becomes a brake for the innovative development in organization. There was named the principal responsibility for timely and qualitative formation of the new competencies, required for realizing the labor potential, in terms of updated reality.

Keywords: labor potential, talent management, loyalty, trustworthiness, engagement, labor activity.

1 Introduction

According to the long-term economic development strategy of the domestic economy in the period up to 2020-2025-2030, the task to modernize traditional sectors of economic growth and to form a digital economy focused onhigh technology was stated [1; 2].

The understandingof the importance to measure the potential employees'capacityis coming, while the new socio-economic relations in society, the introduction of modern digital technologies in production and training, the actualization of new professional competencies, in accordance with the development strategy of updated organizations and their business ideas, are forming. Solving of these issues leads to a change from the personnel management concept to the human resources managementapproach, to the formation and accumulation of intellectual capital in organizations.

The assessment of labor potential has already been led to a transition from knowledge management to talent management. Similarly, all components of labor potential, their essence and measurement technology, using advanced cloud solutions,have become more complex.

2 Problem Statement

Recent years have been marked by a sharp decline in the efficiency of the labor potentialuse, low personnel motivation to highly productive work, lack of experienced specialists, a large amount of unstructured data, errors in the financial forecasting of HR strategy. In this regard, there is an increase in employee dissatisfaction with work, staff turnover and, as a consequence, a dramatic decrease in the loyalty of the organization's staff.

The process of forming the information and innovation economy obliges organizations to keep the most accurate account of the quantity and quality of available resources, as well as to redistribute them with maximum efficiency. In this regard, special attention is paid to the assessment of the labor potential and the effectiveness of its use in technologically new conditions.

3 Research Questions

Today, it can be noted that the task of providing the organization with highly professional personnel has failed. Objective reason is that on the labor market there is no suitably qualifiedstaff with appropriate skills to the modern cloud (digital) technology. That is why supportingthe organization with skilled employees, at the right time and in the right place, is essential for the creation of a coherent and logical talent management system. In this regard, it is required to have innovative approach to the management of thelabor potential and ensuring the flexibility of the entire HR system in accordance with objectives of the business. To do this, it is necessary to identify a modern view at an issue of developing the employees' potential, so it will be possible to manage all HR-processes from one system and improve the efficiency of the organization's staff. In addition, a new approach is needed to identify the problems that cause a decrease in personnel activity in the implementation of the labor potential in the renewed workplaces.

4 Purpose of the Study

Creating such a holistic, transparent talent management strategy, integrated into all elements of the organization's HR management system, is impossible without terminological certainty. For example, you should understand who is a"high-potential"HiPo-employee, and whether it can be included in the"talent pool"? How trustworthy will he be, and what level of loyalty should we expect from him? How is it different from professional HiPro employee? Who has a higher degree of labor activity and / or more consistent with the modern business process strategy of the organization?

The question of loyalty and reliability of the staff arose not by chance. The level of staff loyalty is closely related to the turnover rate. Loyal staff should not be mistaken with trustworthy labor. Trustworthiness of personnel reveals the degree of normative behavior in relation to the company. Itseeks to comply with all the rules and regulations established in the company. But in the case of unfair treatment by the management or mismatch of his personal interests and the interests of the company, is able to commit an illegal act against the company [3].

Staff loyalty is characterized by a positive attitude of the employee to the activities, policies, management of the company. The basis of theloyal employee's behavior is their desire to benefit the company and avoid actions that can harm it. The highest degree of employee loyalty is when it perceives the company's affairs as its own business. A loyal employee, on the one hand, deeply accepts some of the requirements of the company, and on the other hand, is ready to put up with others that do not quite suit him [4].

In fact, trustworthy personnel are passive in the manifestation of their labor potential. They take a "wait-and-see" attitude in assessing the benefits that provided to them in the workplace by leaders of the organization. Such employee can be referred to a "highly professional"HiPro-employee. However, in terms of technological change, when it is necessary to search for innovative approaches and working methods, to operate semi structured data, to master modern tools and technologies, requiring a proactive attitude to perform assigned tasks, just to comply with the requirements and regulations of labor discipline, positive attitude to company policy – it means 'to stand still', and, in fact, inhibit innovation. A highly professional HiPro-employee uses past experience, which, at best, may not fit into a new relationship, and at worst – to spoil innovative initiatives.

Loyal staff may not fall into the category of "highly professional"HiPro-employees, however, due to high motivation, they are ready for any tasks in the updated reality. They have an inquisitive mind and ready to search for non-standard solutions, because they are not burdened with experience and imposed decisions (rules, outdated norms). They are proactive in finding answers, for which they are ready to growth rapidly, always up to master and use modern technologies in their work to get a positive result for the organization.

However, high level of motivation is important condition to be a "high-potential" HiPo-employee. To stimulate active and proactive attitude to work, it is necessary to bring HiPo-worker to the level of partnership, co-management of the organization's business, to increase the level of his involvement. He should understand his personal benefit, and not in the form of a one-time compensation/premium, butin the prospects for his personal development. It should not be a nominal, but a real "personnel reserve" in accordance with the performance evaluation, which reflects the level of achievement of the employee's business objectives. He is personally responsible for the formation of his competencies needful to work in a renewed environment.

Today, the task to motivate employees is shifting towards greater opportunities for providing self-realization, professional growth and career building within the company. According to KPMG research, improving the efficiency of organizations in Russia is becoming a higher priority, which is achieved by increasing employee engagement (55%) and talent management (51%) [5].

Betting on HiPo-workers, employers develop internal training in the format of distance e-learning courses, and are willing to invest in this process, despite the general trend of reducing the HR budget. The main requirement, at the same time, is compliance with two sources: job success profile and skill-matrix, which give a detailed description of the professional skills for particular position.

Following the annual planning system of employee development (progress development guide, PDG), companies use this electronic document (on SAP platform) in which the employee independently and/or with the support of the leader conducts self-assessments (approved by the company's competency model) and is working on a plan of his improvement. Assessing of the labor potential usually follow two criteria: sustained performance (stability of results) and potential level (potential level from 1-level inconsistencies up to 3 maximum). As a result, the employee comes to a certain category in terms of talent development.

However, it is important to strike a balance, engaging an employee in management and self-evaluation processes. The activity of the employee in the process of involvement can dramatically decrease in the state of overload, accumulation of fatigue, not getting an assessment of the results that were achieved, loss of the prospects for their own growth. An employee may experience professional burnout, develop dissatisfaction with work. A loyal employee, losing activity, once again goes into the category of trustworthy, soothing, passive. If the burnout process has gone too far, such an employee is dismissed, the turnover rate increases. The employee is either looking for a new place to use their accumulated abilities, and the company becomes a "forge of personnel", or goes into the underground (changes the scope of professional activity).

5 Research Methods

The developed technology, innovative approaches in the formation and use of labor potential in the updated environment are based on the analysis and comparison of research results published for open use in the Internet environment. The basis for tasks-solving has been put the author's studies either [6-11], that were carried out with the aim to identify specific competencies HiPo (high potential) and HiPro (high professional), relevant digital technologies. The author's method of competence assessment [6] is used to measure the level of labor activity, identify new trends and patterns of competence formation of 'end-to-end' technologies in the implementation of the labor potential.

6 Findings

Based on the above, we can highlight several important points.

Referring to the labor potential, it is necessary to bear in mind not a set of certain knowledge and even performanceexperience of professional duties, which in itself is important, but the degree of active and proactive use of them by the employee in solving the tasks.

The level of activity is achieved through the provision of opportunities to engage in self-development, using advanced cloud technologies (in the format of remote e-learning courses), through which the employee can be involved in the process of co-management of the organization. At the same time, he must be extremely goal-oriented, understand his role and level of participation in the business process, which is determined by the formation of relevant competencies. The type and list of modern competencies is determined and approved by the management of the organization for each specific business task.

The required competency level is assessed according to the HiPo "formula" proposed by SHL for capacity assessment. It includes only three terms [4]:

1. Striving for achievements (potential to take a managerial position),
2. Abilities (questionnaires, Hogan, Talent Q),
3. Involvement (potential assessment - test PiF, Potential in Focus),

At the same time, it is much more important not just to assess the labor potential, but, following the results of the evaluation, to bring information about the enrollment of the person in a particular group at the HiPo level. This is an indicator of forming the employee's labor potential, the effectiveness recognition of his work and success in the personal trajectory implementation of its development within the organization. Enrollment in a certain HiPo group gives an idea of the formed personnel reserve in the organization and the personal benefits that, in this regard, may be available to a particular employee. In this case, the factor of motivational interest determines the level of personnel responsibility in the development of their own labor potential.

The main problem in the functioning of this system is that both the management and the employee, due to different circumstances, can force this process. This has a negative impact on the health of the employee. He is overloaded with work, authority and, sooner or later, there is a professional burnout. This leads to disappointment, dissatisfaction, reduced loyalty of the organization. That is, the employee loses valuable quality HiPro its qualitative characteristics, while remaining highly professional.

7 Conclusion

Thus, it is important for managers not only to provide a cumulative system of forming professional qualities among HiPo-employee in accordance with new technologies of personnel management and business environment. But also necessary to define the figures and indicators, by which it is possible to determine the moment of loss-professional burnout, reduction of employee loyalty, its initiative activity in solving tasks. All this will contribute to the disclosure of the labor potential of staff.

References

1. Federal law "About the independent skills assessment" of 3 July 2016 N238-FZ (2016). URL: http://www.consultant.ru/document/cons_doc_LAW_200485/. Accessed: 01.09.19. [in Rus.].
2. Passport of the national program "Digital Economy of the Russian Federation" (approved by the Presidium of the Presidential Council for Strategic Development and National Projects on December 24, 2018, N16) (2018). URL: http://base.garant.ru/72190282/. Accessed: 01.09.19. [in Rus.].
3. I.N. Makhmudova, Y.A. Krivoruchko, From unreliability to staff loyalty to ensuring the safety of organization personnel. *Kadrovik,* 7, 93-100 (2019). [in Rus.].

4. A.A. Borzunov, The importance of human resources to ensure the economic security of the company in the transition to the digital economy. *Vector Economy,* **7** (2017). URL: http://www.vectoreconomy.ru/index.php/number7/economicmangment-7-2017. Accessed: 01.09.19. [in Rus.].
5. K. Laktionova, Potential employees: The illusion and reality. *Spravochnik Kadrovika* (2015). URL: https://www.pro-personal.ru/article/1083934-qqe-m12-potentsial-sotrudnikov-prizrachnost-i-realnost. Accessed: 01.09.19. [in Rus.].
6. I.N. Makhmudova, The methodology of competency assessment of staff to determine professional standards. *Scientific Review: Theory and Practice*, **1,** 59-74 (2017). [in Rus.].
7. I. Makhmudova, E. Belonogova, Using digital technologies in staff training. *Slovak International Scientific Journal,* **1**(16), 19-22 (2018).
8. I.N. Makhmudova, S.I. Zhuchenko, Unclaimed employment in the digital economy. In G.Yu. Gulyaev (Ed.), *Proceedings of the V International Scientific and Practical Conference «Economics and Management in the XXI Century: Sustainable Development Strategies»* (pp. 84-87). Penza: ICNS "Science and Education" (2018). [in Rus.].
9. L.A. Ilyukhina, I.V. Bogatyreva, I.N. Makhmudova, J.D. Ermakova, Development of a corporate personnel training system: the experience of Russian enterprises. In: V. Mantulenko (Ed.), *Proceedings of GCPMED 2018 - International Scientific Conference "Global Challenges and Prospects of the Modern Economic Development", The European Proceedings of Social and Behavioural Sciences EpSBS*, **LVII** (pp. 654-665). London: Future Academy (2019). DOI: 10.15405/epsbs.2019.03.65.
10. L.L. Kifa, I.V. Kryuchkova, On the issue of staff development trends. *Azimuth of Scientific Research: Economics and Management,* **4**(17), 175-177 (2016). [in Rus.].
11. L.L. Kifa, Features of certification of senior employees of municipal institutions in the transition to an effective contract. *Vector Science of Togliatti State University. Series: Economics and Management*, **2**(29), 54-59 (2017). [in Rus.].

Human Capital as an Object of Accounting in the Knowledge Economy

O. Kogut[1,*]
[*]Corresponding author: kogut.1108@gmail.com.

[1]Al -Farabi Kazakh National University, Almaty, Kazakhstan

Abstract. The modern economy is called the economy of effective human capital, which emphasizes its main role in the development of the economy and society as a productive factor. Human capital is becoming the most important factor in the country's economic growth, the foundation of sustainable economic development. The article considers approaches to the definition of the concept and theoretical content of the category of human capital, its importance for the development of areas of science and practice. It is revealed that human capital is a necessary independent factor determining economic growth. The necessity of reflecting human capital in the system of accounting and reporting of firms is substantiated. The main problems of recognition of this most important factor of production are revealed, the directions of their further solution are determined. The scientific and practical necessity of introducing a new object of accounting and analysis of human capital is substantiated. The article contains information and conclusions that are of practical interest to firms.

Keywords: human capital, intangible asset, investments, accounting, financial statements.

1 Introduction

Throughout almost the entire history of mankind, until the beginning of the 1960s, the main role in the production process was played by physical capital: means of production, material conditions, etc. The requirements for labor were minimal: it was enough to have a muscular, "animal", natural force, which was used as one of the resources, as well as an appendage to technology. The creative abilities of man have not yet played a decisive role in economic development, and the problems related to the formation of labor did not cause much interest among economists.

It was traditionally believed that the labor market has an unlimited supply of workers. If there is a shortage of them in any of the sectors of the economy, it is enough to increase wage rates in it. This will ensure the overflow of labor from other sectors and thereby obtain the required number of workers of any skill level. In the world, the trend was gaining strength, within which the winner in the XXI century. it turns out to be not the one who has more natural resources or basic production assets, but the one who has higher quality of human capital. Therefore, the main strategic directions of the long-term development of Kazakhstan until 2050 is the development of human capital.

Unlike the means of production, human capital is not the property of the company. Human capital is embodied in the identity of its owner. The peculiarity of human capital is that it is inseparable from the person himself. Human capital cannot be the subject of sale, it can only be leased by entering into a contract of employment - an employment contract. For the period of employment, the employer acquires the right to take advantage of the opportunity to work as an employee, and not the labor itself, the owner of which is still the employee. Therefore, when developing plans for the strategic development of the company, it is necessary to keep in mind the structure and value of investments in human capital. From the position of the company, investments in human capital represent the costs that it incurs for the purpose of searching, selecting and training personnel, as well as their subsequent professional retraining.

However, in modern accounting practice, human capital is not taken into account in the company's balance sheet. Disputes surrounding the recognition and measurement of human capital as an asset are based on a subjective interpretation of the traditional foundations of accounting theory, not taking into account the main thing - business needs, which historically predetermined the development of accounting practices. Against this background, there is an underestimation of the human factor in the activities of modern firms, including an underestimation of the need to create transparent accounting and analytical support in relation to human potential. From the point of view of developing the theoretical foundations of accounting and analysis of human capital, this explains the interest in the human factor and the results of its intellectual work.

2 Problem Statement

The task of setting up this study is to consider the effectiveness of the use and development of human capital in modern conditions, in the accounting and information support of accounting for human capital management and the creation of an appropriate financial reporting system in the organization. The approaches to the classification of existing methods in Kazakhstan and foreign practice for assessing human capital will be considered. The main classification criteria will be determined by the economic level and the purpose of assessing human capital.

3 Research Questions

The main research question is the development of practical solutions to the problems of accounting for human capital as part of the implementation of the long-term socio-economic development strategy of Kazakhstan. The purpose of the study is to consider approaches to determining the concept and theoretical content of the category of human capital, its importance for the development of areas of science and practice based on the knowledge economy.

4 Materials and Methods

In the study of human capital and the problems of its accounting, the article used the fundamental works of classics of economic and philosophical thought, monographs, scientific papers, articles of domestic and foreign scientists on the analysis of human capital as a source and basis for increasing and effective activity in the economy. Achieving the goal was based on the methods of neoclassical and institutional economic theory, system analysis, the laws of dialectical development of economic systems, general scientific methods of cognition: analysis and synthesis, induction and deduction; scientific abstraction and concretization, expert assessments, sociological surveys, comparative and socio-institutional analysis.

5 Results and Reasoning

The theory of human capital was created by Nobel Prize winners in economics T. Schulz (1979) and G. Becker (1992). They calculated for the first time showed the advantages of people with special education over people with higher education in terms of income and increased education costs by the level of productivity investments, and education itself by the level of the most important development factor [1].

In the economic literature you can find many definitions of the concept of "human capital". The most common definitions: G. Becker "Human capital is a combination of innate abilities and acquired knowledge, skills and motivations, the effective use of which helps to increase income and other benefits" [1]; A. Dobrynin, S. Dyatlov, E. Tsyrenova "Human capital is a stock of health, knowledge and skills that contribute to the growth of labor productivity and affect income growth" [2]; V. Maksimovich V. "Human capital is a combination of available, accumulated and developed human abilities and qualities that are used in the production and consumption of economic goods, bring profit to its owner, as well as the effect at the place of use and contribute to the growth of the state" [3].

E. Flamholtz, the founder of the concept of accounting for human resources, believed that human capital is the most valuable tool of the company and should be taken into account as an asset of the company. Accounting for human resources is essential in making strategic decisions. He identified three main criteria for recognizing human capital as an asset: potential future economic benefits, ownership or control rights of an entity, and measurability in monetary terms [4].

J. Fitz-enz considered human capital as an asset, a powerful financial lever, the organization's most valuable and productive resource [5]. M. Arm-Strong in his definition of the concept of human capital spoke of him as a key element of the market value of the organization. In his opinion, "its price should be included in the calculation as an indicator for investors or those considering the possibility of a merger or acquisition of an organization" [6, p.848]. R. Likert, who was involved in the theory of human capital management ", also proposed introducing the article" Personnel investments "into the balance sheet as a company asset and reflecting these investments as an asset of the company" [7].

According to T. Schulz, "education is one of the forms of capital, it is called human because this form becomes part of a person, and capital is due to the fact that it is a source of future satisfaction or future earnings, or both." Later, T. Schulz supplemented his theory as follows: "Consider all human abilities either as innate or as acquired properties ... that are valuable and that can be developed with appropriate investments, will be human capital" [8, p.82].

The main characteristic features of human capital as part of human resources and as an object of accounting are the following [9]:

- the dependence of the functioning of human capital on the life expectancy of its carrier;
- the possibility of transforming the potential of human capital into specific scientific and technological developments;
- in many ways, the desire of human capital to integrate with production as confirming its innovative character;
- dependence of human capital on physical and moral depreciation;
- the complexity of calculating the economic benefits of investing in human capital until the actual income.

Traditional accounting science and practice do not consider human capital as an object of accounting, therefore this universally recognized economic category is not subject to registration in the system of accounting accounts and disclosure in financial statements. The main argument of OS Yasinskaya believes that the company does not have ownership of the carrier of human human capital, and traditional accounting does not recognize assets and does not reflect resources in accounting registers if ownership does not apply to them [10]. This is also noted by American economists L. Edwinsson and M. Melone, that the basis of the so-called knowledge economy is huge investments in human capital and information technology, but they do not find a positive reflection in the framework of traditional accounting [11, p.164].

Hence, at present, the idea of recognizing human capital as an asset of a firm seems problematic. Among the most significant arguments against the recognition of human capital as an asset, there is a lack of ownership of the use or possession of it.

Property as an economic category is manifested in the form of relations between people over various means, products of production, material goods through unhindered use, possession and disposal. As an economic category, this is not the means of production or their products per se, not the attitude of an individual person or group of people to the means of production and their products, but the relationship between people about the means of production and the material goods they produce.

Economic ownership requires ownership of the economic benefits of using an asset in one's business, rather than physical ownership. "Assets," noted B. Needles, H. Anderson, and D. Cowell, "are potential revenues that are possible as a result of the use by this legal entity of acquired or controlled values received by the organization as a result of previous transactions or past events" [12, p.496].

In the practice of international accounting, determining the presence of an asset for the right of ownership is not dominant. Most of the assets of the company are present under the financial control of the company when the assets are managed not on the basis of legal ownership, but in terms of ownership of the economic benefits received. According to Ya.V. Sokolova "... if there is a contradiction between the legal and economic interpretation of the fact of economic life, the accountant should choose an economic interpretation of the fact of economic life" [13, p. 62 – 67]. This position is consistent with the accounting principle "priority of content over form", according to which transactions and other events should be reflected in accordance with their economic reality, and not just their legal form.

This approach is implemented in IAS 17 "Leases", according to which the leased asset acquired under financial leasing rights is reflected on the balance sheet of the lessee [17]. At the same time, an entity acquires financial benefits from the use of a leased asset for most of its economic life. Moreover, in order to objectively reflect the level of economic resources and liabilities, financial leasing in the lessee's balance sheet should be reflected as an obligation to pay rental payments in the future.

International financial reporting standards treat assets as organization-controlled resources from which the firm expects economic benefits in the future. Attributes of an asset [14]:

a) the asset brings economic benefits. Labor is a factor of production, allowing the employer to produce and sell products (work, services) produced by workers. The company has the opportunity to receive future economic benefits from the use of human capital as an asset throughout the term of the employment contract;

b) the presence of control over the asset. The organization's acquisition of labor carries the hiring process, is formalized in an employment contract and a work order. The firm has authority over human capital as an asset and has the ability to influence revenue through the exercise of its authority. The employer has the right to make requirements for the level of education, work experience, work experience in the specialty and other requirements. The date of commencement of work, place of work, mode of working time and rest time, size and procedure for the payment of wages are established. The employer entrusts the employee with a specific job, monitors its implementation, in case of non-fulfillment, brings the employee to disciplinary action. Thus, the employer controls human capital;

c) asset valuation. At the stage when labor activity has not yet begun, a person directs his efforts to getting upbringing and education. In order to adapt to the constantly changing conditions of the labor process, the employee repeatedly increases qualifications, improves skills. Assessment of human capital is a monetary expression of the accumulated knowledge of the skills of the individual to obtain generalized data on the organization in the preparation of financial statements.

If human capital is able to bring economic benefits in the future, then the company has full control, and the initial cost is determined, then it can be considered as an asset and can be allocated as an object of accounting. The study of human capital as an accounting category in business management allows you to pay due attention to the development of concepts of accounting and analysis of the company, to use the hidden potentials of intangible resources.

Human capital is attributed to intangible assets, which does not have a physical form, but at the same time has a certain value for the company. Human capital is transformed into company assets. Human capital is not interchangeable. In the company, individual human capital forms a corporate culture, environment. Human capital is inherent in people and cannot be owned by the company.

Intangible assets are currently a problem in the accounting methodology for both Kazakhstan and foreign companies, which is primarily due to the complexity of the subject itself, the definition of criteria for the classification and recognition of intangible assets. If you turn to IAS 38 "Intangible Assets" [17] and a number of provisions of IFRS 3 "Business Combinations" [18], IFRS 36 "Impairment of Assets" [19] and others, you can see that the international approach to the definition of intangible assets is based on a list of criteria for their identification from the rest of the assets, based on asset forms and research objectives.

However, human capital in the conditions of development of innovations, having become, undoubtedly, the main economic resource, should be provided with the accounting methodology and the methodology of its accounting assessment, methods of transformation of accounting results in financial statements. The complexity of updating the theory of accounting for human capital is enhanced by the fact that not only the path to the formation of human capital as a carrier of knowledge is long, but also the process of its transformation into technological innovations. It is human potential, in contrast to other forms of assets, that is distinguished by the property that its creation takes a long time and is based on human mental activity, which personifies the time and resources spent, but which is difficult to account for and evaluate in the subject's information system [15].

The American Association of Accountants determines that, "the process of accounting for human capital consists in identifying, measuring, and transmitting information to interested parties" [16, p. 170 – 174]. The need for accounting and analysis of human capital for a company is determined by the fact that people are value, as they provide services and produce material values that can be quantified. However, this value of people cannot be owned by anyone, unlike other types of assets. The company cannot assign it. In addition, this value of the employee, as a production resource, largely depends on how relations between employees are formed in the company and how the production process is organized. Relations between workers are not taken into account by the classical microeconomic theory. He was ignored by the traditional accounting and analytical practice of economic activity. In the discussion of the qualitative and quantitative presentation of information in the annual reports of the company, the question of how appropriate and to what extent to present information in the financial statements of human capital, which, in essence, is capital in the form of intellectual abilities and practical skills acquired in the process, will be relevant education and implementation of human activities [16].

The only regulatory document for accounting for intangible assets is IAS 38 "Intangible Assets" [17]. But the international standard does not contain special emphasis on reporting such an indicator as human capital. Therefore, there is a need to develop new classification items for account 2700 "Intangible assets". As one of the options, we recommend the following scheme: to account 2700 open a sub-account 2701 "The right to use the knowledge and skills of employees" (analytical accounting is organized for each employee).

At the same time, personnel information can be presented in non-financial reporting. According to IFRS 1 "Presentation of financial statements", a company may provide additional information accompanying the annual financial statements if the governing body considers it useful and reliable for interested users of these statements when making the right economic and managerial decisions [20]. In modern conditions, such additional information may be information about human capital.

6 Conclusion

Functioning like physical capital, human capital has some fundamental differences, the main one of which is its inseparability from the face of its carrier. As a result, the market only sets prices for the "rent" of human capital (in the form of wage rates), while prices for its assets are absent. This seriously complicates the analysis. Secondly, human capital is able to increase the efficiency of activities both in the market and non-market sectors of the economy. Moreover, income from it can take both monetary and non-monetary forms. Human capital is the most important moment in the formation of a modern innovation environment. However, this object of economic reality is not reflected in the traditional accounting system. Therefore, not all facts of modern economic life are subject to registration, which violates the principle of objectivity.

There is an urgent need to take into account human capital in accounting, reflecting investments in this asset, and also analyze their effectiveness. The emergence of a new facility requires changes and additions to the standard Chart of Accounts and the development of methods for reflecting in the accounting of business transactions related to investing in human capital.

References

1. Yu.A. Korchagin, *Development prospects for Russia. Human capital and innovative economy* (CIRE, Voronezh, 2012). URL: http://www.lerc.ru/?part=articles&art=25&page=14,.- page=14. Accessed: 01.09.2019.
2. A.I. Dobryinin, S.A. Dyatlov, E.D. Tsyirenova, *Human capital in the transitive economy: Formation, evaluation, efficiency of use* (Science, Sankt Petersburg, 1999). [in Rus.].
3. A.A. Krugovaya, M.M. Pankova, Accounting and evaluation of the effectiveness of investments in human capital. *Problems of Material Culture - Economic Science,* 35-38 (2013). URL: http://dspace.nbuv.gov.ua/bitstream/handle/123456789/92335/09-Krugovaia.pdf?sequence=1. Accessed: 04.09.2019. [in Rus.].
4. E.G. Flamholtz, *Human resource accounting: Advances in concepts, methods and applications* (Springer, Boston, MA, 1999).
5. Ya. Fitts-ents, *Profitability of investments into personnel: Measuring the economic value of personnel* (Vershina, Moscow, 2006). [in Rus.].
6. M. Armstrong, *Practice of human resources management*, 10 edition (Piter, St. Petersburg, 2010). [in Rus.].
7. A.A. Kulikov, Human resources as a registration category. *Siberian Financial School*, **1**, 64-67 (2007). [in Rus.].
8. A.G. Mokronosov, Y.V. Krutin, Human capital or human potential. *Ideas and Ideals*, **2**(32/2), 80-89 (2017). DOI: 10.17212/2075-0862-2017-2.2-80-89. [in Rus.].
9. M.A. Kuznetsova, Cost approach to human resources management in the innovation economy. *Fundamental Research*, **2**(3), 578-582. [in Rus.].
10. O.S. Yasinskaya, Theoretical justification of economic essence of human capital as a new object of accounting and analysis, *Bulletin of Polotsk State University, Series "Economic and Legal Sciences"*, **5**, 112-118 (2015). [in Rus.].

11. L. Edvinsson, M.S. Malone, *Intellectual capital: Realizing your company's true value by finding its hidden brainpower* (Harper Business, New York, 1997).
12. B.E. Needles, H.R. Anderson, J.C. Caldwell, *Principles of accounting* (Finance and statistics, Moscow, 2004). [in Rus.].
13. Ya.V. Sokolov, *Fundamentals of accounting theory* (Finance and statistics, Moscow, 2003). [in Rus.].
14. T.N. Obushenko, Accounting human capital, *Science of Science*, **8**(3), 11-12 (2016). [in Rus.].
15. R.S. Kaplan, D.P. Norton, *Balanced system of indicators, from strategy to action* (Olimp-Biznes, Moscow, 2016). [in Rus.].
16. O.A. Kuzmenko, N.A. Masyuk, Human capital as object of accounting and reporting, *Economics and Law Issues*, **12**, 170-174 (2013).
17. IAS 38 "Intangible Assets" (2004). URL: https://www.iasplus.com/en/standards/ias/ias38. Accessed: 01.09.2019.
18. IFRS 3 "Business Combinations" (2008). URL: https://www.iasplus.com/en/standards/ifrs/ifrs3. Accessed: 01.09.2019.
19. IFRS 36 "Impairment of Assets" (2017). URL: https://www.ifrs.org/issued-standards/list-of-standards/ias-36-impairment-of-assets/. Accessed: 01.09.2019.
20. IFRS 1 "Presentation of financial statements" (2017). URL: https://www.ifrs.org/issued-standards/list-of-standards/ias-1-presentation-of-financial-statements/. Accessed: 01.09.2019.

The Strategy for the Development of Digital Education in Russia

E.B. Kolbachev[1], *A.A.* Pakhomova[1*], *T.I.* Lomachenko[2], and *T.A.* Kokodey [3]
*Corresponding author: tivano@yandex.ru.

[1] South-Russian State Polytechnic University (NPI), Russia
[2] Sevastopol Branch of Plekhanov University of Economics, Russia
[3] Sevastopol State University, Russia

Abstract. In a wide range of works of domestic and foreign scientists only some aspects of the digital economy are touched upon, which in the long run should develop into broader research. The authors' researches are aimed at forming a strategy for the development of digital education in Russia as the basis for implementing the national concept of "digital economy". To meet the requirements, the authors of the MBIC should pay more attention to the principles of management based on a cost approach, as well as to the interests of stakeholders in the planning and control of the achievement of goals. Stakeholders may be individuals, groups of individuals, institutions and organizations that influence the achievement of SSC goals and the implementation of their own goals, which in turn depend on the success of each SSCC author's activities.

Keywords: digitalization; education; strategy; strategy; network; national interest.

1 Introduction

At the present time, there is no common understanding of the concept of "digital economy" in scientific literature, so Doctor of Economics, corresponding member of the Russian Academy of Sciences Vladimir Ivanov suggests the widest definition of digital economy as a virtual environment that complements the existing reality [1]. R. Meshcheriakova, distinguishes between two approaches to the term "digital economy". The first approach is "classical": the digital economy is a digital economy and it is more appropriate to characterize only the field of electronic goods and services. The second approach is an extended one: "digital economy" is economic production using digital technologies [2]. The current work on the digital economy covers a wide variety of areas, such as the role of big data and application economics, standards for Smart Cities and the Internet of Things, vertical applications (e.g. workflow management and insurance telematics), and so forth.

2 Problem Statement

In addition, many domestic scientists, as well as foreign researchers consider cyberphysical systems as a key element of the digital economy.The essence of cyber-physical systems is that they link physical processes, such as manufacturing, with software and electronic systems. A.N. Raikov's contribution to the study of this issue is extremely great, and in his outstanding article [3] he pointed out the limitations of classical methods in solving the problem of education of schoolchildren and students, training of specialists of higher qualification taking into account the future needs of society (education for the future). This scientist proved that such a situation is connected with unreliable forecasting of needs, because the market can fluctuate in an unpredictable way. Specialist training, starting with kindergarten, is carried out in an environment of very high uncertainty. However, it is mainly based on retrospective knowledge and methods.

3 Research Questions

According to the study [4], the authors prove that the most stable and optimal structure of the education management system is a pyramidal three-level structure. At the same time, taking into account the development of the processes of horizontalization of relations between the subjects of the education system developing in the networked reflexive-active environment, it is possible to increase the levels of management. However, it is advisable to ensure proper control over the functioning of the system, monitoring and forecasting of market needs with appropriate modeling and theoretical justification.

If this study had not been carried out, it could have been argued that the existing works on digitalization of education cover only isolated aspects of individual areas of the digital economy and do not contain the development of comprehensive tools for effective management of the digital economy. However, the above work [4] considers ways and directions of further development of the system of strategic management of digital education development in the Russian Federation (SSMDEDIRF).

4 Purpose of the Study

The aim of the study is to develop the strategy of digital education of the Russian Federation as the basis for the implementation of the concept of "Digital economy". The authors of the article in their work explore the risks and threats facing the education system. They consider it necessary to increase the competitiveness of the country's economy through the tools of the educational system.

5 Research Methods

In order to create a conceptual model and management mechanism for the formation and operation of the system of strategic management of digital education development in the Russian Federation, it is necessary to clearly define the goals of digitalization in Russia in general and the digitalization of education, in particular, the formation of development trajectories leading to the achievement of such goals. These trajectories are formed by selecting a set of expedient management decisions of organizational, technical and other nature, ensuring a synergetic effect in the practical implementation of the decisions made.

Many possible solutions (initially of a chaotic nature) serve as an information base for the formation of alternative options for the trajectories of innovative development of SSMDEDIRF. One (or several) of these alternative trajectories in the process of competitive (quasi-competitive) procedures becomes/become dominating/dominant. This is a concrete embodiment of the rational choice of authors, carried out in specific conditions and under certain preferences.

In the context of our study, the issue of the goals of digitalization in general and innovation related to digitalization (including the digitalization of education) is important. This issue can be solved by considering the axiological aspects of digitalization - aspects related to the nature of the values of education and innovation, the interaction of different values among themselves.

Taking into account these values [5, 14, 15], it can be argued that the formation of a strategic management system for the development of digital education (SMSDDE) at the national level should solve the following problems: the formation of the conceptual provisions of the SSD; the formation of the composition of subjects of the digital education system; the formation of a system of interaction between the subjects of the digital education system; the formation of a system of resource support of SSD activities :in this case, the subjects of the digital education system include: public educational institutions of general secondary education; public educational institutions of secondary professional education; public educational institutions of higher education; individuals - developers and rightholders; state educational authorities of different levels; elements of the digital education system infrastructure; Federal Analytical and Project Center for Digital Education (SSC); Distributed Regional Situation Centers for Digitalization of Education (RSCFDE); Smart-Centers of educational organizations;

The nature of interaction between subjects of the digital education system is determined by the following features: a) organizational (including legal) independence of subjects; b) competitive nature of relations between developers and other subjects of SSC; c) the presence of common goals and objectives of activity determined by the interests of the Russian state and society as a whole. In this case, point "c" seems to be extremely significant, as it is directly related to the above axiological aspects of the digitalization of education.

On the basis of the above, it can also be stated that the quantitative assessment of the development of the socio-economic system (SES), for the sake of which digital education is created, can be made on the basis of determining the volume of information contained in it, and the entropy of SES. With regard to socio-economic systems as a whole, a similar conclusion was made by V. Maevsky in his famous work [6]. In particular, he found that economic evolution can be presented as a process of increasing negative entropy of the economic system. At the same time there are processes of complication of economic relations and ordering of the internal structure of the system.

6 Findings

The above described nature of interaction between the subjects of SSC allows to assert that the set of subjects of the digital education system forms a business network (BN), the specifics of which should be taken into account in the formation of SSCS. The information stage of economic and social development was marked, among other things, by the strengthening of the authors' integration, which made it possible to speak about the formation by the end of the nineties of a new economic reality - the global SES, which, in turn, consists of numerous networks based on new methods of integration in the implementation of joint projects related, in particular, to the introduction of computer (digital - in modern terminology) technologies [6, 7, 8].

Such BN factors as risk sharing and optimization of information management are very important for understanding possible goals and objectives of domestic SSC. In fact, participation in the business network may become the most important way for the Russian subject of SSC to increase the stability of functioning, especially in the conditions of modernization of educational processes (part of the risks is removed and functioning conditions are stabilized).

Conventionally, the situation in the network organization can be represented in the form of "abstract of the socio-economic world", which received further development in works by Y. V. Shishkov, M. Shpagina. [9,10]. The main feature of this model is the hypothesis that there are differences in the parameters of information exchange between participants of joint activity. According to this model, the social space in which the participants of the joint activity are distributed has a non-zero information permeability. The specific values of information permeability are determined by the level of development of information and communication technologies and mean that information from point A to point B of the space reaches a certain time, which is the greater the "distance" between these points in accordance with the topology of information dissemination in it. Accordingly, the longer the time that information is disseminated, the more likely it is to become outdated due to constant changes in the state of the economic space at each point. The first group is located within a radius of RO equal to the maximum distance from point A, at which it is still possible to exchange information intensively enough to use the community form of management. The second group is outside the radius of the RO, but inside the border with a large radius of RE. The RE distance is the maximum, at which the information exchange, sufficient for the functioning of the hierarchical form of control, is possible. All participants of the second group can potentially establish hierarchical relations with the "observer", but they can no longer establish communitarian relations. The third group is beyond the RE radius, and therefore the "observer" can interact with them on the basis of an even less information-intensive form of control.

With the development of digitalization, there is an increase in the speed of information dissemination, which leads to an increase in the length of the RO and RE radii and, consequently, the boundaries of the zone where community-based and/or hierarchical forms of control can be applied are extended. Theoretically, the network structure has no clear boundaries. In the literature [11] only the main subsystems forming the network organization are singled out: the corporate center; the subsystem of service organizations; members of the network. In the context of SSC, their roles are performed by the entities that support the activity of SSC (Federal Analytical and Project Center of SSC (FAPCSSC); distributed regional situational centers of digitalization of education (DRSCDE); Smart-Centers of educational organizations), as the infrastructure of SSC (ISSC).

It is advisable to entrust the Corporate Center (FAPCSSC) with the functions of coordinating the interests of all network members, representing them in relations with third parties, coordinating and supporting the corporate image of the network. In addition, the Corporate Centre FAPCSSC ensures interaction between the members of the network (it is an information and communication center).

In our opinion, such an approach can be adopted as a basis with the reservation that the allocation of FAPCSSC plays a functional rather than a structural role: the same organizations can be members of the network and at the same time perform infrastructural service functions for the whole BN.

Members of the network in terms of access and participation in the work of the network can be subdivided into: valid, i.e. having access to all functionalities of the network and participating in the work of the network directly; associated, i.e. having access not to all functionalities of the network (access to the network can be organized through one of the full members).

The place and nature of the links vary: members of the first level network who are in constant interaction with other members of the network as well as with FAPCSSC; members of the second and subsequent levels who interact with the group of members of the network and have almost no interaction with FAPCSSC.

The degree of interaction in the network organization can be characterized by the density of interaction, defined as the ratio of the number of real interactions between members of the network to their theoretically possible number:

$$P(t)= Ir(t)/ Im(t)*100 \qquad (1)$$

where P(t) is the network density at time t;

Ir(t) - number of real interactions at the moment t;

Im(t) - number of mathematically possible interactions at the moment [12].

The study of L. Hirschhorn describes the advantages of a network organization as follows: «Network organizations are characterized by the creation of better communications than the market can do, network forms of organization provide better coordination in the face of changes, the significance of which cannot be fully conveyed or understood through price signals. The last seems to be the most important in the context of our research. As has been shown above, commercial criteria for the efficiency of modernization projects are not objective enough. This is one of the examples of inefficiency of "price signals" discussed in the above paper» [12, p.56].

By their very nature, classic networks are close to coalitions - voluntary associations formed to achieve common goals and objectives. Such a view of networks as polycentric informalized structures is characteristic of the European research tradition. The American authors, on the contrary, tend to interpret networks in a more formal way, emphasizing the moment of deep interdependence of participants and noting the possibility of the presence in the networks of a single control center in the person of the company - network integrator. In general, the scientific literature presents a wide range of definitions of networks in the absence of a common interpretation.

There is an opinion that the closest to the practical understanding of networks is the definition of F. Kotler and R.S. Achrol, as it indicates the lack of hierarchical control in the network organizations, but at the same time it notes the presence of common goals among the participants. Involvement, the authors summed up, «is the main difference between the network concept and the traditional theory of organization» [13, p.10].

The natural interest to the network form of SSC is connected with revealing of the mechanism coordinating network interactions as, without assuming the presence of administrative apparatus of compulsion, simultaneously provides joint work of subjects. Unlike hierarchies based on internalization (assignment) as a way to ensure integration processes, the network is an organizational and management form based on externalization (alienation).

7 Conclusion

Strategic organizational challenges related to the competitive nature of the relationship between developers and other SSC entities appear to be quite complex and important, given that competitiveness (CS) is a complex, multi-level concept whose analysis and evaluation should be linked to the type and geographical scope of the competitive field. Therefore, the formation and analysis of the system of factors and indicators characterizing SSC in the market of educational services, it is advisable to consider in the context of competitive processes at the macro and meso-levels.

It is obvious that the factors determining the CS of the SSC authors at macro- and meso-levels are external to them. It should be noted that the existing interrelation between the factors is bilateral in nature. On the one hand, a favorable socio-economic and political climate, formed at the macro level, is extremely important for the CS of SSC authors. On the other hand, the CS of education in the country depends on the ability of all subjects to build effective chains of intellectual products, their ability to create and develop competitive advantages and influence the formation of socio-economic policy of the state.

Acknowledgments

The work is executed at financial support of Russian Foundation for basic research under research project 19-010-00377 "Designing the strategic management system development digital education of the Russian Federation".

References

1. The World Bank, *Development of the digital economy in Russia. World Report 20 December 2016* (2016). URL: https://www.vsemirnyjbank.org/ru/events/2016/12/20/developing-the-digital-economy-in-russia-international-seminar-1. Accessed: 13.10.2019. [in Rus.].
2. RIA NOVOSTI, *Digital economy: How experts understand this term* (2017). URL: https://ria.ru/20170616/1496663946.html. Accessed: 02.09.2019. [in Rus.].
3. A. Raikov, Accelerating technology for self-organising networked democracy. *Futures*, **103**, 17-26 (2018). DOI: 10.1016/j.futures.2018.03.015. [in Rus.].
4. T.I. Lomachenko, T.A. Kokodey, E.B. Kolbachev, A.A. Pakhomova, Development of the pyramidal conception of the Russian Federation digital education development strategy formation. *Druckerovsky Vesnik*, **3**, 134-141 (2019). DOI: 10.17213/2312-6469-2019-3-134-141. [in Rus.].
5. E.B. Kolbachev, Social effectiveness of organizational and economic decisions affecting the development of higher education in the regions of Russia. *Bulletin of the South Russian State Technical University (NPI). Series: Socio-Economic Sciences*, **1**, 118-124 (2012). [in Rus.].
6. V. Maevsky, Economic evolution and economic genetics. *Economics*, **5**, 4-21 (1994). [in Rus.].
7. L.A. Volovik, *Fundamentals of foreign economic activity* (KSU, Kaliningrad, 1998). [in Rus.].
8. N.V. Khmel'kova, Network management in marketing. Practice of domestic metallurgical companies. In S.S. Chernova (Ed.), *Problems of Economics and Management of Enterprises, Industries, Complexes* (pp. 198-207). Novosibirsk: ZRNS (2009). [in Rus.].
9. Yu.V. Shishkov, *Globalization of economy – A natural product of industrialization and informatization of* society (IMEMO RAN, Mocsow, 2016). [in Rus.].
10. M. Shpagina, Revision. Should we reject the values of the new economy. *Expert*, 10.09.2001 (2001). [in Rus.].
11. B.L. Kuznetsov, A.A.Galiakberova, Synergetic effect of the technopolis development of the region. *Economic Synergetics and Innovation Process (abstracts of Interuniversity scientific and practical conference)* (pp. 42-43). Nabereznye Chelny: Kampi publishing house (1998). [in Rus.].
12. L. Hirschhorn, Beyond mechanization: Work and technology in a postindustrial age (MIT Press, Cambridge, MA, 1986).
13. F. Kotler, R.S. Achrol, Marketing in the network economy. *Journal of Marketing*, 63, 146-163 (1999). DOI: 10.2307/1252108. [in Rus.].
14. E.B. Kolbachev, A.A. Pakhomova, Designing the system of strategic management of digital education development in the Russian Federation: Some methodological features. In T.I. Lomachenko (Ed.), *Proceedings of the III Interdisciplinary All-Russian Scientific Conference: Development of methodology of modern economic science, management and education in the conditions of information-digital trends* (pp. 136-145). Sevastopol: Sevastopol branch of the "Russian Economic University named after G.V. Plekhanov" (2019). [in Rus.].
15. E.B. Kolbachev, A.A. Pakhomova, Development of digital education in Russia: Some strategy issues. *Druckerovsky Vestnik*, **4**, 162-172 (2019). DOI: 10.17213/2312-6469-2019-4-162-171. [in Rus.].

Modern Teacher's Worldview and Ethics in the Context of Technological Paradigms (Case Study of a National Region of Russia)

A.D. Nikolaeva[1,*], *S.A.* Filin[2], and *D.* Davaasurengiyn[3]
*Corresponding author: allanikol@list.ru.

[1]Ammosov North-Eastern Federal University, Pedagogy Chair, Yakutsk, Russia
[2]Plehanov Russian Economics University, Moscow, Russia
[2]Mongolia State University of Culture and Arts, Science and Innovation Chair, Ulaanbaatar, Mongolia

Abstract. The authors discuss the process of formation of worldview and ethics among modern teachers in the context of establishment of a new system of values due to changing technological paradigms. Based on historical-pedagogic and moral values, which are the foundation of pedagogic education, prospects for the development of worldview ethics of future teachers have been identified in the context of changing technological paradigms. The authors emphasize the influence on the worldview and ethics of teachers of, on the one hand, natural-climatic factors, connected with the implementation of ethno-pedagogic origins of education, and, on the other hand, scientific-technical revolution, connected with intensive economic innovative development processes. The significance of environmental awareness in forming new ethical outlooks, connected with the understanding of existing environmental issues, is confirmed by empirical evidence.

Keywords: teacher, environmental awareness, ethics, innovative worldview, technological paradigm, educational space.

1 Introduction

The relevance and originality of the research topic is attributed to the idea that the processes of globalization, ongoing in the world and affecting various spheres of life in society, are characterized by accelerated changes in the existing ways of life, establishment of a new system of values and re-evaluation of the global goals of human existence. In this context, the highest significance is gained by the role of a teacher, as an instrumental influence in forming the personalities, way of life and innovative worldview, consciousness and mentality of students. The diversity of natural-geographic, climatic, ethnic, demographic, sociocultural, educational and other features of Russian regions, including the national region of the Sakha Republic (Yakutia), brings about the need for multicultural cooperation in order to, among other things, form individual personal qualities and personalities in general. A special role in modern conditions of societal development is played by the influence of worldview and ethics of a contemporary teacher in the upbringing of young students.

2 Methods

The methodological framework of the research is based on the concept of interrelation and interdependence of social, culturological, pedagogical, objective and subjective factors in the development of pedagogical education on the foundation of interdisciplinary connections: economics, philosophy, culturology, pedagogics. To reach the goals of our research a number of theoretical methods were used, based on the requirements of expert-analytics work: reflective analysis of literature on philosophy, economics, psychology and pedagogics; methods of theoretical analysis, synthesis of pedagogical approaches, comparison, classification, systematization, surveys, observation, questionnaires.

3 Results

N.U. Yarychev [1] identifies two levels of socio-cultural environment: "macro-environment (economic conditions, social relations, culture of society)" and "microenvironment" (family, friends, specific organizations, homes, specific natural landscape)", which are constantly interacting and influencing each other. As the ethics and worldview of a teacher are influenced by socio-cultural and ecological environments, the diversity of training models of modern teachers, including in national regions of Russia, has unique socio-cultural, ecological and other features. Each nationality combines both universal human traditions and original features in forming the culture of teachers, which are all organically integrated into the single socio-cultural standard of forming the personality and professional qualities of Russian teachers. The evolution of ethics and worldview of a personality in Russia, including in Yakutia, "as a leading socio-cultural activity, has largely been a historic process of developing ethno-cultural education and its influence on personalities" [2]. According to U.S. Borisova [3], during the process of ethno-cultural education future teachers assimilate ethno-cultural values: language, culture, historic and spiritual heritage. Correspondingly, the foundation for the evolution of worldview ethics of a teacher is the preservation and reproduction of spiritual traditions of peoples and folk pedagogic. This has been a long process of evolution of ethno-pedagogical education, starting from the first schools of Kievan Rus, through the school education reforms of the 17th-19th centuries, to the revival and development on the cusp of the 20th and 21st centuries.

Principal factors that influenced the formation and evolution of folk pedagogic were the natural-climatic environment and factors connected with the implementation of ethno-pedagogical origins of upbringing. Analysis showed that in thousands of years specific means of regulating the relations of people and the environment have formed in Russia and Yakutia during the process of societal and natural evolution: natural-geographic conditions in Yakutia are closely connected with traditional economies and affect the development of peoples and formation of their ethical values and general worldview. The richness of traditions and experience of folk pedagogic "contributed to the development of a worldview ethics of teachers" [4]. Thus, the specific feature of the evolution of worldview ethics of teachers in Yakutia is the fact that its formation has been influenced by various conflicting factors, due to which the qualitative changes that took place during this process have significant scientific, historic and practical value for modern pedagogy.

The analysis of worldview of modern teachers in the context of establishment of a new system of values during shifting technological paradigms (TP) shows that it is influenced by many modern processes, in particular, the ever-increasing demand of humanity for energy, primarily in economically developed countries, due to which there are significant changes in the natural-geographic and climatic conditions of Russian regions: thawing of permafrost, melting of ice sheets in the Arctic part of the country, warmer and shorter winters, increasing mean annual air temperatures in Russia and the world. The number and intensity of natural cataclysms is rising, and the environment is worsening. Shifting TPs inevitably lead to changes in the system of values of social groups and societies in general, as well as adaptation of people to shifting TPs due to the fact that it is human capital which is the driving force behind the economic development of countries. Without characterizing all the TPs we would like to point out that in conditions of the sixth TP (1985-2035) (digitalization of economy, homestead urbanization of a "cloth" type, new medicine, new types of communication) enhanced capabilities are expected from organizations and people, as well as investment into them, planning of future and its management, resolution of problems within the frames of coevolution, presence of a new level of education system, artificial intelligence, high humanities and environmental technologies (new environmental management), and technologies of assembly and destruction of social subjects. The potential for development of this paradigm allows us to expect by the year 2020 radical changes in economic and social spheres: improvements in the quality of life and life expectancy, changes in the character of labour. Convergence of nano-, bio-, cognitive and information and communication technologies that form the sixth TP [5] is oriented towards cognitive processes. This is cognitive science that provides for new rationality and vision of the world. However, if by 2040 the humanity does not fully transition to the sixth TP, new global issues are unavoidable, connected with the provision of energy and resources, which can lead to a severe global crisis [6, p. 21].

The analysis of worldview of modern teachers in the context of establishment of a new system of values during shifting TPs shows that at present there are two principal global trends that have the highest chance in influencing the direction of development of pedagogical school, people and human civilization in general: 1. humanity is concerned with issues connected with the shrinking and coorganization of sociohumanitarian and socio-cultural diversity due to globalization processes; 2. humanity is carrying over to the future the determinant of the environment: the crisis in the organization of life of humanity in general, which for the most part causes financial-economic and other types of crises, due to which humanity will be actively searching for ways of organizing its life in the future and for new ways of using intellectual capital.

At present, Russia is systematically introducing new sociohumanitarian technologies on a large scale, which can improve quality of life. This is why there are opportunities in the country to develop and effectively use sociohumanitarian and socio-cultural technologies with a positive synergetic effect in order to accelerate its innovative development based on original traditions and ways of thinking in these spheres, as well as intellectual, creative and scientific capacity of the population and the involvement of Russians in innovative activities. According to O. Bakhtiyarov [7] "the source of human activity is shifting from external (social, cultural, enforced) stimulation to internal consciousness, to active, strong-willed, creative layers." This signifies a change in the concept of a human and another project to create a new human, oriented towards technologies, the basic characteristic of whom will be the ability to generate new socio-cultural and technological realities and the possibility of "coding" it. Possible programming of a part of human thinking, when it will stop being a purely biological phenomenon, and the human brain will become an analogue of a hard drive, will increase human efficiency: if some knowledge is forgotten or disappears with a human, it will be possible to restore it by downloading the relevant information. That is, capabilities of human memory will be expanded. Thus, the worldview of a teacher is influenced by natural-climatic environment and factors connected with the implementation of ethno-pedagogical origins of education on one hand, and the scientific-technological revolution and connected processes of intensive economic innovative development on the other [8]. Russian and foreign researchers agree that in modern conditions environmental ethics as the foundation of environmental awareness becomes extremely relevant, forming ethical-environmental views in every region of Russia: "the topic of ecology and environmental ethics is firmly ingrained in the social consciousness. The fragility and vulnerability of nature in the Russian North, the delicate nature of the balance of the "human – nature – society" system are now fundamental truths in social consciousness" [9-11]. Environmental awareness is crucial in forming new ethical values connected with the understanding of existing environmental issues, rational and careful attitude towards nature and the environment, new ways of developing the world, information civilization, actualization of the creative potential of a personality, complex understanding of the modern world image, consolidation of social activities and establishment of various social environmental movements and, correspondingly, new (innovative) type of worldview [12-14]. The above point is confirmed by the results of a survey held by the authors among the teachers and students of Russian universities (sampling): environmental issues have been shown as significant by 100% of respondents, including the issues of air quality, drinking water, safety of food products, declining health of population etc. Global climate change was a point of major concern, signifying the spread of environmental awareness and the level of environmental education. Russian students answered that the main factor in formation of ethical culture is the protection of the environment and its influence on the formation of worldview ethics (84%). Most respondents (teachers and postgraduate students – future teachers selected in the sampling) include among the factors that influence the worldview ethics of modern teachers such points as the evolution of society, state and political system of a country, the level of moral and cultural development of society, historic and cultural features, natural and socio-cultural environment, development of science and education. 82.9% of the Ammosov North-Eastern Federal University students responded favourably to the preservation of traditional, national way of life, reasoning that this will preserve a unique civilization living in harmony with nature. Teachers pointed out that the system of conditions and activities for forming ethics and worldview of a modern teacher can be based on the integration of personality-oriented, competence, culturological and axiological approaches, the foundation for which is the modelling of results of education in university on the basis of forming a value-worldview culture of personalities of future teachers [15]; orientation towards the motives and goals of a teacher's personality; translation of societal values onto the personality and education process, including the values formed by the modern scientific-technological revolution and connected processes of intensive innovative development etc.

Questioning students on the ethical worldview of future teachers also showed the indisputable influence of family (38%), mass media (27.6%) and educational institutions (schools – 16.2%, university – 10.1%) that comprise the socio-cultural environment, which builds pedagogical activity that recreates in teachers ethical, worldview and ethno-cultural values. The leading aspect that actualizes the significance of worldview and ethics of the younger generation for social development is that based on their ethical and worldview understanding young people define for themselves which directions of development are progressive at the current point of time and which are not. Additionally, under the influence of professional activities of teachers, the cultural-educational environment of an educational institution can change, and subsequently, under that influence, the socio-cultural environment can change in the educational institution at first, and then, possibly, in the society at a corresponding level. While studying the evolution of socio-cultural educational environment in the context of historical stages of societal development, closely connected with traditional economies, specific means of regulating relations of people and nature, as well as the diversity of cultural-labour and spiritual-moral orientations and traditions, we found that this environment forms ethical views of modern humans, including teachers: in the aspects of establishment of a new system of values, formation and development of ethical views of a teacher both in the educational process and the open environment of a university (school), the bearers or subjects of which are teachers as personalities and systems, theories and technologies of education and pedagogical upbringing introduced by them; in the socio-cultural aspect of forming an innovative worldview, consciousness and way of thinking of young people as personalities (politics, culture, creativity, traditions, family, education, socially useful work, sport, health etc); in the "human – society – nature" system (civic activities, self-organization, positive or negative synergetic effect depending on the person's attitude towards nature, society, own life). Russian questionnaire participants place equal emphasis on such social phenomena that affect the formation of a personality as understanding of the value of nature (40.1%), attitude towards political and social life of a country and participation in it (47.4%), considering the traditional way of life as a unique civilization (82.9%).

4 Conclusion

From the point of view of establishment of a new system of values in the context of shifting technological paradigms, the worldview of a modern teacher is characterized by its relation to the teacher's attitude towards nature, society, different facets of the environment, self and own life, which is expressed in convictions, beliefs, ideals, values that characterize the spiritual-moral sphere of the teacher's personality, direction of worldview, consciousness and way of thinking. Formation of innovative worldview, consciousness and way of thinking of a teacher is largely determined by, on the one hand, understanding of the priority of environmental awareness that projects ethical views connected with the recognition of existing environmental issues, rational and gentle attitude towards the environment, and, on the other hand, scientific-technological revolution and connected processes of intensive development of innovative production activities that result in a digital economy that promotes the actualization of a creative potential of a personality, complex understanding of the modern world image and, subsequently, establishment of this type of worldview. Socio-cultural environment and open socio-cultural educational space are factors that define the regional features of historic and modern educational practices, significantly influencing pedagogical processes and ethics formation, innovative worldview, consciousness and way of thinking of both teachers and, under their professional influence, socio-cultural environment in an educational institution and society at the corresponding level. Beliefs and convictions of teachers transform into motives that compel them to change and develop their moral worlds that affect through pedagogical upbringing and educational process the formation of ethics and innovative worldview, consciousness and way of thinking of students, who are the future human capital of innovative economies in new technological paradigms.

Acknowledgements

The article was written with the financial support of the Russian Foundation for Basic Research, Project #17-26-03003.

References

1. N.U. Yarychev, *The concept of developing conflict resolution culture of a teacher in a self-educating organization*. Ph.D. thesis. Chelyabinsk: Chelyabinsk Institute of Retraining and Continuing Education of Educators (2011). [in Rus.].
2. A.D. Nikolaeva, S.A. Filin, D. Davaasurengiyn, Origins of the formation of the world outlook ethics of the teacher in Russia (on the example of the national region of the Republic of Sakha (Yakutia). *Discussion*, **7**, 90-95 (2017). [in Rus.].
3. U.S. Borisova, *Ethnocultural education: Historical-sociological analysis: A case study of the Sakha Republic (Yakutia.)*. Ph.D. thesis. St. Petersburg: St. Petersburg University (2006). [in Rus.].
4. A.D. Nikolaeva, O.P. Osipova, The formation of ethics and world outlook of the future teacher. In V. Chernyavskaya, H. Kuße (Eds.), *Proceedings of the 18th PCSF 2018 - Professional Culture of the Specialist of the Future. The European Proceedings of Social & Behavioural Sciences EpSBS*, **51** (pp. 615-662). London: Future Academy (2018). DOI: 10.15405/epsbs.2018.12.02.67.
5. Yu.V. Yakovets, *Landmark innovations of the 21st century* (Ekonomika, Moscow, 2004). [in Rus.].
6. I.A. Shumpeter, *Theory of economic development. Capitalism, socialism, democracy* (Eksmo, Moscow, 2007). [in Rus.].
7. O. Bahtiyarov, People of new willpower: Social-humanitarian paradigm and its creators. *Development and Economics*, **3**, 150-163 (2012). [in Rus.].
8. A.Zh. Yakushev, Formation of innovative consciousness as one of the key aspects of ensuring the innovative scenario of economic development. *Science Journal of Volgograd State University. Global Economic System*, **4**(33), 150-160 (2015). [in Rus.].
9. F.M. Giorgett, C. Campbell, A. Arslan, Culture and education: Looking back to culture through education. *Pedagogica Historica*, **53**(1-2), 1-6 (2017).
10. A.D. Nikolaeva, V.M. Savvinov, Multi-ethnic school in the Russian Federation: The preconditions of formation and development (a case study of a national region). *International Electronic Journal of Mathematics Education*, **11**(10) (2016).
11. H. Ivon, D. Kuscevic, School and the cultural-heritage environment: Pedagogical, creative and artistic aspects. *Center for Educational Policy Studies Journal*, **3**(2), 29-50 (2013).
12. Z. Komar, Free time as time of true education. *Filozofska istraživanja*, **2**(114), 297-302 (2009).
13. N.D. Neustroev, A.D. Nikolaeva, A.N. Neustroeva, A.V. Ivanova, Problems and modernization trends of ungraded schools of the Russian North. *International Electronic Journal of Mathematics Education*, **11**(10) (2016).
14. F. Trede, Role of work-integrated learning in developing professionalism and professional identity. *Asia-Pacific Journal of Cooperative Education*, **13**(3), 159-167 (2012).
15. S.A. Filin, A.D. Nikolaeva, A.Z. Yakushev, I.S. Alexeeva, Training of specialists for "digital" and innovative economies. In V. Chernyavskaya, H. Kuße (Eds.), *Proceedings of the 18th PCSF 2018 - Professional Culture of the Specialist of the Future. The European Proceedings of Social & Behavioural Sciences EpSBS*, **51** (pp. 680-688). London: Future Academy (2018). DOI: 10.15405/epsbs.2018.12.02.74.

Developing Student's Key Competencies through Cooperation with Enterprises

R. Dušek[1,*]
*Corresponding author: dusek@mail.vstecb.cz.

[1]VŠTE, České Budějovice, Czech Republic

Abstract. The aim of this paper is to describe proposal of the two semester course inspired by project-based learning methods that could be applied to develop students' key competencies. Proposed method should be considered especially by university for business and economics, because its purpose is to integrate and develop competencies of final-year university students through working on specific projects in the areas of trade, marketing, economics or tourism. Project topics should be specified by various enterprises that are interested in cooperation with universities, so the course topics will be connected to business practice.

Keywords: competence approach, high education, cooperation with enterprises.

1 Introduction

The actual labour market requirements are no longer based only on theoretical or expert knowledge, but especially on higher level of key competencies. They can be defined as a generic complex of knowledge, values and attitudes, but primarily life skills such as [1, 2]:

a) problem solving,
b) ability to communicate effectively,
c) collecting and analysing information,
d) ability to work independently,
e) leadership,
f) creativity,
g) ability to cooperate effectively,
h) lifelong learning.

These competencies allow individuals to act adequately and efficiently within the various professional and personal life situations. They represent and incorporate a wide range of long term qualifications and requirements for a flexible response to the changing labour market demands and they also ensure maximum performance in the defined tasks [3].

Key competencies are behavioural, so if students have certain basic individual capabilities, their competencies can be developed throughout higher education. So, their further well managed development should be in the interest of those universities that want their graduates to be successful on the labour market, which should be very important especially for university for business and economics [4, 5].

The next step is to find and implement suitable approach enabling students to develop such key competencies. There are plenty of methods that have been proven to help in that, for example case studies [6], business simulation games [7, 8] or project-based learning methods [9]. The last mentioned options is widely considered as the most suitable method for students' key competencies development. This approach can be characterized by creating the students' research teams to work individually on long-term projects, which are connected to business practice and are designed to solve the actual business problems and cases [10, 11].

2 Methods

The aim of this paper is to present draft of the course that allows to develop students' key competencies through cooperation with enterprises. The course structure use as a basis the project-based learning methods. It is intended mainly for business and economics universities.

The structure of the course was inspired by various studies focused on implementing of effective competency and project-based learning methods [12-15], but it was adapted to the local conditions of the Czech labour market requirements.

Based on the papers about implementing project-based learning courses mentioned above, the main objectives that should be fulfilled by the proposed course have been defined as follows:

1. Integrating wide range of skills and knowledge.
2. Connecting the students' skills and knowledge with the experience, that will help to increase effectiveness of the learning process.

3. Activating students for lifelong learning.
4. Developing students' ability to communicate and cooperate in a team.
5. Developing students' organizational and time management skills.

3 Results

The proposed two semester course purpose is to integrate and develop key competencies of university students through working on various projects in the areas of trade, marketing, economy, tourism etc. As it is inspired by project-based learning methods, it is important to mention that all project topics should be specified by several enterprises that are interested in this kind of cooperation with university and its students.

According to course structure as shown in fig. 1, students should at first obtain enough relevant theoretical sources in the reseach area and then start to work on concrete solutions or ideas to a given problem by an cooperating enterprise's manager, that expects relevant recommendations suggested by students at the end of the second semester.

Due to the necessary skills and knowledge level of students participating on this projects, the course is recommended especially for students that are in their final year of study. The initial expectation is that this course should effectively combine both theory and practice, so that, in effect, the students can be more enthusiastic about trying something new during their studies. Simultaneously it can result in the enteprises' opportunity to test students as potential employees, so managers of the cooperating enterprises will be able to choose the right candidates for theirs actual graduate jobs more easily. The whole process flow of the proposed course is shown in fig. 1.

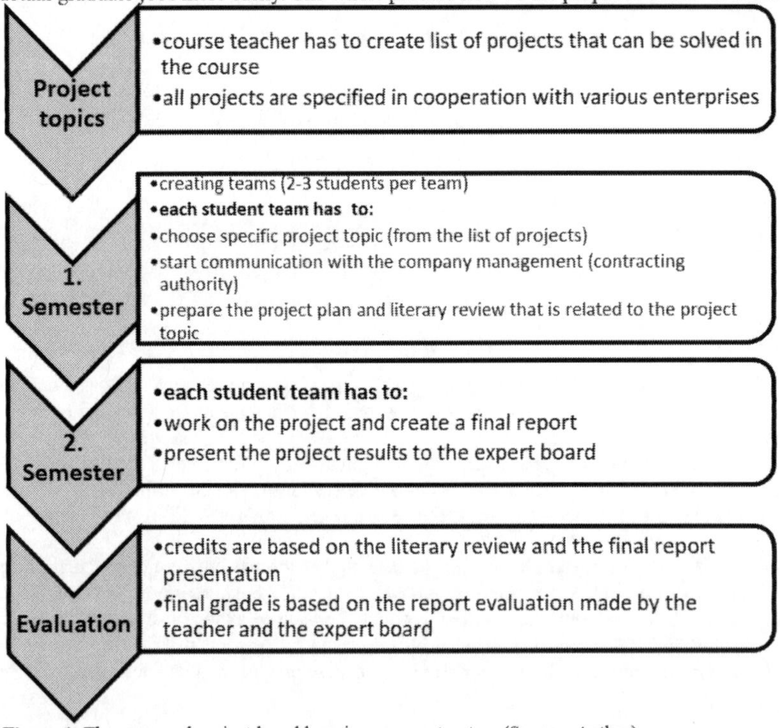

Figure 1. The proposed project-based learning course structure (Source: Author).

3.1 Recommended project topics

Projects solving should develop the students' ability to apply their theoretical knowledge on specific, practical and actual problems from their fields of study, so in the case of business and economics schools, the projects can be focused on the following topics:

1. Marketing research,
2. Economic analysis,
3. Data mining,
4. Financial analysis,
5. Event management,
6. Marketing mix,
7. Marketing communication.

3.2 Projects evaluation

As proposed in fig. 1, the first semester of the course would be evaluated with a credit (based on quality project plan and literary review made by all teams), while the second semester is evaluated with both credit and examination. Students will receive credits for presenting the final report to the expert board (which consists for example of managers from the cooperating enterprises). Final report, respectively project documentation, which is the main course output, should contain relevant recommendations for cooperating enterprises how to solve concrete problems in accordance with the aim of the project.

Final grade for each team should be based on criteria defined by course teacher at the beginning of the first semester. The main criteria may be defined as follows:

1. Level of target achievement,
2. Level of the final presentation,
3. Level of literary review,
4. Level of final report,
5. The cooperating enterprise's satisfaction with the final report and results.

4 Conclusions

The aim of this paper is to describe the project-based learning method draft, that could be relevant factor in developing students' key competencies such as problem solving, communication, creativity or leadership. Formulated method should be considered especially by university for business and economics.

The main purpose of the proposed two semesters course is to integrate and develop competencies of each student through working on specific projects in the various areas, for example trade, economics, marketing or tourism. Student teams must prepare theoretical and practical solution of the given project, so in the process of working on the topic they can develop their practical skills in a s area. Defined course has been inspired by project-based learning methods, so all projects for students' teams should be specified in cooperation with various enterprises that are interested in this kind of cooperation with concrete university and its students.

As stated in the proposed course structure, students should at first obtain enough relevant theoretical sources in the project area and then start to work on concrete solutions to a given problem or case study by an specific enterprise. At the end of the course, managers involved in the cooperation on the project will expect students' relevant recommendations or solution in the final report. The main expectation is that this course should combine both theory and practice, so that, in effect, the students' skills could be improved more faster and effectively than by regular teaching methods.

References

1. M. Armstrong, *Armstrong's handbook of human resource management practice. 13th Edition* (Kogan Page, Philadelphia, 2011).
2. H. Belz, M. Siegrist, *Klíčové kompetence a jejich rozvíjení* (Portál, Praha, 2001).
3. T Weiss, S. Kolberg, *Coaching competencies and corporate leadership* (CRC Press, Florida, 2003).
4. G. Debling, The employment department/training agency standards programme and NVQs: Implications for education. In J.W. Bruce (Ed.), *Competency Based Education and Training* (The Falmer Press, London, 1989).
5. M. Hay, Business schools: A new sense of purpose. *Journal of Management Development*, **27**(4), 371-378 (2008).
6. E. Anderson, B. Schiano, *Teaching with cases: A practical guide* (Harvard Business School Publishing, Boston, 2014).
7. A. Avramenko, Enhancing students' employability through business simulation. *Education and Training*, **54**(5), 355-367 (2012).
8. S.R. Tiwari, L. Nafees, O. Krishnan, Simulation as a pedagogical tool: Measurement of impact on perceived effective learning. *The International Journal of Management Education*, **12**(3), 260-270 (2014).
9. W.N. Bender, *Project-based learning: Differentiating instruction for the 21st centrury* (Corwin Press, Thousand Oaks, 2012).
10. M. Slavík, *Vysokoškolská pedagogika pro odborné vzdělávání* (Grada, Praha, 2012).
11. J. Atkinson, *Developing teams through project-based learning* (Gower Publishing, Hampshire, 2001).
12. J. Allen, Measuring competencies of higher education graduates. *New Directions for Institutional Research*, **126**, 49-59 (2005).
13. W.M. Krompf, *Identify core competencies for job success* (Association for Talent Development, Alexandria, VA, USA, 2007).
14. R. Maltese, *Project based learning: 25 projects for 21st century learning* (Dog Ear Publishing, Indianapolis, 2012).
15. M.R. Weber, D.A. Finley, A. Crawford, D. Rivera Jr., An exploratory study identifying soft skill competencies in entry-level managers. *Tourism & Hospitality Research*, **9**(4), 353-361 (2009).

Current Trends in Innovative Development of Higher Education

M.O. Suraeva[1,*], and *O.S.* Andreev[1]
*Corresponding author: marusyasuraeva@mail.ru.

[1] Samara State University of Economics. Samara. Russia

Abstract. The relevance of the studied problem is due to questions of improving the trends in the innovative development of higher education, since the methodological level of its development does not fully reveal their essence. The analysis of conceptual views on this issue points to many arguments, which indicates a divergence of theoretical positions. The purpose of the article is to study the trends in the innovative development of higher education, the influence of higher education factors and other factors of innovative development in terms of economic growth.

Keywords: knowledge management, innovative economy, knowledge economy, commercialization, entrepreneurial university.

1 Introduction

The manifestation of increased attention to the problem under study is based on an understanding of the dominant role of higher education and the manifestation of its specific features in market conditions. The main results of the article will contribute to the formation of tools for testing the most progressive innovations aimed at improving the effectiveness of higher education. The main components of scientific novelty are areas that have been further developed and contain elements of a real increase in scientific knowledge, their market positions in the domestic and international markets. Article materials can be used to improve public policy in the field of higher education, as well as in teaching courses in economic theory, management theory and innovation management.

A great contribution to the development of the theory and methodology of higher education system management was made by such scientists as J.G. Backhause [1], D.A. Garvin [2], C. Freeman [3], R.M. Solow [4] and others.

Various issues of higher education management are considered in the writings of such scientists as M.I. Stolbov [5], M. Sergeev [6], V.Yu. Pashkus [7]. G.A. Orlova and A.K. Markov [8] and other.

Despite the presence of a large number of works by scientists, many issues related to the functioning and development of higher education institutions in the context of Russia's integration into the global educational space are still not sufficiently developed.

2 Methods

The main research tasks were to study actual trends in the higher education innovative development and the impact of different factors on the development of higher education. To solve the research tasks the authors used methods of analysis and generalization of scientific literature on the topic.

3 Results

The process of globalization has a significant impact on the socio-economic system of modern society, including the higher education system. The higher education system is a complex of institutions that influence the social, legal and financial nature.

Currently, the main vector of transformations taking place in the system of higher education is its improvement. The changes affected all aspects of the functioning of the higher education system: both the organizational and legal base, and the institutional basis of higher education. Along with this, among them should be attributed changes in the requirements for training personnel, thereby placing its qualitative characteristics at the forefront.

The reforms carried out in the system of higher education are caused not only by global processes of globalization, but also by the need to become part of the global educational space. At present, it is important to perceive the country's position at the present stage, taking into account the quality of human capital, the peculiarity of development and the state of education and the level of use of technology in production.

The higher education system has been undergoing reforms for about 20 years. The direct borrowing of Western standards and the strengthening of the regulatory functions of the state leads to an increase in the number of reporting documents at the teaching staff of universities to the detriment of research.

Scientists believe that the higher education system should play a leading role in effectively ensuring the transition to a knowledge economy. A review of the economic literature shows that the transfer of innovation through cooperation between universities and industry has become a dynamic problem. Universities and firms have different perceptions of cooperation, which often leads to a lack of trust and communication problems throughout the innovation process, and is transformed into the problem of susceptibility of enterprises to innovations in general. The low level of university-industrial cooperation can be explained by the fact that there is a mismatch between the needs of business and the motives of professors.

The identified problems of university-industrial cooperation are associated with internal factors that the university administration is able to develop independently with the full understanding of the business or research mission of the university. The development of science and technology is the main accelerator and an important factor for the development of the country's competitive advantage. In this regard, the economy, which is based on the use of knowledge and technology, is a key factor for the effective indicators of qualitatively new transformations in building a qualitatively new society with high-tech characteristics.

In the modern world, universities as generators of innovation should play an increasingly important role in the transition to a knowledge-based economy. Since it is universities that should determine the development of modern science and subsequently become an integrator of the main components of an innovative economy: education-science-production.

Higher education allows countries to use effectively available technologies in the direction of economic growth. In the works of domestic and foreign scientists are provided various definitions of the concept "education". C. Freeman consider education a specific type of material-spiritual production, which develops under the influence of moral and spiritual values [3]. The authors of this definition give an interpretation of the concept of "education" from the standpoint of the theory of the economy knowledge, a feature of which is a rational combination of spiritual, informational and material production with the decisive role of spiritual production.

We consider such an interpretation to be very relevant, since in the modern world new knowledge is one of the most important sources of competitive advantage at the micro, macro and global levels. However, C. Freeman demonstrate a limited approach to the disclosure of the essence of education, since they do not name either its purpose or its subjects.

Highlighting various aspects of education makes it possible to study it from different points of view, as well as from the perspective of a systematic approach, that is, in the aggregate and the relationship of these aspects. We agree with the position of C. Freeman who attribute education to the sphere of spiritual production [3].

Author considered the interaction processes of the innovation system and the learning economy, according to the results of studies of which he concluded that innovation is not a static indicator, but a dynamic and, moreover, cumulative process.

Higher education institutions do not produce anything except knowledge. The products of their activities are research results and educational services, between which there is a relationship. At the same time, estimating the cost of such products is no less difficult than optimizing industrial production with unknown technology.

Higher education, scientific research have a direct impact on economic growth.

Among the new approaches in the innovative development of education can be identified:

- paradigm approach;
- systematic prognostic approach;
- phenomenological and synergetic approaches;
- personality-activity approach.

All these approaches in general determine the innovative elements of the higher education system. Today, there is no clearly defined model of innovative higher education aimed directly at the market needs of relevant specialists. Each of the models has its positive and negative points. But, it should be noted that not one of them is functioning in its pure form at present and cannot be called innovative.

Currently, they are increasingly leaning toward models of universities directly, because in the conditions of competition in the educational market, as well as the possibility of independent choice of the learning path, universities themselves determine innovative learning parameters.

The Kneller and Stevens model uses the stochastic boundaries of higher education analysis. The basic model is associated with the production function defined by the Solow equation, but with the introduction of two stochastic conditions [4]:

$$Y_{it} = f(K_{it}L_{it}H_{it})\exp(-n_{it})\exp(\varepsilon_{it}), \text{ where} \qquad (1)$$

Y_it– Country Produced Product (GDP);
K_it- capital assets;
L_it – labor intensity (number of working hours per week);
H_it – human capital;
n_it- reflects economic inefficiency (0 < n_it< 1).

It is assumed that the production function will be determined using factors of production, labor, capital and human capital.

In our model, in our opinion, the definition of the boundary is less satisfactory. The effect of human capital is small, but incorrectly described, which may be due to the presence of h in the part of the model that explains the efficiency. The capital efficiency ratio is much larger, which is consistent with the observed capital factor, while labor efficiency is correspondingly lower.

R.G. Ehrenberg [9] considers the behavior models of higher education institutions based on the principles of maximizing utility, which were first used by D.A. Garwin [2] to explain the decision-making processes at universities.

From our point of view, the models do not fully cover issues that consider the effects of higher education on economic growth. In models, microeconomic analysis provides estimates of the impact of education on individual incomes, and authors usually supplement their models with macroeconomic studies that provide estimates of the rate of return, similar to those found in microeconomic studies, in the range of 6-12% per annum.

J.G. Backhause believes that higher education institutions can avoid the inefficiency of nonprofit organizations [1].We agree with the position of B.F. Rogge and P.F. Goodrich, who came to the conclusion that the recommendation of A. Smith to pay most of the salary from the funds received as payment for tuition would be correct if education was the only product of the university. However, research (both basic and applied) is an important product of higher education institutions, along with training [10].

Despite the many different models that reflect the impact of education on the economic development of the country, they do not take into account factors that are essential. Namely, regional aspects, features of investments in education as costs, etc. are not taken into account.

As a result, when assessing the impact of education on the innovative development of a country, the number of indicators is constantly increasing. Under these conditions, it is advisable to study foreign experience both in the field of organizing innovative education and in the sphere of its influence on the economic growth of the state as a whole.

The following arguments can be made in favor of the need for state support for higher education:

- presence of external effects in the form of innovative technologies and the diffusion of scientific and technological achievements that contribute to increasing the efficiency of social production;
- equality (including financial) in the access of all citizens to education;
- function of higher education as a pillar of basic and secondary education.

The state and development of higher education in Russia today is largely determined by international integration processes that have occurred and are taking place in the global system of higher education.

Modern humanitarian discourse is characterized by various contexts of the concept of "globalization of education", appealing to global trends in the formation of global markets, the development of modern information technologies and the transformation of knowledge into a direct productive force [11].

New technologies for creating information flows, practically unlimited possibilities of the Internet and the growing relevance of education in an environment where knowledge becomes obsolete at a high speed, have led to completely new relationships in national educational systems that are gradually transforming into global markets for the provision of educational services. The influence of globalization processes on the education sector is quite mixed and associated with the emergence of certain problems and contradictions. Due to the effect of globalization, modern higher education institutions take on numerous social functions, which are not only reduced to the training of specialists. The determining directions for the development of higher education in the future will be the individualization of education, its internationalization and interdisciplinary relations. Globalization processes lead to fierce competition in the educational market. This fact is a reflection of the requirements of our time, when new information technologies contribute to the global level of knowledge. Progressive movement towards a continuous process of improving the quality of the provision of educational services, which in the current situation is a guarantee of the university's stable position in the market. International partnership is one of the main components of the reform of the education system of the Russian Federation, as it helps to determine the ultimate goals and objectives of the reform through taking into account generally recognized world trends and universally recognized educational standards. The implementation and development of international relations attracts additional financial resources in the form of grants and technical assistance, which has an impact on increasing the attractiveness of a higher educational institution, improving its image, and as a result, increasing its competitiveness. On a practical level, it is quite difficult to assess separately the impact of international partnerships on changing the competitiveness of the entire university. It is possible rather to establish a connection between international educational programs and the organizational development of the university.

The problem of ensuring the quality of education is global in nature and is due to global trends in approaches to its solution. It is important to note that the concept of ensuring the quality of education has a broader understanding than the quality management system. One of the main requirements of stakeholders when consuming educational services is the level of their quality, which must meet the criteria for both the world and domestic labor markets.

The most objective form of assessing the quality of the organization and implementation of educational activities of a higher educational institution is an external assessment carried out through the international accreditation procedure. The compliance of the organization of the educational, scientific, and educational processes of the university with the standards of a foreign accreditation body, confirmed by the corresponding certificate, is a significant indicator of the university's competitive advantages in both international cooperation and competitiveness in the educational services market as a whole. The integration processes in the global system of higher education and the reforms implemented in the system of higher education, aimed at training the staff needed by the modern economy, depend on the adequacy of the volumes and coordination of financing mechanisms.

The systemic principles of economic transformation and the formation of a new way of thinking and behavior of business entities in the conditions of economic transformation can be served by the theory of human capital, which defines the key role of education, science and intellectual potential with the characteristic features of public-private partnership. Undoubtedly, a significant factor has appeared in post-industrial society, which must be taken into account. This factor is the intellectual capabilities of the person himself.

Researchers at various levels note the existence of certain difficulties in conducting a comparative assessment of the cost characteristics of the human capital of countries with varying degrees of socio-economic development, thereby affecting the coefficient values of human capital efficiency, which naturally manifests itself in their labor productivity.

The role of the financing mechanism for higher education is the effective use of tools and methods of funding and use of financial resources of universities, the final result of which is its quality and accessibility. At the same time, the sequence and effectiveness of the goals set directly depends on the respective sources of financing, the adequacy of not only their combination among themselves, but also the costs for which they are provided.

In determining the forms and methods of financing the higher education system for public universities, the key is the regulatory method of financing. For non-state universities, financing methods have the nature of mixed financing, based on a combination of the costly method with the mandatory use of components of the normative method. In this case, there is a need to clarify the application of the normative method in the formation and use of various costs for the areas of social direction, including the higher education system. The quality of human capital is one of the key success factors in the commercialization of technology.

Education forms a qualitative side of human capital, thereby defining such criteria-specific characteristics as the level of health, material values and the psychological component of human capital. It is this significance that makes it possible to confirm the arguments in favor of the conclusion that education should be a key priority in the development of human capital. We consider education as the basis for the formation of human capital, namely as a factor, tool, control and indicator of its condition.

As a factor, education includes all stages of education, from primary to higher, from school to retirement, on the principle of "education throughout life." The factor in this case will be considered as the accumulation of knowledge and skills throughout the entire cycle of the formation of human capital. As a factor, education is the starting point in the formation of the quality potential of human capital, thereby combining all periods of knowledge: at home, at school, college, university, company.

As a tool, we consider education through a financial mechanism. In our opinion, the effect of the return on investment in education is more obvious, since a person who has invested in education has a higher chance of receiving higher fees in the future than a person with a lower level of education.

As an indicator it is advisable to consider education as a qualitative component of human capital, since education is a key criterion in determining the well-being of a country, shaping national wealth, affecting diversified all elements of the formation of human capital.

Investing, in our opinion, improves human capital. Investing is the most important condition for the reproduction of human capital. Although it is undoubted that investment in human capital is not the main factor affecting its qualitative characteristics.

What is needed is not just investment, but real, meaningful activity that will become the "engine" for the reproduction of human capital. Therefore, effective investment in human capital creates sufficient conditions in order to increase the competitiveness and efficiency of the state economy, which, in turn, is due to the strengthening of the intellectual characteristics of human capital and the efficient use of labor resources.

Some studies dispute the role of scientists in the commercialization of scientific research, highlighting the "market" and "expert" orientation of scientists. At the same time, US scientists recognize and accept the role of the entrepreneur, but put in the first place the role of the researcher, used in the financial support of the higher education system. An important condition for the commercialization of university technology is to ensure its information transparency. Current trends in the economic development of Russia require the integration of components and their systematic work.

Currently, a strategic partnership between universities and business is needed. Most universities are at the initial stage of the development of entrepreneurial infrastructure or, as foreign scientists call it, the entrepreneurial ecosystem. A new focus on commercialization and knowledge creates a certain degree of uncertainty role of universities. However, an entrepreneurial university is a command of the time, and if a university declares this mission, then the above factors can achieve the goal with less loss.

4 Conclusion

The development of the open circuit problem "education - science - production" is facilitated by such factors as the lack of regulation of the relationship between higher education and employers, the underdevelopment of social partnerships, the inadequate legislative framework for attracting employers to training personnel, the low efficiency and degree of practical use of scientific research and applied developments. The creation of innovative resources and the formation of innovative infrastructure involves providing the market with scientific personnel and creating favorable conditions and incentives for research and development. It is also necessary to create appropriate levers not only to improve the quality of higher education, but also to stimulate universities to research and development. In this perspective, the state creates an appropriate innovative infrastructure, i.e. the mechanism of functioning of specialized institutions that promote the effective implementation of innovative products and create demand for innovation.

References

1. J.G. Backhaus, The university as an economic institution: The political economy of the Althoff system. *Journal of Economic Studies*, **20**(4/5) (1993). URL: https://www.emerald.com/insight/content/doi/10.1108/EUM0000000000169/full/html. Accessed: 11.09.2019.
2. D.A. Garvin, *The economics of university behavior* (Academic Press, New York, 1980).
3. C. Freeman, The national systems of innovation in historical perspective. *Cambridge Journal of Economics*, **19**(1), 5-24 (1995).
4. R.M. Solow, A contribution to the theory of economic growth. *The Quarterly Journal of Economics*, **70**(1), 65-94 (1956).
5. M.I. Stolbov, *The influence of the financial market on economic growth*. Ph.D. thesis. Moscow: Moscow State Institute of International Relations (2007). [in Rus.].
6. M. Sergeev, How to overcome obstacles to the knowledge economy. *Politkom* (2005). URL: http://politcom.ru/1151.html. Accessed: 11.09.2019. [in Rus.].
7. V.Yu. Pashkus, "New economy": Features and problems. In A.I. Yuriev (Ed.), *Proceedings of the IV International scientific conference "Real and virtual world of the new millennium"* (pp. 100-103). St. Petersburg: Red dragon (2002). [in Rus.].
8. G.A. Orlova, A.K. Markov, Economic growth: Theory and practice. *Russian Foreign Economic Bulletin*, **7**, 3-22 (2014). [in Rus.]
9. R.G. Ehrenberg, Adam Smith goes to college: An economist becomes an academic administrator. *The Journal of Economic Perspectives*, **13**(1), 99-116 (1999).
10. B.A. Rogge, P.F. Goodrich, *Education in a free society*. In A.H. Burleigh (Ed.) (Liberty Fund Inc, Indianoplis, 1973).
11. G.D. Hansen, E.C. Prescott, Malthus to Solow. *American Economic Review*, **92**(4), 1205-1217 (2002).

Innovations in the Management of Marketing Activities of the Higher Education System

M.O. Surayeva[1], *I.A.* Grigoryants[1], *N.V.* Kochetkova[1,*], *and A.N.* Evdokimov[1]
*Corresponding author: nvkochetkova@bk.ru.

[1] Samara State University of Economics, Samara, Russia

Abstract. Modern educational organization is actually a full member of market relations. Universities now have the right to make money and manage it. On the education market there are many factors that affect the stability of the educational organization. The authors made an attempt to analyze the factors and risks in the education market. Analyzing the set of factors operating in the education market, the authors identify the most dangerous of them, which currently affect the sustainable functioning of universities. It is proposed to solve the problem of sustainable functioning of the University with the use of marketing management as one of the components of management studying market processes.

Keywords: marketing, sustainability, University, education system, management.

1 Introduction

The relationship between the participants of higher education in the Russian Federation has undergone significant changes in recent decades due to the change of socio-economic foundations of society. In relation to higher education, this has caused a change in the system and mechanisms of regulation by the state, restructuring of sources of funding due to public underfunding, increased competition, reform in this area, the transition to new standards of education.

The education sector is under considerable pressure from environmental factors, which negatively affects the sustainability of the functioning of universities:

- unfavorable demographic situation in the Russian Federation, which takes place today, the consequence of which is the disproportion between the increasing supply of universities and the decreasing number of applicants;
- the deterioration of economic development, which has led to the inability of a large number of people to pay for increasingly expensive educational services.

In this regard, there is a need to improve the management system of universities, which is recognized by both the scientific community and society.

2 Methods

The problem of unstable functioning of universities can be solved by modeling the marketing management system. It is the marketing approach that will allow universities to adapt to the dynamics of the external environment. The basis of marketing management is the satisfaction and formation of the population's demand for educational services. Active marketing policy is necessary for the development of educational programs that will prepare qualified personnel that meet the interests of the labor market.

Many aspects of marketing management of sustainability of higher education functioning are not sufficiently developed and need to be clarified taking into account the changes. Stability and risk of the social system are interrelated concepts. At the heart of the categorical structure of risk theory is the concept of "the danger of losing something." Under the risk can be understood the threats that are fraught with the external environment identified in the SWOT analysis. The threat is what the university risks when it comes to negative events. This is an objective regularity that determines the processes of quantitative and qualitative changes in the systems perceived as a threat to the vital interests of people.

From the point of view of the theory of systemic risks are considered as a property inherent in any type of expedient activities, manifested as the uncertainty of the implementation of the objective functions, the nature, content, direction and conditions of achievement of which are not fully clear to the decision-maker. Considering the risks of systems of any level of complexity allows us to classify them by forms of manifestation and investigate them as relatively stable and repetitive.

The complexity of the system of public relations, the growth of overall productivity, the creation of a system of mass consumption leads to the fact that an increasing number of individuals who are excluded from the production process and become passive consumers. In these conditions, there are growing risks associated with the fact that resources aimed at education, professional training of large groups of people will never pay off.

Therefore, risk assessment through a system of quantitative indicators will minimize the costs of the company to divert resources to various reserve funds. Destructive risks, as a rule, cannot be accurately quantified, information about them is mostly probabilistic and is not available to most business entities. This is due to the fact that the qualitative parameters of possible losses are so great that the company is not able to create reserve funds to eliminate the damage from the occurrence of destructive risks. Production risks are risks associated with the possibility of shortfall in profits or losses as a result of inefficient cost management, personnel and as a result of errors in the choice of strategies to increase productivity, product differentiation.

3 Results

In the scientific literature there are various approaches to the definition of risk parameters [1, 2]. Some authors consider risk as one of the types of danger associated with political, social and economic activities of people [3]. Other do bias in the direction of the man-made nature of the risk [4]. The third group of authors examines risks as a form of uncertainty of the result, which is associated with a special type of economic activity [5]. Within the framework of this direction, the regularities of the risks manifestation are investigated, which in the conditions of the system functioning have a set of maxima, each of which violates different equilibrium States. Particular attention is paid to such characteristics of risks as entropy, which is a measure of uncertainty of a certain state of the system, which violates its equilibrium.

The main indicator of risk is the probability of occurrence in time of events leading to changes in the equilibrium stability of socio-economic systems. Their source is the conditions and factors that cause uncertainty in the state of the system. In management theory, there are many types of uncertainty. Most of them are related not to the effect of direct impact on the targets, but to the time parameter of this influence. Uncertainty is considered relative to the time when a negative event occurs [6].

Along with the above types of risk should be identified the concept of marketing risk. Marketing risk is the danger of mismatch of needs and opportunities to meet them. Achieving a result is always multivariate, therefore, there is a multiple opportunity to minimize the needs and costs to ensure the stability of the system. Marketing risks are largely determined by SWOT analysis.

Management of marketing risks should be carried out minimizing them by matching the needs and opportunities in the socio-economic system.

It is possible to allocate following characteristics of risks of the marketing:

- mismatch of goals and results of expedient activities in social, political and economic spheres;
- deviation of the actual result from the possible one, which is defined as the base (threshold) value.

Marketing risks can be classified on such methodological grounds as the nature and form of manifestation. The nature refers to the essential features of economic risks, the manifestation of which causes damage, leads to losses and reduces the level of rationality and efficiency of the use of limited resources and factors of production.

The risk of any system, including marketing, is characterized by resource (material) and spiritual contradictions in the system, determined by the transition of the system from one state (when the system has a disaster – partial performance of the system's functions) to another (in which the system does not have a disaster - the full performance of the system's functions).

Universities education institutions are part of the socio-economic environment and all changes in it are reflected in universities. Universities are full participants of market relations. Increased competition in the field of educational services requires a marketing approach to the management of the University.

Modern educational organization interacts with a huge number of different objects, which is reflected in figure 1.

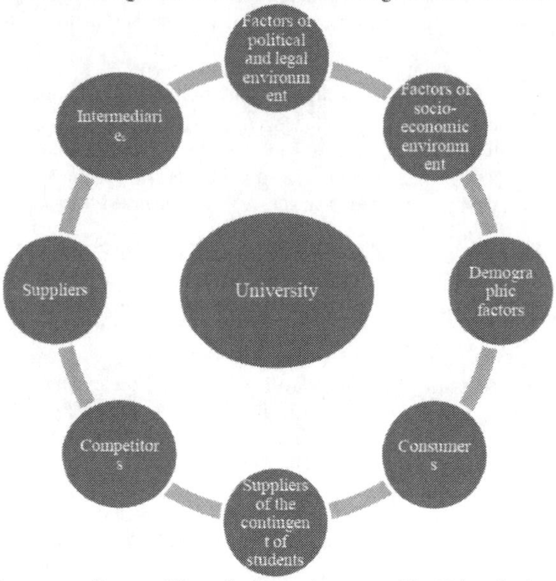

Fig. 1. The structure factors of the marketing environment of the University (Source: Authors.)

The author's analysis of the factors of the external environment of the University showed that the activities of educational institutions have a pronounced social character. The social orientation of educational institutions creates conditions for solving a wide range of socially significant problems. Therefore, the concept of socially responsible marketing, according to the author, is fundamental in the activities of the University.

Elements of this concept contribute to the establishment of social responsibility of educational institutions and are key factors in the formation of the image of the University.

Since the University is a participant of market relations, its ability to adapt to changes in the external environment is of great importance.

A feature of a large University is a significant inertia, which can be compensated only through marketing research activities, which will allow to timely identify threats from the external environment and develop measures to avoid or reduce the negative consequences of their occurrence.

4 Conclusion

The University is an open system that has the ability to transform the external environment, that is, the system is able not only to respond to changes in the dynamics of the external environment, but also to provoke them. In the process of such interaction, the structure and functioning of the University are changing in order to ensure its sustainability.

Adaptation to changes in the environment and the structuring of its elements are made possible through the use of a marketing approach to the management of the University. The right strategy and tactics of marketing management will lead to the sustainability of the educational institution.

References

1. E.P. Pecherskaya, L.V. Averina, L.G Karanatova, S.A. Kozhevnikova, Innovation educational institutions as a point of growth in digital economy. In V.V. Mantulenko (Ed.), *Proceedings of the International Scientific Conference "Global Challenges and Prospects of the Modern Economic Development"*, GCPMED-2019. *European Proceedings of Social and Behavioural Sciences*, **57** (pp. 365-372). London: Future Academy (2019). DOI: 10.15405/epsbs.2019.03.36.
2. I.K. Belyaevsky, G.D. Kulagina, L.A. Dancheno, A.V. Korotkov, A.A. Romanov, V.V. Usov, *Statistics of market of goods and services* (Finances and Statistics, Moscow, 2002). [in Rus.].
3. E.A. Kandrashina, A.M. Izmailov, V.V. Mantulenko, N.S. Mirzayev, Theoretical aspects of gaps formation in the system of Russian higher education, In V.V. Mantulenko (Ed.) *Proceedings of the International Scientific Conference "Global Challenges and Prospects of the Modern Economic Development"*, GCPMED-2019. *European Proceedings of Social and Behavioural Sciences*, **57** (pp. 1135-1142). London: Future Academy (2019). DOI: 10.15405/epsbs.2019.03.114.
4. N.N. Pavlova, Features of the service and its marketing. *Librarium* (2009). URL: http://www.elitarium.ru/2009/01/23/osobennosti_marketinga_uslugi.html. Accessed 12.09.2019. [in Rus.].
5. S.I. Ashmarina, E.A. Kandrashina, I.A. Plaksina, Interest alignment model in the higher education system. In V. Mantulenko (Ed.), *Proceedings of the 17th International Scientific Conference "Problems of Enterprise Development: Theory and Practice"* (November 26-27, 2018, Samara, Russia). *SHS Web of Conferences*, **62**, 12001 Les Ulis: EDP Sciences (2019). DOI: 10.1051/shsconf/20196212001.
6. V.K. Chertykovcev, A.N. Bogusonov, To the question of stability of economic systems. *Bulletin of Saratov State Socio-Economic University*, **2**(13), 78-79 (2006). [in Rus.].

"Mobilization of Ethnicity" as Europe-Asian Project: Constructing a Higher Education Policy

N.F. Tagirova[1], *E.I.* Sumburova[1], *Yu.A.* Zherdeva[1], and *A.S.* Zotova[1,*]
*Corresponding author: azotova@mail.ru.

[1] Samara State University of Economics, Samara, Russia

Abstract. The article discusses the ethnic policy in the field of higher education in Russia between the First and Second World Wars. The "mobilization of ethnicity" in the educational policy of the first years of Soviet power is presented in the study as the "European-Asian project" of Soviet Russia - an attempt to create a single supranational economic, political and cultural space in a significant part of Eurasia. Based on the materials of the multicultural region of the Middle Volga region, the authors analyze the process of integration of the Volga peoples (Tatars, Mordovians and Chuvash) into the higher education system. The study showed that at the first stage of designing the new policy (1920s), the key direction was to provide ethnic groups with relative cultural autonomy (preserving and supporting national languages, religious customs and social traditions) in exchange for recognizing the legitimacy of the new government. At the second stage (1930s), the national educational policy of the state became more straightforward, integration processes intensified and, as a result, the mutual economic dependence of the capital and regions.

Keywords: higher education, ethnic educational policy, industrialization, «homo soveticus», Europe-Asian education, Russia.

1 Introduction

The process of changing technological structures providing the next technological transition is accompanied by the adaptation of old and the emergence of new institutions. The analysis of the institutional forms of this transition shows the general and special trajectories of the economic development of states included in the transformation process. At the stage of the modern fourth industrial revolution, called digitalization, the Russian version of the second industrial revolution, which took place a century ago, can be useful, despite the temporary remoteness. It solved many "attendant" problems, was determined by the territorial extent and multinational diversity. In the Russian Empire and then in the USSR, it took place over a considerable space from the Baltic Sea to the Pacific Ocean, and in this regard, the experience of the second industrial revolution in Russia can be regarded as a kind of "Eurasian model" for the development of economic design. Recent studies of the Russian Public Opinion Research Center have shown that more than a quarter of Russians (28%) would like to see "the USSR in a new form" in the modern Eurasian Economic Union [1]. In this context, the study of the experience of economic, political and cultural integration of the peoples of Eurasia in the conditions of the Soviet model of the economy is relevant. Then, a century ago, the creation of an industrial economy, the training of highly qualified personnel for it, the solution of the tasks of creating a classless society merged into one large social project, conditionally called "Homo Soveticus".

2 Problem Statement

In the scientific literature, the social and socio-political mechanisms of imperial and Soviet industrialization, of course, successful, are not considered from the point of view of its educational and especially ethnic component. Various aspects of the formation of ethnic education are covered mainly in the context of studying the national policies of the Russian Empire and the Soviet state, regional history, and certain areas of ethnocultural research. The first to note the connection between national and economic issues was E. Lohr [2]. F. Hirsch, analyzing the experience of building the Soviet state, examined in detail the scientific and inter-party discussions on administrative-economic and national construction in the 1920s, linking economic, national and administrative-territorial subjects [3]. In turn, J. Cadiot also noted the relationship between the national policies of the late Russian Empire and the early USSR, which allows us to consider them as a single process in this study [4].

At the same time, the transformation of educational policy and its connection with national and economic construction eludes the attention of researchers. Only in the works of R. Suny, T. Martin and T. Krasovitskaya showed the influence of ethnic education on the projects of political and economic modernization of the country [5; 6; 7]. In particular, R. Suny shows that "education was one of those areas in which nationality began to signify the differentiation of Russian society" [5, p. 65].

3 Research Questions

In this paper we try to fill in the gaps in understanding economic and educational modernization and analyze the national educational experience of the USSR through the factors of the technological revolution (socialist industrialization). The key question of the study is to find out how ethnic policy was embedded in the program of technical and economic reorganization. Answering this question, we examine the integration practices and mechanisms of "embedding" the Eurasian peoples inhabiting the multinational Soviet Union, using the example of one of its regions. As an object of study, we took the region of the Middle Volga. Until the mid-19th century, Middle Volga and Trans-Volga Region, as modern Russian historians P. Kabytov, E. Dubman and O. Leontieva consider, could be described as "a frontier of the Russian state" [8, p. 118].

4 Purpose of the Study

The purpose of the study is to trace the relationship between centrifugal trends (the concept of "the right of nations to self-determination" and the diversity of ethnic schools) and centripetal trends (building a single ethos "homo soveticus" and unification of education) in national educational policy in the USSR of the 1920s-1930s from the perspective of solving the state goals of the country's industrialization (technological revolution). Diversity and unification are differently directed vectors, and the historical experience of their coordination, according to the authors, can be useful for creating a modern model of Eurasian education. The logic of our study is based on three main subjects. The first section examines the ideological and conceptual sources, the conceptual apparatus and state organizational structures related to the national question. In the second section, we analyze the ethnic politics of Soviet authorities in the 1920s and 1930s. The third section, based on the factual material of the Middle Volga region, traces the specific implementation of these steps in the field of higher education.

5 Research Methods

The theoretical basis of the study is the concept of "Big Waves" by K. Peres, proposed for the analysis of technological revolutions. According to this concept, technological transitions are accompanied by institutional and sociocultural changes. Often they are associated with political changes in the country. The system of higher education in Russia was transformed under the influence of all these changes. We consider the sphere of higher education as the deployment element of the Great Wave of the industrial revolution - a kind of "social project" of industrialization.

In the project for the reconstruction of society according to the Soviet version, the key role was given to education of the population and ethnic politics. Those were the factors which were supposed to form a society loyal to the new government and sharing its ideals. The authors' methodological approach consists in recognizing the structural interdependence of all these elements.

6 Findings

6.1 "Mobilization of Ethnicity": from idea to organizational structures

The success of industrialization in the USSR in the first half of the twentieth century was indisputable. The average annual rate of economic growth during the years 1870-1950 was quite comparable and even ahead of the European ones (Table 1). The average GDP per capita exceeded the Asian one and only approached Western European, but in the rate of increase this indicator in the USSR was clearly faster.

Table 1. The GDP changes in the regions of the world in 1870-1950.

	The average annual rate of economic growth, %		**The average GDP per capita, $**	
years	1870-1913	1913-1950	1913	1950
Western Europe, totally 12 countries	1,33	0,84	3688	5018
The USSR (Russia)	1,06	1,76	1488	2841
Asia (except Japan)	0,43	- 0,08	658	639
The whole world	1,31	0,88	1526	2113

Source: Authors based on [9].

Economic success was the result of the active involvement in the technological revolution of the labor potential of all the peoples of the country mobilized to solve this problem. The term "mobilization of ethnicity" in relation to the Russian history of the civil war was first used by Mark von Hagen and quickly entered the research practice devoted to the period of the war, revolution and the first years of Soviet power [2; 3]. We use this concept in a wider historical context, identifying the conceptual sources, the conceptual apparatus and organizational structures created by the state to involve different peoples in solving national problems.

The imperial authority recognized the ethnic group as a separate, different from the main part of the country's population living on the territory of the Russian Empire, accounted for the late XVIII - early XIX centuries. It was associated with the study of territories relatively recently included in the state and economic space of the country. First, the military [10] and only then did geographic scientists compose the first descriptions of the provinces and the cultural and historical regions of the Russian Empire [11]. The military-strategic tasks predetermined the methodological approach: the "people" became the "population", which needed to be counted and which needed to be managed. In this case, the criterion of difference was not ethnicity, but religion, which was shown by the first general census of the population conducted in the Russian Empire in 1897.

In 1917, Soviet power proclaimed the equality and equal rights of all peoples. Ambitious plans were developed for the technical, economic, socio-cultural and ideological reconstruction of society. They were planned to be implemented by the newly formed state bodies (people's commissariats) - the People's Commissariat for Ethnic Affairs (People's Commissariat), the People's Commissariat of Education (People's Commissariat for Education) and the State Planning Commission (Gosplan).

6.2 Ethnic Policy of Russia: from "non-Russian" to "national minority"

In the years 1917-1923. the Bolsheviks worked out policies aimed at integrating the entire multinational population of the country based on the recognition of the equal rights of all nationalities [4]. Outwardly, this looked like a direct contradiction to the politics of the late Russian Empire, which sought to establish "state nationalism" and limited the rights of nationalities. However, closer attention to the content of the ethnic policy of the Soviet government shows that it professed the same idea of "linear evolution" of peoples, which considered the "indigenous peoples" of Russia (Volga, Siberia, the Caucasus and the Caucasus, etc.) as "the most undeveloped" and therefore requiring special custody of the central administration, as imperial Russia. And if in the XIX century. the term "foreigners" was used to designate the non-Russian population of the empire, after the revolution of 1917 the Bolsheviks began to use the concept of "national minorities".

The People's Commissariat considered its main task the economic, cultural and political revival of the "backward peoples" (meaning nomadic peoples) of Russia, as well as the inclusion of nationalities in the mainstream of Soviet construction [12]. Having abolished class division, the Soviet government equalized all citizens of the country, but granted special status to "national minorities" as ethnic groups that were oppressed by the imperial administration and needed special state guardianship for their equal accession to Soviet society. At the same time, there was a restructuring of the territories based on the ethnic principle and measures were taken to attract indigenous representatives to the local administration. As noted by J. Cadio, such a policy encouraged people to use ethnicity as an important identification feature [4, p. 117].

In the 1920s Soviet policy in the field of nationalities was aimed at the development of national languages, elites, cultures and territories. In fact, the Soviet regime institutionalized nationalities: changes in society were accompanied by large-scale mobilization of national identities and created nationalities even where they were still poorly aware of themselves.

However, since the beginning of the 1930s, when the legality of power desired by the Soviet government was achieved and the ethnic groups of the state supported the establishment of the Soviet regime, "indigenization" was curtailed, and the government headed for the unification of nationalities. The Soviet government, in fact, returned to the imperial model of unification and Russification, and the status of "national minorities" from privileged turned into marginal.

6.3 The practices of designing the new nation by the means of higher education as part of social project "homo soveticus"

The Soviet government saw in education the most important tool for influencing the masses and one of the means of rebuilding society. The implementation of these principles is visible in the opening of national schools and professional educational institutions, in the creation of writing for peoples who did not have it before, in the publication of national literature and the training of national teaching staff.

The issue of enlightening non-Russian peoples remained particularly relevant for regions with a multinational composition, which included the Middle Volga region. The average percentage of literacy in the region from 1926–1927. in general, it was 36.4%, while among the Mordovian population - only 22.7% (for women - 8.4%), and for the Chuvash - 32.4 (for women - 17.3%) [13].

Since the mid-1920s, the country's leadership has established a quota for the training of representatives of national minorities in higher educational institutions. In this way, the authorities tried, on the one hand, to level the situation with the illiteracy of the non-Russian population, and, on the other, to attract "foreigners" with higher education in the economic, cultural and socio-political life of the region. First, the local authorities were interested in attracting representatives of those folks, who made up a significant share among the population of the region. For the Middle

Volga, these are the Tatars, Mordovians and Chuvash, who, in the terminology of the 1920s and early 1930s. were the "national minorities" of the region.

The country's party leadership has declaratively announced to universities what percentage of representatives of indigenous peoples they should accept. For example, in 1925, when admitting to the Samara Agricultural Institute, a quota was established: 5 Tatars, 5 Chuvash and 15 Mordvinians, a total of 21.4% of the total number of applicants [14, f. R-81, op. 1, d. 705, l. 23]. In 1929, at the Samara Pedagogical Institute, representatives of national minorities should have been at least 10% [14, f. R-2304, op. 1, d. 1212a, l. 4.6]. The largest percentage of indigenous students in 1930 was at a medical institute - almost a third of freshmen (28%), the dominant ethnic group among whom were Mordovians (more than a third) and Chuvashs (13%). However, the local party leadership also considered such an indicator to be low, requiring decisive measures from the leadership of universities.

After 1936, the curtailment of national components in the education system began, and the model began to take shape of the so called "Soviet education" - a unified, international, aimed at educating the "Soviet man".

7 Conclusion

In a multinational country located in the Eurasian space, the national educational state policy was, from the point of view of the authors, an important component of the socio-economic, socio-political and socio-ethnic project "homo soveticus". During the implementation of this project at the first stage (1920s), the national theme was a key one. The peoples were to become the social pillar of the new government and for the support of its legitimacy they received benefits: the preservation and support of national languages, religious customs and social traditions. The study of peoples, their statistical accounting and the identification of the development potential of the territory in which they lived, became the basis for the development of plans for the administrative and economic regionalization of the country, and, ultimately, for industrialization and the technological revolution. With the strengthening of Soviet power and the success of industrialization already in the late 1920s. a hierarchy of relations between the center and the regions was built. At the second stage (from the beginning of the 1930s), the national educational policy became more straightforward, integration processes intensified and, as a result, the mutual economic dependence of the capital and regions was achieved. The government has relied on the formation of an ethnic Soviet elite, capable of leading the processes of technological reorganization of society in national regions. Such institutionalization of peoples did not become a policy of equality, but a "leveling" of peoples.

Acknowledgments

The authors express their gratitude for support given by Russian Found for Basic Research (RFBR) and the Government of Samara Region to the project № 18-49-630008 "Experience of social design in the sphere of the higher economic education of the Samara region in the first half of the 20th century".

References

1. Russian Public Opinion Research Center (WCIOM), *Eurasian integration: Form, goals and consequences*, 3862 (2019). URL: https://wciom.ru/index.php?id=236&uid=9525. Accessed: 03.09.2019. [in Rus.].
2. E. Lohr, *Nationalizing the Russian Empire: The campaign against the enemy aliens during World War I* (Harvard University Press, Cambridge, 2003).
3. F. Hirsch, *Empire of nations: Ethnographic knowledge and the making of the Soviet Union* (Cornell University Press, Ithaca & London, 2005).
4. J. Cadiot, *Le laboratoire imperial: Russie – URSS, 1890-194* (CNRS Éditions, Paris, 2007).
5. R.G. Suny, The empire strikes out: Imperial Russia, "National" identity, and theories of empire. *Ab Imperio*, **1-2**, 9-72 (2001). [in Rus.].
6. T. Martin, *The affirmative action empire: Nations and nationalism in the Soviet Union, 1923-1939* (Cornell University Press, Ithaca & London, 2001).
7. T.Yu. Krasovitskaya, *Modernization of Russian educational space: From Stolypin to Stalin (the end of XIX century – the beginning of XX century (1920s)* (New Chronograph, Moscow, 2010). [in Rus.].
8. P. Kabytov, E. Dubman, O. Leontieva, The Middle Volga and Trans-Volga region in the context of development of the Russian state system: A modern concept. *Quaestio Rossica*, **2**, 117-137 (2015). [in Rus.].
9. A. Maddison, *Contours of the World economy, 1-2003 AD: Essays in micro-economic history* (Oxford University Press: Oxford, 2007).
10. P. Holquist, *To count, to extract, and to exterminate: Population statistics and population politics in late Imperial and Soviet Russia*. In R.G. Suny, T. Martin (Eds.), *A state of nation. Empire and nation-making in the age of Lenin and Stalin* (pp. 111-114). New York, N.Y.: Oxford University Press (2001).

11. V.P. Semyonov-Tyan-Shansky, P.P. Semyonov-Tyan-Shansky, V.I. Lamansky (Eds.), *Russia: The full geographical description of our motherland: Bed-side and road book for Russian people.* In 19 volumes (A.F. Devrien, St. Petersburg, 1899-1914). [in Rus.].
12. RSFSR, *People's Commissariat for nationalities. Six years of national policy of Soviet power and People's Commissariat in 1917-1923.* Report (Edition of information and publishing department of People's Commissariat for nationalities, Moscow, 1924). [in Rus.].
13. Krajplan, *The Middle-Volga region (economic and socio – cultural review)* (State publishing house, Middle-Volga regional branch, Moscow, Samara, 1930). URL: http://ulrgo.ru/upload/iblock/557/sredne-volzh_krai.pdf. (Accessed: 03.09.2019). [in Rus.].
14. Central State Archive of the Samara Region, *Information on the institute history,* f. R-2304, op. 1, d. 1212a, l. 4.6. Tractor course manager's memo. f. R-81, op. 1, d. 705, l. 23 (1929). [in Rus.].

Language Education in the Age of Globalization and Innovation

V. Dobrova[1,*], *P.* Labzina[1], *N.* Ageenko[1], and *S.* Menshenina[1]
[*]Corresponding author: victoria_dob@mail.ru.

[1]Samara State Technical University, Samara, Russia

Abstract. Globalization and innovation have recently resulted in the extensive use of the latest technological products practically everywhere, and in education especially. Various technologies are now employed in different spheres of education. Virtual Reality (VR) is a global innovative technology with great potentials and enormous pedagogical possibilities that offers new methods and techniques for education. The main features of it are visibility, security, involvement, presence and focusing. It enables to combine the computer-generated virtual information and the real environment in real time. The presented VR language program is based on the concept of 3D modeling and semantic frame method.

Keywords: language education, globalization, innovation, virtual reality.

1 Introduction

Globalization has become a widely discussed idea in national and international context in recent years. Yet the term "globalization" is relatively new, and its meaning and impact remain essentially contested within the research community.

R. Robertson [1], whose name is closely associated with globalization theory, considers globalization as an accelerated compression of the modern world and also as the intensification of world consciousness as a singular entity. Some researches believe that globalization is the process of the world changing so that individuals can communicate to each other immediately through the use of various technologies and especially the internet. Others consider globalization to be a social change at which physical presence of place on the social, educational and cultural aspects are diminishing, and the general public know that it is diminishing [2]. Therefore, it can be said that globalization is the weakening of the physical boundaries associated with countries and nations that affects various spheres of human life such as economy, politics, education, social interaction, culture and many other aspects, and promotes technological progress.

2 Globalization and Education

2.1 Impact of globalization on education

It is becoming obvious that processes of globalization have impact on education around the world in many important ways. As education plays a great role in shaping a society, so globalization has to be associated with education thus having a deep impact on it.

The brief categorization of globalization show that some authors mention the following effects of globalization in education: commodification, internationalization, citizenship and accessibility of knowledge [2]. World globalization aims at increasing attention to the internationalization of the academic studies subjects thus creating the opportunities for new international partnerships in teaching and research with companies and institutions across the world [3]. Another impact of globalization on education is the general accessibility of knowledge. With the use of the internet knowledge is readily accessible, that leads to the faster acquisition of knowledge by individuals as they have a wide range of information and knowledge available for them there. This makes possible that teachers and learners can create a global understanding of education on the basis of their communication with different people from around the whole world.

The effects of globalization lead to important changes in education by bringing rapid developments in communication and technology that foresees changes within education systems across the world in terms of values, ideas and knowledge, that also change the roles of teachers and students, and what's more produce a shift from the industrial society towards the information-based one. With such changes the distance barriers are easily being broken down due to the key aspect of globalization. People of all ages got an opportunity to learn in every suitable way, to attend live and virtual classes, and they don't need to be present in a class or an educational institution physically in order to learn. It allows for individuals to explore new horizons of thinking and learning that could not be done in the traditional educational environment with pen and paper and a teacher that dominates the classroom setting. Now knowledge can be gained through experimentation and inquiry.

2.2 Use of innovative technologies

Globalization and innovation have recently resulted in the extensive use of the latest technological products practically everywhere, and in education especially. Various technologies are now employed in different spheres of education. With the beginning of today's Digital Era the information and communication technology has been developed allowing real-time communication across space and time zones in one common environment reachable from everywhere around the globe - the virtual reality (VR). Thus, virtual environment enables the access of broad public to civilization knowledge, allows to share and acquire expertize in various fields across borders without geographical limitations [4].

These advantages are continuously being integrated into the educational sector. The use of VR in education is considered to be one of the global natural evolutions of IT- and computer-based training. The conducted analysis proved that the virtual environment stimulates learning, comprehension and reproduction, as it provides a tight and unique combination of experiential and symbolic information. VR is changing the way people learn in global and interesting ways. Rather than looking out the window, students could potentially see exactly what their teacher is talking about.

Immersing students in what they're learning has the potential to truly engage and increase their overall comprehension, no matter what type of learning method they prefer. Thus, VR in education provides the following benefits:

- Prevailing of the active experience over the passive one;
- Gaining of unforgettable great experience with no distractions;
- Getting immediate engagement;
- Helping trainees to understand complex phenomenon and subjects;
- Time compression.

Today information technology is widely used everywhere from educational establishments to the work places all over the world [5]. In schools in Japan, China and the USA such VR technologies and VR environments are commonly used, and VR lessons take a great part of their curriculum. In Russia it hasn't become general practice yet. So far VR lessons and materials are not included in the compulsory educational program, the legal, physiological and psychological aspects of VR usage have not been studied, though several cases of VR classes implementation have already taken place. For the Russian reality it means that the supplemented reality in learning and teaching is applied only by adding interactive materials into the ordinary textbooks. For example, when a student points a camera, tablet or phone on a picture in a textbook, he can see a molecule structure or a Maxican pyramid in 3D, study it from different angles, move it or interact with it primitively on the device screen. Another example is about implementation of vidoe materials such as a video of a biological process or a chemical reaction.

Some researcher determine the problem of modern higher education as the lack of new and effective approaches to learning and teaching [6]. He criticizes the use of lecture rooms with rows of seats as the main physical learning space for studying and teaching in higher education. Thus, he suggests to use such learning spaces as blended, physical/virtual, mobile, formal/informal, professional and to consider adaptability, flexibility, and time. Such learning spaces will reflect up-to-date learning and teaching techniques and strategies thus emphasizing individual and peer-based learning both in physical and virtual learning environments. In the educational sector in Russia, the use of VR technologies is a rare practice and the preference is still given to the older and more traditional methodologies.

While education institutions of highly developed countries are adopting technology rather effectively, less developed societies are left behind, hardly implementing development programs for education. A lack of funding and infrastructure hardly makes possible the implementation of any communication or technological development. However, notwithstanding the differences in politics, economy, culture and education, advanced and underdeveloped economies have adopted the best educational ideals and are anxious to rise education and promote it as a symbol of modernity and development, as a result of globalization.

3 Language Education in VR

3.1 VR in teaching

The application of VR was originally designed for aviation to visualize flights for pilots training, healthcare to visualize surgical operations, way finding and architecture to visualize large-scale projects. Now it became obvious that it can be widely implemented to teaching due to its unlimited possibilities and universal character.

The most general features of VR educational environment are:

- the use of 3D vector geometry when modelling the environment,
- unlimited movement possibilities for the users,
- updated views of the users as they move through the environment,
- the objects respond to the user action within the virtual environment,
- 3D audio.

Many researchers [7, 8] emphasize such a possibility of VR as to operate and act in the environments that are hardly or not accessible to individuals due to distance, scale, time and security.

The following possibilities of VR in education were singled out:

1) Visibility – as the use of 3D graphics makes it possible to detail various phenomena and processes [9, 10]. VR is able to provide information as well as to demonstrate the phenomenon with various details.

2) Security – as VR can immerse the person in any circumstance without any threat to life.

3) Involvement - as VR makes it possible to influence the reality and interact with it. For example, one can change the course of an experiment, drive a spaceship, etc.

4) Focusing – as the virtual world is designed specially to allow the viewer to concentrate entirely on the material and not distract on external stimuli like in the real world.

5) Presence – as an individual has a first person view and a sense of presence in VR world.

3.2 Innovative language training system

The team of Samara State Technical University, Russia designed and implemented the innovative educational VR course of English that creates a real situation of interaction allowing the student to make contacts, to give and receive certain information, to practice scientific research and educational techniques, to be engaged in project activities [11]. By introducing this course the students got the opportunity to understand the processes and situations that can hardly be explained in the traditional settings, to become a part of the real world through VR simulation.

It's obvious that the total immersion of a student into the real language environment can give results that both teachers and students are looking for. The educational process in classrooms can't replace this real experience of learning, no matter how well-organized, creative and communicative it is. Educators will agree that effective language learning depends on the motivation and the right language environment. As the majority of people who study a foreign language cannot afford to do it abroad, educators and teachers are searching for the alternative methods and inventing training models that are close to the natural and real language environment.

Today VR combines the two main factors for effective learning – immersion and motivation. VR allows students to immerse into another nation's culture for little or no cost. Students will definitely be interested in learning with VR technologies as it offers unique learning experiences by providing of real-time three-dimensional visualization and by introducing of various interactivity types within virtual learning environment.

With the help of VR a foreign language teacher can easily immerse a student into a real situational interaction, for example, a job interview, the situation in a shop, street, etc. VR space can be developed and adjusted according to the study objectives, time limits, number of participants, language level, animated or real characters, preferable devices, etc. It is flexible and convenient in use - that is a key factor for the students to use this technology in learning. Besides it helps to solve many problems of traditional learning. The new designed generational course of English «VR Customs» is based on three-dimensional VR technology. The minimum set necessary for using the program is a laptop or PC and VR glasses.

The language material was selected and organized on the basis of the frame technology. This is a convenient tool for carrying out the task of structuring a situation. If it is necessary to structure a dynamic situation, and this is exactly our case, it is customary to use a frame script. This is a dynamic structure in which elements are scanned, "run" by the mind's eye in a certain sequence [12, 13]. Scripts are characterized by the fact that they describe not objects and not static states of affairs, but events unfolding in time, which add up to the situation by forming a sequence. Thus, on the basis of script and frame description methodology, the structure of various classes of situations were characterized and described [14].

The virtual environment where the student receives information and makes a virtual inspection is the international airport. The main goal is learning and training professional English for customs officers. We have three locations – the main hall of the airport, red channel and green channel. The student has to interact with passengers, workers of the airport and do their tasks to find a smuggler. They give the student tasks like bringing documents they have lost somewhere, or choosing necessary bags from the porter's list of luggage. After doing exercises passengers give some hints and features of the smuggler the student is looking for. So at the end he does the final test and helps the customs officer find the smuggler. The next level is the red and green channel, where customs officers work. Students can learn the names of prohibited items and goods customs officers work with every day. Students listen to the dialogues that contain the most used expressions and terms. The customs officer interacts with the passenger, asks about his declaration, baggage, presence of prohibited items, valued things. Students can communicate with customs officers and get the necessary information about the Customs, job description and main responsibilities through the dialogue. Then the student should do lexical or grammar exercises to test the learned material.

So, the integrated learning system "Virtual Customs" is the universal innovative method of studing a foreign language as it uses the latest VR technologies. It also allows to train foreign language teachers from higher educational institutions to conduct lessons with the help of VR technologies taking as example the training system "Virtual Customs".

4 Conclusions

The information and communication technology has been evolutionary changing the known way of working and living through interconnecting people across time and space and enabling globalization. The revolutionary development of information and communication technology enabled the global information exchange to increase in volume and speed rapidly and continuously during last decades [15]. It created a new environment globally accessible for individuals for mutual interaction in one common place that is reachable from everywhere - the virtual world.

It should be mentioned that global education using VR allows you to conduct the training visually, to present all the aspects of the real processes and objects. As a whole it is highly effective, inhances the quality and the speed of the educational process, reducing its cost. With the help of VR technology a person can fully use the information from the outside world as 80% of it an individual perceives visually, 20% an individual remembers if only sees, 40% if sees and hears, and 70% is he sees, hears and does.

In general, VR can be considered an ideal learning environment. Its possibilities for teaching and studing have very high potential for application in a global sense. The full immersion in the professional environment allows students to build an individual way of studying the discipline alone or use it as an additional source.

References

1. R. Robertson, *Globalization: Social theory and global culture* (SAGE, Thousand Oaks, 1992).
2. Essays, UK, *The definition of globalization education essay* (2018). URL: https://www.ukessays.com/essays/education/the-definition-of-globalization-education-essay.php?vref=1. Accessed: 01.10.2018.
3. C.A, Twiggs, D.G. Oblinger, *The virtual university*. A report from a joint Educom/ IBM round table. Washington, Washington D.C. (1996). URL: https://www.educause.edu/ir/library/html/nli0003.html. Accessed: 22.11.2018.
4. M. Dávideková, J. Hvorecký, Collaboration tools for virtual teams in terms of the SECI model. *Advances in Intelligent Systems and Computing*, **544**, 97-111 (2016).
5. P. Standen, D. Brown, Mobile learning and games in special education. *The SAGE handbook of special education*, **2**, 719-730 (2014).
6. M. Keppell, K. Souter, M. Riddle, *Physical and virtual learning spaces in higher education: Concepts for the modern learning environment* (IGI Publishing, Hershey, PA, United States, 2011).
7. M. Bricken, Virtual reality learning environments: Potentials and challenges, computer graphics. *ACM SIGGRAPH Computer Graphics*, **25**(3), 178-184 (1991).
8. J. Cromby, P. Standen, D. Brown, Using virtual environments in special education. *VR in the Schools*, **1**(3), 1-4 (1995).
9. S.-Y. Yoon, J. Laffey, H. Oh, Understanding usability and user experience of web-based 3D graphics technology. *International Journal of Human-Computer Interaction*, **24**(3), 288-306 (2008). DOI: doi:10.1080/10447310801920516.
10. K.M. Stanney, M. Mollaghasemi, L. Reeves, R. Breaux, D.A. Graeber, Usability engineering of virtual environments (VEs): Identifying multiple criteria that drive effective VE system design. *International Journal of Human-Computer Studies*, **58**(4), 447-481 (2003).
11. V.V. Dobrova, P.G. Labzina, Virtual reality in teaching foreign languages. *Bulletin of Samara State Technical University, Ser. Psychological and Pedagogical Sciences*, **4**(32), 13-20 (2016).
12. R.C. Schank, R.P. Abelson, *Scripts, plans, goals and understanding: An inquiry into human knowledge structures* (Hillsdale, Psychology Press, 1977).
13. G. Mushayabasa, *Translation techniques in the Pershtta to Ezekiel: A frame semantics approach*. Studia Semitica Neerlandica, **63** (Brill, Leiden, 2015).
14. O. Kistanova, V. Dobrova, V. Savitskiy, E.Gashimov, L. Nurtdinova, Modeling as a method of describing the lexical meaning structure. In E.Yu. Malushko, N.L Schamne, N.V. Martyushev (Eds.), *Proceedings of the 7th International Scientific and Practical Conference "Current Issues of Linguistics and Didactics: The Interdisciplinary Approach in Humanities". Advances in Social Science, Education and Humanities Research*, **97**, 125-130 (2017).
15. M. Dávideková, M. Greguš, The impact of globalization on team composition: Virtual and local team. In T.Kliestik (Ed.), *Proceedings of the 16th International Scientific Conference Globalization and Its Socio-Economic Consequences* (pp. 354-361). Rajecke Teplice: UNIV ZILINA (2016).

How University Acts in the Development of "Smart Cities"

M. Afanasiev[1], and *M.* Lysenkova[1, *]
[*] Corresponding author: lysenkovam@gmail.com.

[1]Central Economics and Mathematics Institute of Russian Academy of Sciences, Moscow, Russia

Abstract. "Smart city" is a new model of territorial development, taking into account the growing importance of information, innovation and human capital. One of the main elements of the "smart city" is a developed system of higher education. The aim of this work was to study the impact of higher education on "smart" and innovative cities. The objectives of this work were to obtain quantitative characteristics of the impact of the University on the smart city. Approach to solving this problem was to build indicators based on indicators that characterize the quality of life, and ratings to compare cities on their basis. The hypothesis of the study is that the higher education system has a positive impact on the development of smart and innovative cities. A theoretically justified method of constructing an indicator of a certain direction of socio-economic development is a component analysis of indicators characterizing this direction. As a result, the rankings of Russian and foreign cities based on the characteristics of quality of life, which prove that education is a key indicator of the development of "smart" and innovative cities.

Keywords: "smart city", University, territorial development, innovation.

1 Introduction

The growing role of the knowledge economy in strengthening the competitiveness of cities and improving the quality of life of the population is based on the development of innovation. The problem is that innovation processes, although global in nature, can take place outside the borders of a particular region. Therefore, the potential of cities and regions largely depends on the level of development of the regional innovation system.

2 Methods

Analyzing regional innovation systems Asheim and Isaksen focus on the role of innovation space consisting of business, universities, research centers and institutional conditions that ensure their interaction [1]. The authors also emphasize that regional development largely depends on external interactions, access to global channels of reproduction of advanced achievements of world science, their dissemination and implementation in innovative products and services.

Islankina E.A. and Fiyaksel E.A. emphasize that the regional innovation system includes the subsystems of production and consumption of knowledge which are in interaction under the influence of supply and demand for the results of intellectual activity, the prevailing institutional conditions and practices of state regulation. At the same time, the regional innovation system is open to interaction with the external environment through global research networks, in which knowledge is disseminated and trained, as well as global production networks, through which investments are attracted to the region, and local companies enter the markets [2].

In the field of research of the regional innovation system, a model of a "smart city" has been formed and received wide worldwide distribution. The idea of a "smart city" implies the presence of a localized infrastructure facility that combines the interaction of such structures as the state, business and science. The city uses innovative technologies, provides high quality of life of people on the basis of development of technologies and improvement of ecology of environment. The concept of "smart city" involves the formation of an institutional environment conducive to the creation and dissemination of innovations in all spheres of life.

An important role for a smart city is the interaction and connection of the efforts of government, business and science. A special role for the development of innovative economy, achieving economic growth and accelerating scientific and technological progress in the interaction of the three structures (government, business and science) has already been noted in his works by Henry Itzkowitz [3]. He calls this interaction a triple helix, noting that universities have a leading role in building such cooperation. Itzkowitz believes that cities with several universities can become generators of innovation-smart cities. Thus, the aim of this work is to quantify the impact of higher education on the development of smart cities.

https://doi.org/10.1051/shsconf/20197105011

3 Analysis of international experience in "smart city" development

At the moment, there are about 143 smart city projects in the world [4] – implemented and still in the project stage. Most of these cities are now present in North America and Western Europe, but many projects have been created in Eastern Europe, Latin America, the middle East and even in Russia.

The aim of this work was to obtain quantitative characteristics of the impact of the University on the smart city. To achieve this goal, Spearman's comparative analysis, ranking and rank correlation coefficient were used. International experience is the most pronounced and amenable to analysis. The most striking examples of smart cities can be called only 20 in the world, they are worthy of the status of "smart city".

The World Council on City Data (WCCD) on its website provides a fairly wide range of innovative standardized indicators of cities [5]. For each innovative city, information is provided for several years (2014-2017), the list of indicators follows the ISO standard 37120:2014 "Regime of Sustainable development of the Russian community. The indicators of the city services and quality of life" [6]. 11 cities were selected for the study, among which 4 are considered "smart" cities (Melbourne, London, Toronto and Amsterdam), and the remaining 7 are developing smart cities (Boston, Dubai, Buenos Aires, Makati, Guadalajara, Amman, Mecca). Data for 11 cities for 2014 and 2015 were selected for the analysis. The collected data make it possible to make a ranked rating of a sample of cities for the whole set of indicators, for this purpose calculations for all blocks of collected indicators are carried out. Ranks are defined for each indicator in the work. These ranks are averaged over a set of indicators. Using the values of the averaged ranks, the rating of the sample of cities is constructed (table 1).

Table 1. List of smart and developing cities in the world

City	Country
Melbourne	Australia
London	Great Britain
Amsterdam	Netherlands
Boston	USA
Toronto	Canada
Buenos Aires	Argentina
Dubai	UAE
Guadalajara	Mexico
Amman	Jordan
Makati	Philippines
Mecca	Saudi Arabia

Source: Authors.

A natural approach to solving this problem is to build indicators based on indicators that characterize the quality of life, and ratings for comparing cities on their basis. A theoretically substantiated method of constructing an indicator of a certain direction of socio-economic development is a component analysis of indicators characterizing this direction. The most complete methodology of its application and the results of testing in assessing the quality of life are presented in [7]. Further indicators are based on 106 indicators of quality of life standard ISO 37120: 2014. All these indicators are relative and comparable for cities of different scales and different countries. Therefore, a simplified heuristic approach is used to construct the indicator for the entire set of indicators. At the first stage on the basis of the economic theory indicators for which the desirable direction of their change is known are revealed. For example, an increase in the proportion of the urban population provided with wastewater services or a decrease in the mortality rate of children under 5 years per 1,000 live births indicate an improvement in the quality of life. For each of these indicators, the ranks of the cities selected for comparison are determined. The indicator is plotted as the sum of ranks for all indicators. If for some indicator, for example, "number of firefighters per 100 000 inhabitants", the desired direction of its change is not known a priori, the correlation coefficients of the indicator with the already built indicator are calculated to identify it. After that, the ranks of cities are determined by this indicator and included in the indicator. On the basis of the constructed indicator the rating of cities on set of indicators is constructed.

In order to build ratings that take into account Russian cities, it is necessary to select from a total of 106 indicators such as those presented in the statistical reports of the Russian Federation. At the first stage of solving this problem, out of 106 indicators were selected those that are most correlated with the indicator. Indicators with correlation coefficients above 0.7 list further.

Spearman rank correlation coefficient:

20.1 Share of urban population provided with wastewater disposal services – 0,9146
12.1 Life expectancy – 0,9035
16.4 Proportion of municipal solid waste disposed of in organized landfills – 0,8875
6.7 Tertiary education rate per 100,000 inhabitants – 0,8712
18.1 High-capacity public transport system mileage per 100,000 inhabitants – 0,8444
13.2 Area of outdoor public recreation areas per capita – 0,8425
8.2 Suspended particle concentration (PM10) – 0,8293

12.4 Under-5 mortality rate per 1,000 live births – 0,8276
6.5 Percentage of male school-age population enrolled in educational institutions – 0,8161
4.7 GDP per capita – 0,8147
21.3 Proportion of urban population with access to quality facilities – 0,8127
12.3 Number of doctors per 100,000 inhabitants – 0,7961
6.6 Percentage of school-age population enrolled in educational institutions – 0,7878
14.3 Crimes against property per 100,000 – 0,7797
15.2 Number of homeless persons per 100,000 inhabitants – 0,7728
12.2 Number of hospital beds per 100,000 inhabitants – 0,7696
16.3 Proportion of municipal solid waste that is recycled – 0,7623
15.1 Percentage of urban population living in slums – 0,7523
7.2 Percentage of urban population with authorized electricity connections – 0,7477
5.6 Number of business entities per 100 000 people – 0,7454
21.1 Proportion of urban population connected to drinking water supply – 0,7394
10.1 Number of firefighters per 100,000 inhabitants – 0,7203
Further, five of these indicators were selected to build a combined rating of foreign and Russian cities. The main selection criterion was the availability of data on these indicators for all the cities under consideration for 2014 and 2015.
5.6. number of business entities per 100 000 people;
12.1 life expectancy;
12.2 number of hospital beds per 100,000 population;
12.3 number of doctors per 100,000 inhabitants;
6.7 tertiary education rate per 100,000 inhabitants

To check the informativeness of the selected indicators, ratings of foreign cities were built on four indicators (5.6, 12.1, 12.2, 12.3) according to 2014 (see column 3 of table 2) and 2015 (column 5 of table 2). The rank correlation coefficient of the 2014 ratings for 106 indicators (column 2 of table 2) and 4 indicators (column 3 of table 2) is 0.9273. The rank correlation coefficient of the 2015 ratings for 106 indicators (column 4) and 4 indicators (column 5) is 0.9364. For the period under consideration, the selected indicators are sufficiently informative to be used for building ratings.

Table 2. Ranks of foreign "smart" cities by indicators of ISO 37120:2014

№	Rating 2014 (106 indicators)	Rating 2014 (4 indicators)	Rating 2015 (106 indicators)	Rating 2015 (4 indicators)
(1)	(2)	(3)	(4)	(5)
1	Melbourne	Melbourne	Melbourne	Melbourne
2	London	Boston	Amsterdam	Boston
3	Amsterdam	London	London	London
4	Boston	Amsterdam	Boston	Amsterdam
5	Toronto	Buenos Aires	Toronto	Buenos Aires
6	Buenos Aires	Toronto	Dubai	Toronto
7	Dubai	Dubai	Buenos Aires	Dubai
8	Guadalajara	Makati	Amman	Makati
9	Amman	Amman	Mecca	Amman
10	Makati	Guadalajara	Guadalajara	Guadalajara
11	Mecca	Mecca	Makati	Mecca

Source: Authors.

0.9273 - Spearman rank correlation coefficient rankings 2014
0.9364 - Spearman rank correlation coefficient rankings 2015

Then there was built the combined ratings for the selected Russian and foreign cities on 4 indicators: column 2 of table 3 for 2014. and column 4 of table 3 for 2015. In is normal that the relative position of foreign cities in the ratings for 4 indicators in tables 2 and 3 coincides. In the 2014 ranking of four indicators, all four Russian cities occupy adjacent positions in the middle part, taking into account indicator 6.7 "share of higher education per 100 000 inhabitants". In the 2015 ranking, Moscow and St. Petersburg maintain their positions on four indicators. Kazan and Tomsk slightly worsen them. In the rankings for 5 of the indicators of higher education regarding the state of Russia's cities is improving for 2014 and 2015 In the last row of table 4 shows the sum of ranks of four Russian cities in each of the ratings. Indicator 6.7 improves the position of Russian cities in the ratings of 2014 by an average of 0.5 points, in the ratings of 2015-by 0.25 points.

Table 3. Ranks of foreign and Russian "smart" cities
by 4 selected indicators without taking into account the characteristics of education.

№	Rating 2014 (4 indicators)	Rating 2014 (5 indicators)	Rating 2015 (4 indicators)	Rating 2015 (5 indicators)
(1)	(2)	(3)	(4)	(5)
1	Melbourne	Melbourne	Melbourne	Melbourne
2	Amsterdam	Boston	Amsterdam	Boston
3	Boston	Moscow	Boston	Amsterdam
4	London	Saint-Petersburg	London	Moscow
5	Moscow	Amsterdam	Moscow	London
6	Saint-Petersburg	London	Saint-Petersburg	Saint-Petersburg
7	Kazan	Kazan	BuenosAaires	Toronto
8	Tomsk	Toronto	Kazan	Kazan
9	BuenosAaires	BuenosAaires	Toronto	BuenosAaires
10	Toronto	Tomsk	Tomsk	Tomsk
11	Dubais	Dubai	Dubais	Dubai
12	Makati	Makati	Makati	Makati
13	Amman	Guadalajara	Amman	Guadalajara
14	Guadalajara	Amman	Guadalajara	Amman
15	Mecca	Mecca	Mecca	Mecca
The sum of the ranks of cities of the Russian Federation	26	24	29	28

Source: Authors.

Improvement in 2014 on (26-24)/4 =0.5
Improvement in 2015 on (29-28)/4 =0.25

4 Conclusion

1. The results of the constructed ranks, led to the conclusion that education is a key indicator of the development of not only smart cities but also innovative cities.
2. Spearman's rank correlation coefficient showed that in order to compile the rating according to the ISO standard 37120:2014, it is possible to use not the whole set of 106 indicators, but a number of indicators that were obtained with maximum values.
3. In the considered sample there is a positive relative effect of the higher education index influence on the rating of the city development level.
4. The positive influence of the University on the development of Russian "smart" cities is traced.

References

1. B.T. Asheim, A. Isaksen, Regional innovation systems: The integration of local 'sticky' and global 'ubiquitous' knowledge. *Journal of Technology Transfer*, **27**(1), 77-86 (2002).
2. E.A. Islankina, E.A. Fiyaksel, Glocalization of innovation: The role of clusters and the international context in regional development. *Innovation*, **11**(205), 64-74 (2015). [in Rus.].
3. H. Etzkowitz, *Triple helix universities-enterprises-state. Innovations in action* (Routledge, New York, N.Y., 2008).
4. V.I. Drogginov, From the city of the mad to the half-mad and from it to the smart. Round table report "Smart cities: Potential and prospects of development in the regions of Russia", April 11, 2014 (2014). URL: https://issuu.com/epliseckij/docs/ bc9fac678b9405/2?ff&e=7773934/74 747. Accessed: 25.09.2019.
5. World Council on City Data, *Open City Data* (2016). URL: https://open.dataforcities.org/. Accessed: 25.09.2019.
6. ISO/IEC AWI 30145, *Information technology – Smart City ICT reference framework – Part 1: Smart city business process framework* (2016). URL: https://www.iso.org/ru/standard/76371.html?browse=tc. Accessed: 25.09.2019.
7. V.L. Makarov, S.H. Aivazyan, M.Y. Afanasiev, A.R. Bakhtizin, A.M. Nanavyan, The estimation of the regions' efficiency of the russian federation including the intellectual capital, the characteristics of readiness for innovation, level of well-being, and quality of life. *Economy of Region*, **4**, 9-30 (2014).

Methods for Overcoming Psychological Barrier by Adults while Studying Information Technology

V.V. Andreeva[1*], A.A. Makarov[2], and T.K. Makarova[2]
*Corresponding author: AndreevaVV-IT@mail.ru.

[1] Samara State University of Economics, Samara, Russia
[2] International Institute of the Market, Samara, Russia

Abstract. The article considers the problems of teaching information technology to the adult population of Russia in accordance with the strategic direction of the country's development until 2025 - the digital economy, which requires improving the process of teaching information technology (IT), developing new pedagogical technologies and methods focused on modern conditions. Evaluation of developed methods of teaching IT and overcoming the psychological barrier in a large sample of the country's population for a long time shows their replicability and effectiveness.

Keywords: information technology, training, methods, psychological barrier, adult population, digital economy.

1 Introduction

The digital economy is included in the list of the main directions of the strategic development of Russia until 2025. Despite the fact that the strategy involves the creation of new jobs, overcoming poverty and social inequality, there is a high probability of the so-called “digital divide” - a gap in digital education, in terms of access to digital services and products, and, as a result, the gap in well-being of people [1-9]. Especially strongly negative manifestations will affect people of the middle and older age groups of the population, most of whom do not own computer technologies.

Consequently, there is a need to create a system for teaching computer technology to the adult population, the goal of which is to introduce the population to new realities of the digital economy and eliminate “computer illiteracy”. In accordance with these goals, the main task of the system is to develop computer or information technology (IT) training methods for the adult population of the country, which will minimize the risk of a digital divide and overcome the psychological barrier in mastering IT [2, 4, 5].

2 Methods

To develop methods of overcoming the psychological barrier in the study of IT for different age categories of trainees, the authors use a long-term analysis of their activities during theoretical and practical classes [2, 4, 5].

The study involved:

- Schoolchildren studying engineering at schools;
- Students studying various IT courses at colleges;
- Students studying various IT courses at higher educational institutions;
- Undergraduates, graduate students and applicants for the scientific degree of candidate of sciences (to a limited extent);
- Participants in various advanced training courses.

We performed modeling of various non-standard situations that arise when working with a computer. We used the methods of system analysis and methods of “brainstorming” to find a way out of these situations, so that students can subsequently cope with problems when working with a computer on their own.

3 Results

The effectiveness of developed methods of IT teaching and overcoming the psychological barrier was tested (2000-2019) in different categories of students:

- Teachers of various specialties of higher and secondary professional educational institutions of the city of Samara and the Samara region (from humanitarian disciplines to natural science disciplines);
- Employees of the financial and economic sphere (accountants and economists of schools in the city of Samara and the Samara region);
- Workers of industrial enterprises of Samara;
- Executives of the education department of the Samara region.

The study of the behavior of these categories of trainees showed that their main problem is a psychological barrier that arises due to the complexity of the computer and lack of faith in ability to successfully master IT. This barrier leads to errors of various kinds, incorrect reaction of the participant to arising situations and inadequate actions, negatively affecting training.

Due to the age characteristics of schoolchildren and most students, the process of overcoming the psychological barrier occurs less painfully and more quickly than in adults who are not associated with information technology before this training. In addition, the length of study for schoolchildren (2 years) and students (4 years) allows them to gain sufficient IT experience and believe in their strength [2, 4].

The duration of IT training in advanced training courses is, as a rule, from 36 to 72 hours, which is insufficient for studying them.

Thus, the tasks of adult training are the most difficult and can be solved by:

- Overcoming the psychological barrier;
- Learning the basic methods of working on a computer;
- Preparing for independent mastering of a computer after the completion of classes. It means that adults develop research behavior.

The problems of the emergence of the psychological barrier in teaching various disciplines have been studied for many years, but most of these works are focused on the study of foreign languages (in particular English) by students of universities [6-10].

The problems of human interaction and computing machines have been studied since 70-80s of the last century. A lot of materials are also devoted to IT. As for the problems of adult education, they were not considered or were considered rather superficially. Pedagogical technologies, which allow relieving psychological tension in adults and make them confident in computer skills, were not developed [6, 8, 10].

Let us consider methods of overcoming the psychological barrier, since overcoming the natural distrust of participants to a complex technique, their fear of accidentally spoiling it is an important condition for their effective training, successful mastering the techniques of working on a computer.

The psychological barrier often occurs in people of middle and older age, as well as people with education in the humanities. Therefore, it is advisable to form groups in accordance with the age and professional characteristics of students and apply teaching methods adapted to these categories.

Consequently, it is necessary to build a pedagogical process so that, in addition to studying IT, students will learn to overcome the fear of technology and strengthen the faith in their own strengths and capabilities.

The removal of psychological tension and mobilization of participant's abilities are specific compared with the usual psychological counseling [2, 4]. In order to successfully solve the problem, the teacher teaching the class must improve his own communication skills. In addition, he must clearly know the causes of psychological tension and how to relieve it. As a result of studying the typical mistakes of inexperienced users, the main, most common causes of tension were identified, their classification was carried out and methods for their elimination were developed (Table 01). The developed methods are primarily aimed at teaching participants how to overcome critical situations that arise when working on a computer.

Table 01. Reasons of the psychological barrier and ways to remove it.

№	Main reasons	Ways to remove psychological barrier
1	Complexity of the computing system	The structure and work of the system is considered simplified. All concepts and devices of the computer are considered in a close logical relationship. In the future, with the emergence of basic skills, knowledge of the system is deepened and supplemented by new, more complex concepts.
2	Fear of spoiling expensive equipment	The computer has protection against incorrect user actions (Fool Proof). Therefore the system, in general, will not allow incorrect actions by the participant or input of incorrect data.
3	Unfamiliar situations, inability to find a way out of them without a teacher	Different situations and ways out of them are considered. Actions that lead to unexpected results are performed and ways out of such situations are shown. Problems that may occur in a particular situation are noted and solutions are considered.
4	Inadequate response of the system to participant's correct actions	You must learn to analyze the situation that has arisen, checking actions accuracy: - Actions seem to be correct, but in fact they are wrong. You need to carefully repeat all actions from the very beginning; - Software error (occurs relatively often) or computer failure (rarely), you need reset.

5	Loss of orientation in a large number of new terms, operations and algorithms for their implementation	Before learning new concepts, you need briefly revise the basic terms and operations studied previously. Thus, definitions, concepts, operations are fixed, and only after you study new elements. It is recommended to write down concepts, operations and methods of working on computer in notebook. It is proposed to work independently with these notes, since even IT-experts work this way.
6	Lack of experience	The functions of the help system - help, reference, hints, etc. are considered. It is proposed to develop your own help system in notebooks, since a relatively small number of techniques, elements, etc. are used in practical work. All IT specialists keep similar records at the initial stage of their work because of the large amount of new information and peculiarities of human perception.

Source: Authors.

4 Conclusion

Based on the considered features and techniques of IT training, pedagogical methods for teaching non-specialists to work on a computer, the authors develop the curricula and technology for conducting classes:

- The main attention is paid to the logical understanding of the process of working on a computer. It is viewed in dynamics, and the participant, together with the teacher, moves from simple operations to complex computer procedures;
- Under the guidance of the teacher, the participant gains experience in researching Windows applications, which allows him to independently study and use new applications in the future;
- The emphasis is not on cramming. It is on developing the ability to generate new methods based on the acquired basic knowledge and skills;
- The participant learns skills to overcome situations associated with erroneous actions of man and computer.

The developed methods were tested on various categories of participants in IT advanced training courses - accountants of schools in Samara and Samara region, workers of industrial enterprises of Samara, executives of the education department of Samara region, teachers of many Samara and Togliatti colleges, schools and universities, and these methods showed good results [2, 4, 5].

References

1. Fingramota, *What is the digital economy* (2017). URL: http://www.fingramota.org/teoriya-finansov/item/2198-chto-takoe-tsifrovaya-ekonomika. Accessed: 09.09.2019. [in Rus.].
2. V.V. Andreeva, The system of multi-level training of specialists in the field of information technology (Samara State Academy of Culture and Arts, Samara, 2005). [in Rus.].
3. V.A. Sukhomlin, Results of the reform of higher education: Movement for the revival of domestic science (2011). URL: http://www.za-nauku.ru/index.php?option=com_content&task=view&id=3851&Itemid=31 Accessed:09.09.2019. [in Rus.].
4. V.V. Andreeva, Problems of developing working programs on information technologies, In G.R. Khasaev, S.I. Ashmarina (Eds.), *Proceedings of the 16 International Scientific-Practical Conference* (pp. 191-194). Samara: Samara State University of Economics (2017). [in Rus.].
5. V.V. Andreeva, Improving the quality of teaching bachelor-computer science in digitalization. *The Portal of the Teacher* (2019). URL: https://portalpedagoga.ru/servisy/sbornik_konferencii/doklady_uchastnikov_konferencii/1369.doc. Accessed: 09.09.2019. [in Rus.].
6. O.V. Doronina, Fear of a computer: Nature, prevention, overcoming. *Questions of Psychology*, **1**, 68-78 (1993). [in Rus.].
7. T.J. Blayone, O. Mykhailenko, R.Vanoostveen, W. Barber, Ready for digital learning? A mixed-methods exploration of surveyed technology competencies and authentic performance activity. *Journal Education and Information Technologies*, **23** (3), 1377-1402 (2018).
8. L. Skvortsova, Psychological problems of using computer technologies. *E-learning Portal* (2010). URL: http://www.e-learning.by/Article/psichProblems/ELearning.html. Accessed: 09.09.2019. [in Rus.].
9. E.R. Korzhavina, Overcoming the psychological difficulties of mastering modern computer technologies. *Pedagogics Online* (2013). URL: http://aneks.spb.ru/srednee-spetcialnoe-obrazovanie/preodolenie-psikhologicheskikh-trudnostei-osvoeniia-sovremennykh-kompiuternykh-tekhnologii.html. Accessed: 09.09.2019. [in Rus.].
10. C. Iaosanurak, S. Chanchalor, E. Murphy, Social and emotional learning around technology in a cross-cultural, elementary classroom. *Education and Information Technologies*, **21** (6), 1639-1662 (2016).

Some Aspects of Digitalization in the Study of Criminal Law Disciplines

I.E. Milova[1,*], *E.A.* Milova[1], *D.D.* Osipov[1], and *T.A.* Kalenteva[1]
*Corresponding author: irina.milova@ro.ru.

[1] Samara State University of Economics, Samara, Russia

Abstract. Digitalization is a part of modern reality. It covers all areas of activity, including the educational process. For the training of lawyers is extremely important transition to the figure in the study of disciplines of criminal law orientation. It is in this area of knowledge that the use of artificial intelligence can give breakthrough results. Electronic formats of training not only develop the logic of thinking of lawyers, giving them a sum of theoretical ideas, but also prepare them for independent practical work, giving them the opportunity to gain professional applied skills. At the same time training simulators, business games, help to identify and develop the ability of students to implement specific law enforcement functions. At the same time, future lawyers develop a non-formal approach to obtaining information, with the transition from its simple perception to a logically oriented search, generalization and application, the ability to identify and identify the problem, to develop an algorithm for its step-by-step solution, to obtain optimally informed conclusions.

Keywords: modern education, computer technologies, University education, criminal law Sciences, modern technologies in education.

1 Introduction

The modern period of civilization is characterized by the spread of digital technologies. In Russia, public policy is focused on setting priorities in the field of information technology. It is emphasized that the transition to the figure in various fields of activity is one of the strategic objectives that determine the development of the country in the near future [1].

They are quite actively implemented in the educational process [2], including the high school, including the study of the disciplines of criminal law orientation [3]. In fairness, we note that the trend was ambiguously evaluated by the scientific community, because some scientists do not consider this vector measure scientific progress, noting there are a number of serious risks [4].

At the same time, most authors accept this phenomenon as an objective reality, considering it rather promising [5]. Moreover, it is believed that digitalization can really contribute to the breakthrough development of legal science, bringing it to a new conceptual level [6]. With this understanding and orientation training on the formation of students ' competences after graduation, a graduate is ready for independent work on the chosen specialty [7].

2 Problem Statement

Russia is a very promising country in the field of digitalization development, as there is a real social demand for the use of information technologies in all spheres of activity, including education. At the same time, legal professionals with such knowledge are in great demand in the investigative bodies, the Prosecutor's office, the judicial system, expert institutions and the bar. The problem is that legal education has long been focused on traditional methods of knowledge transfer, with an emphasis on their simple retransmission. As a result, at the output we received a graduate who was not ready to perform a specific job in the field of law, which required internships, qualification trainings. The transition to innovative forms of education is associated with the active involvement of students in the development of new material, with the development of their skills to find information and the possibility of its practical application.

3 Research Questions

We have made an attempt to explore the possibilities of digitalization in the development of students of criminal law disciplines. The role of forensic laboratories, which are a platform for the application of innovative methods of training lawyers, is shown. In addition, some proposals have been made to expand digitalization in this area of information acquisition; the advantages of electronic control methods for checking the residual knowledge slice have been identified.

4 Purpose of the Study

The purpose of this work is to try to summarize the experience accumulated in the Samara state University of Economics in digitizing the educational process of bachelors and masters specializing in this profile. We tried to make specific proposals to expand the use of electronic options for obtaining knowledge in this area, with an emphasis on the formation of students ' competence approach.

5 Research Methods

During the research we used such methods as system analysis, structuring, comparison, sociological and experimental approaches, scientific forecasting and modeling.

The basis for the preparation of the study were the publications of domestic and foreign authors, whose research interests are the issues of digitalization of the educational process in the study of criminal law disciplines by lawyers.

At the first stage, we carried out the selection of scientific materials on the chosen research topic, with their subsequent analysis.

At the second stage, the main conclusions of the study were formulated.

6 Findings

The Samara State University of Economics has a teaching and methodological office "Forensic Laboratory". Using its technical equipment, students, under the guidance of teachers, gain practical skills, having the opportunity to conduct expert research, with the inclusion of digital techniques. So, during fingerprinting, the automated information retrieval system Papilon is actively used, containing data on fingerprints in the database [8]; during ballistic examinations electronic data banks on firearms and cold steel are used. The laboratory has electronic booklet folders summarizing materials on various expert studies, with conclusions on specific criminal cases. Created electronic content allows you to monitor regional investigative, judicial and expert practice.

The problem is the fullness of forensic suitcases, as the equipment inside them becomes outdated quickly enough. At the same time, digitalization would allow them to be virtually supplemented with modern tools, which, of course, would make the study of forensics more relevant.

In addition, in the educational process, when studying a number of courses in the criminal law cycle, digital photo and video recording tools are used. For example, with their help, educational investigative actions are carried out, such as, inspection of the scene of the incident, verification of testimonies on the spot, presentation for identification. The use of digital equipment increases the accuracy of information consolidation, which, in turn, eliminates subjective moments in its assessment [9].

Information necessary for the study of criminal law disciplines, including tests, training simulators, forensic cases, business game mock-ups, educational films with elements of role-playing games, is stored in electronic format [10], which allows you to quickly find and effectively use it.

Interactive forms are very productive in the educational process, such as a business game - a trial in a criminal case, where students have the opportunity to try themselves in various procedural qualities, both on the part of the prosecution and the defense. Preparing to give speeches, the participants of the simulator game use digital technologies, both for developing a mock criminal case, and in order to study related materials; familiarization with the prevailing investigative and judicial practice in this category of cases.

Invaluable benefit in training is provided by various electronic reference systems and open access to resources on which judicial acts are placed, including in criminal cases. The information available in these databases is used during practical exercises, especially when solving problems-incidents.

Attractive is the creation of various kinds of information banks [11], in different areas of educational work within the framework of the courses being studied. Suppose, it seems advisable to create an electronic bank of methods for investigating economic crimes.

Traditionally, it is important to monitor residual knowledge [12]. Digitization of this process makes it possible to optimize the control procedure for students mastering theoretical material [13] and the formation of their practical application skills. Testing results are placed in the electronic environment of SSUE, with a clear indication of the criteria for their assessment.

It seems useful to create a digital forensic laboratory at the university, where it would be possible to concentrate the efforts of the faculty and students on innovative research, with interactive elements (for example, to create a forensic training ground on the basis of which using digital technologies it would be possible to simulate various forensic situations and work out the methodology and tactics of crime investigation). At the same time, such an innovation would contribute to the revival of publication activity; would introduce a modern orientation in the selection of development topics; would activate the applied component when writing master's theses.

The corresponding laboratory can be used as a platform for dialogue, including with remote participants on the most pressing problems [14]. The experience of this kind of virtual communication at the university is accumulated. An illustration is the holding of international, All-Russian conferences and round tables on the basis of the SSUE on the fight against extremism, anti-corruption, anti-drug readings.

The equipment of this laboratory is also advisable to use for career guidance work, with the involvement of potential applicants. In this sense, an interesting experience is the holding by the university teachers with the participation of students of master classes in the program "Young forensic scientist". These demonstration games-releases through online broadcasts are shown simultaneously in a number of schools, with the opportunity to participate in the discussion and laboratory research.

In addition, the digital format should be part of preventive work with the public, especially with young people, to prevent crime. The teenage audience, of course, responds better to the video series, with vivid illustrations, accompanied by small in volume, but memorable in content comments. At the same time, online communication is a more familiar format for a youth audience, which helps to establish trustful contacts and get a real response to the issues discussed.

A digital forensic laboratory could become a starting incubator for modern developments in the field of criminal law disciplines, with simultaneous electronic communication with other universities, including foreign ones [15], and practitioners interested in such studies.

Digital technologies are successfully applied at the SSUE as part of the functioning of the Legal Research Center. On the landscape of this structural unit, students, under the guidance of teachers, advise citizens who apply for various legal issues, including remotely, prepare the necessary procedural documents, electronic extracts from legislation, and reference materials. The materials obtained in the legal clinic are effectively used in the educational process as illustrative material, making it lively and practice-oriented.

We see extremely promising digital formats for interaction between lawyers (in the person of teachers, students, practitioners) and IT specialists, in creating projects that are at the junction of criminal law topics and programming. In particular, some stages of the educational process could be robotic, using artificial intelligence as an assistant, both for the teacher and for students. It is, first of all, about working in a forensic laboratory and using this content to control the knowledge gained.

7 Conclusion

To summarize what has been said, we note that digitalization is an objective reality [16]. Its task is to optimize social processes, including educational. At the same time, digitization will allow you to optimally quickly receive the information necessary for an in-depth study of criminal law disciplines, free teachers and students from routine technical functions, and expand the speed of their communication with other contractors. Switching to a number in law does not mean giving up creativity, a creative approach, individual serious work, which is characteristic of the legal profession.

References

1. Decree of the President of the Russian Federation of December 1, 2016 No. 642 "On the Strategy for Scientific and Technological Development of the Russian Federation" (2016). URL: http://www.consultant.ru/document/cons_doc_LAW_207967/. Accessed: 25.09.2019. [in Rus.].
2. The order of the Government of the Russian Federation of July 28, 2017 No. 1632-r "On approval of the program" Digital Economy of the Russian Federation" (2017). URL: http://www.consultant.ru/document/cons_doc_LAW_221756/. Accessed: 25.09.2019. [in Rus.].
3. M. Goodman, *International dimensions of cybercrime*. In S. Ghosh, E. Turrini (Eds.), *Cybercrimes: A Multidisciplinary Analysis* (Heidelberg, Berlin, 2011).
4. V.N. Voronin, Criminal law risks of digital technologies development: Problem statement and research plan. *Bulletin of the University O.E. Kutafina,* **12**, 73-80 (2018). [in Rus.].
5. M.M. Babayev, Risks as a component of the determination of the complex crime. *Bulletin of the Nizhny Novgorod Academy of the Ministry of Internal Affairs of Russia,* **1**(41), 104-110 (2018). [in Rus.].
6. D.S. Latyshev, A brief review of information technology used in legal activity. *Innovation Science*, **10**, 14-17 (2017). [In Rus.]
7. A.H. Dzhatdoev, Information technology in jurisprudence. *Young Scientist,* **6,** 20-24 (2018). [In Rus.]
8. I.P. Rak, Information technology in the activities of law enforcement agencies. *Innovation Science,* **2-3**, 132–135 (2016). [in Rus.].
9. V.A. Meshcheryakov, Digital forensics. The Forensic Library. *Science Journal,* **4**(15), 231-241 (2014). [in Rus.].
10. Ju.V. Gavrilin, A.V. Pobedkin, Gathering evidences in the form of data on electronic media in criminal justice in Russia: Necessarily to improve the procedural form. *Proceedings of the Academy of the Ministry of Internal Affairs of Russia,* **3**(47), 106-112 (2018). [in Rus.]

11. D. Brown, M. Warschauer, From the university to the elementary classroom: Students' experiences in learning to integrate technology in instruction. *Journal of Technology and Teacher Education*, **14**(3), 599-621 (2006).
12. E. Judson, How teachers integrate technology and their beliefs about learning: Is there a connection? *Journal of Technology and Teacher Education*, **14**(3), 581-597 (2006).
13. C. Pereira, M. Afonso, D. Santos, Educating for entrepreneurship: A new challenge for teacher training institutions. In C.S. Reis, P. Tadeu, T. Paiva (Eds.), *Conference on Enabling Teachers for Entrepreneurship Education, Guarda Polytechnic Institute* (pp. 50-51). Guarda: Guarda Polytechnic Institute (2013).
14. M. Boulet, B. Boudarbat, The economic performance of immigrants with Canadian education. *Regional and Sectoral Economic Studies*, **15**(2), 23-40 (2015).
15. M. Bouhajeb, H. Mefteh, R.B. Ammar, Higher education and economic growth: The importance of innovation. *Atlantic Review of Economics: Revista Atlántica de Economía*, **1**(01) (2018). URL: https://doaj.org/article/ceafcbafcb2647b08908d1d33c294d95. Accessed: 01.10.2019.
16. P. Uetela, Higher education and the challenges for economic growth in Mozambique: Some evidence. *International Journal of Sociology of Education*, **4**(3), 276-294 (2015).

Data Inventory as the Development and Promotion Basis of Universities Educational Programs

P. Zhelev[1], *I. A.* Svetkina[2], *A.S.* Zotova[2], and *A.A.* Chudaeva[2,*]
*Corresponding author: chudaeva@inbox.ru.

[1] University of National and World Economy, Sofia, Bulgaria
[2] Samara State University of Economics, Samara, Russia

Abstract. Digitalization affects universities, including management decision-making processes. Different management decisions require different data sets. The paper studies the issues of making managerial decisions at the university on the development and promotion of basic educational programs at the educational services market based on the results of data inventory. Two approaches are proposed: the study of data belonging to the university, and the search for the necessary data in the external environment. The article identifies the areas that need to be studied in order to make managerial decisions on ways to promote the main educational programs at the university, and analyzes the information accumulated both inside the university and in the external environment. The authors of the article proposed an algorithm for database inventory, which involves the identification of data sources, including digital traces, the analysis of the authority to use databases with the necessary registries of information systems and website directories. The result of data inventory should be the design of updating systems for the development and promotion of a university educational program. These actions will lead to the creation of an information system that will collect and process the diverse information needed to adjust the university's strategy. As a result, the university should expand the functional system that accumulates and systematizes various data required to promote the university educational programs at the market.

Keywords: digital traces, data inventory, university educational programs, data sources, information system catalogues.

1 Introduction

An inventory of university data is carried out considering the digitalization process, which involves new approaches to the organization of educational and related processes in higher education. So, in July 2019, 26 universities in the Russian Federation signed a charter on digitalization of the educational space. The charter contains uniform principles for the formation of the market for IT solutions for universities and will facilitate their networking and the dissemination of best digital practices [1]. Today, the use of digital traces in the educational process is being actively discussed [2, 3]. It is assumed that the data about the previous educational experience, or digital traces, will allow universities to analyze the student development and tailor the curriculum personally for him [4, 2]. Starting in 2021, a digital service based on the student's digital traces will be introduced in several Russian universities: it will evaluate student performance, lecture activity, participation in the university's public life, and overall behavior [5]. However, the organization of the educational process is preceded by the elaboration and development of educational program, the offer of the program at the market of educational services. Russian universities already have experience in applying data on potential consumers aiming to promote educational services on the market more effectively. So, the number of graduate students and those expelled in the first year of Tomsk State University (TSU) was reduced due to the fact that, during the admission campaign, the university began to analyze the social networks of applicants and scored the most motivated ones [6, 7]. Thus, the potential for using digital technologies as a tool for searching, ranking and organizing information for solving specific problems is huge.

Nowadays it is necessary to use digital tools not only as the educational tool [8, 9] but also in the way that allows to have real-time effective management decisions in the organization. Higher education institutions are no exception. For example, state universities operate in a highly competitive educational market and this leads to the need to transform and adapt digital methods of researching the catalog of educational products [10], including basic educational programs, tools for their development and promotion, as well as educational services consumers.

Questions of how modern technologies influence the study of the needs of the educational services market in educational programs are of specific interest within the framework of this study. Another important issue is what tools can be helpful in studying the data that are at the university's disposal and information that influences the adoption of managerial decisions, but which is formed outside the university and independent of it. In order to find the answer to these questions, it is necessary to determine the parameters of the data inventory in the decision-making process on the creation, development and promotion of educational products in the educational services market. In the context of digitalization, the inventory of university data should include the analysis of digital traces that need to be classified for the purposes of this study.

2 Methods

The purpose of the study is structuring the data inventory process as the basis for decision-making on the creation, development and promotion of the university main educational programs at the educational services market. It is necessary to sort out the information available to the university and in open sources and find out whether it would be enough to justify management decisions to promote the main educational programs at the market. In order to perform this efficiently, it is necessary to develop catalogs of the information that is necessary for making managerial decisions in matters of promoting educational programs at the market, add decision models to the database available at the university.

The methods which were applied while performing the research were such as: systemic, activity-oriented, integrated, social and managerial approaches. The authors used theoretical and empirical research methods. Theoretical methods include analysis, synthesis, comparison, generalization, classification, and empirical: observation, analysis of documents and information array.

3 Results

Digitalization processes significantly affect the activities of higher education institutions and the effectiveness of managerial decision-making in them. The basis of management decisions is data, the collection and processing of which in the digitalization environment becomes, on the one hand, simpler and, on the other, a very complex process. To build an inventory algorithm for the data array that allows you to make informed management decisions, you should determine the goal and the tasks necessary to achieve it.

The selection of the tools for the development and promotion of university educational products based on the study of the consumers interests is possible using data inventory that involves two approaches: studying the database formed at the university (UDB) and searching for the necessary data in the external environment (EED). In our opinion, inventory and monitoring [11] of a complete database (UDB + EED) allows not only to form a support system for making managerial decisions, but also find reserves for creating new products and / or services.

To develop and make managerial decisions at the university for its main educational programs' promotion at the market, the following tasks should be solved:

- to determine the characteristics of the target audience for main educational programs promotion (for undergraduate applicants - graduates of secondary vocational education and general educational organizations and their parents, for graduate applicants - graduates from different years of both the region and the country);
- to study the requirements for applicants for employment in the areas of training for which the university is accredited, as well as the requirements of professional standards and other regulatory documents for graduates;
- to study the opinions of educational program students and graduates about the quality of training and compliance with the requirements of employers, professional standards and other regulatory documents;
- to analyze the quality of graduates' employment;
- to explore the capabilities of the university in terms of resources availability for educational programs implementation meeting market requirements;
- to develop / process educational program in accordance with market requirements;
- to identify channels for promoting educational program at the market.

The solution of these tasks is possible by studying information, including digital, which the university accumulates, and searching for the necessary data in sources located in the external environment.

During the process of preparing for the inventory, we determined that the university has:

- the database of the university, which reflects data on applicants and their parents, on applicants for undergraduate studies and applicants submitting documents for graduate studies (the amount of data (gender, age, place of residence, etc.) is determined by the regulatory requirements of the Ministry of Education of the Russian Federation);
- regulatory framework (Consultant +, Guarantor, etc.), which allows to study the requirements for applicants for employment in the areas of study for which the university is accredited, as well as the requirements of professional standards and other regulatory documents for university graduates in areas of education which are accredited by the university;
- the University reports on the employment of graduates;
- the list of educational programs implemented by the university.

In addition to the listed data, which the university has independently of whether there is a request for a specific management decision or not, it is necessary to find / collect the following information:

- opinions of applicants, students of educational programs, their parents about the quality of training and compliance with the requirements of employers, professional standards and other regulatory documents, - various surveys conducted by organizations external to the university;
- the results of universities monitoring, monitoring the employment of graduates;
- the requirements for applicants for employment in the areas of training for which the university is accredited;
- the quality of employment of graduates of regional universities;

- which educational programs that meet the requirements of the market are implemented by universities - competitors (universities are considered to be competitors offering educational programs in the same areas of training as the university that is developing a promotion strategy);
- which channels for promoting services are used at the market and which of them are the most popular and effective.

To identify the necessary information, an inventory of the above data should be carried out. The results can be presented in Table 1.

Table 1. The sources of data necessary for making managerial decisions about educational programs promotion.

Data sources	The form of processing the data	The source for information system inventory	Web-sites catalogues
University data base	Multistage coordination of the right to receive certain information by managers and performers of various levels.	1C, IS «Dean office» UDB	University site
Regulatory base (Consultant+, Guarantor)	Open access to the full package of regulatory information. The problem is in the skills of searching, interpreting information in order to solve the problem, etc	Consultant+, Guarantor	Consultant.ru, https://minobrnauki.gov.ru/
University reports on graduates' employment	Graduates employment office	UDB EED	http://indicators.miccedu.ru/monitoring/
Information about the resources used by university	To access the full volume of data on resources, a decision of the university authorities on granting the access to it if it is necessary	1C, UDB	University site
The list of university educational programs	Information is open - the university website contains information in full	University site	University site
Students and parents feedback	Gathering, analysis and interpreting of information at the level of universities authorities	UDB EED	Social networks
Universities monitoring, monitoring the employment of graduates	Open information	UDB EED	http://indicators.miccedu.ru/monitoring/
The requirements for applicants for employment in the areas of training for which the university is accredited	Identification of the person / group of persons who will collect and process this information from open sources, cooperation with employers	EED	Sites, aggregating the information about vacancies (www.hh.ru, www.avito.ru, SuperJob.ru, etc.)
The quality of regional universities graduates' employment	At the level of university leadership, define the concept of "quality of employment" and those responsible for collecting information that allows you to analyze the quality of employment	EED http://indicators.miccedu.ru/monitoring/	http://indicators.miccedu.ru/monitoring/
Educational programs of universities - competitors	Open information at the universities' sites	EED	http://indicators.miccedu.ru/monitoring/ universities sites
Educational products promotion channels at the market	The analysis of such data for making managerial decisions is not performed.	EED	
Social nets and web search engine as the channels for universities promotion	Analysis of open "digital traces" of users for the promotion of educational programs, as well as for proposals for creating individual learning paths based on data about the applicant / student	EED	www.vk.com, www.instagram.com,

Source: Authors.

The process of inventorying university data involves the analysis of digital traces (or digital footprints) for certain markers, which should be understood as a collection of information about user visits and user contributions during the time he spent in the digital space. For inventory purposes, digital traces should be classified. In our opinion, digital traces can be internal and external.

Internal data should include information which can be directly controlled by the university. Internal digital traces remain in the information and educational environment of the university, it contains documents that ensure the implementation of basic educational programs, information about the university, work programs of disciplines, term papers and student projects, etc.

External digital traces include data from sources external to the university (social networks, search forms, etc.). External digital traces form a "digital shadow" for the university, which is not structured, is constantly expanding and does not obey the administrating procedures of the university. External digital traces are characterized either by the possibility of indirect control by the university, or by the inability to exercise control. External digital traces carry reputational risks for the university, which leads to the need to develop additional measures to promote the main educational programs of the university. One of the ways may be the latent promotion of educational programs through network users, which is a reference to the positive experience of students and graduates of educational programs in the Internet space. At the same time, users must be identified in the network as graduates of a specific educational program.

Also, digital traces are distributed in active and passive ways. In relation to the university and the educational program, passive digital traces are formed on the basis of data unintentionally created by the user - visiting various sites, social networks and resources associated with the university. This kind of information is automatically stored on the servers of providers, in web archives and data centers, and is reflected on site counters. The user leaves active digital traces consciously - for example, in social networks, choosing a university in which a person is studying or studied, participating in thematic groups with a university marker, posts, comments on discussions, etc. Analysis of the Vkontakte social network showed that the number of digital traces about SSEU is huge (and this number is not static, it increases every day): the number of individuals who have noted themselves as affiliated with SSEU is 37641, of groups containing in their name SSEU - 578, mentions of SSEU in the news - 21641, audio records - 61, video records - 1253 (data are given on the date of analysis).

Therefore, we note that an important part of the data inventory process is the quality control of digital traces with university markings. Unfortunately, often the quality of the data is poor, due to several reasons. Among them are the following: users' lack of understanding of the value of "digital traces" for decision-making, incl. strategic, the inability to structure the data in the context necessary for analytical purposes due to a lack of understanding by both developers and users.

Another point that should be considered is the answer to innumerous questions about the ethics and politics of processing such information. If the information is not kept forever, under what conditions should it be erased? Does the information produced through digital platforms impose any obligations on those who have access to it [12]?

When designing future information systems, the listed reasons should be considered in order to envisage measures aimed at eliminating them — this will improve the quality of the developed algorithms and the convenience of using information grouped as necessary.

For a project to promote a university degree, it is necessary to compare the goals and objectives with the available data. As the data in Table 1 show, the data available at the university will not allow us to solve the tasks and achieve the stated goal - there is a need to attract specialized companies and identify those responsible for collecting and analyzing the available data within the university.

4 Conclusion

The result of the inventory should be the design of system updates from tasks. We believe that it is necessary to develop catalogs of the information necessary for making managerial decisions in matters of promoting university educational programs at the educational services market and add decision models to the database available at the university. An IS can also be developed in which the collection and processing of external information necessary for developing a strategy for promoting educational programs will be carried out. The result, in our opinion, should be a target system that accumulates and systematizes the various data arrays needed to promote an institution of higher education on the market.

Acknowledgement

The research is done in the frame of project 574060-EPP-1-2016-KZ-EPPKA2-CBHE-SP "Enhancement of Higher Education and Corporate Sectors Integration in Accordance with New Social Environment" (ENINEDU) co-funded by the Erasmus+ Program.

References

1. D-Russia, *26 universities signed a charter on digitalization of the educational space* (2019). URL: http://d-russia.ru/26-universitetov-podpisali-hartiyu-o-tsifrovizatsii-obrazovatelnogo-prostranstva.html. Accessed: 10.10.2019. [in Rus.].
2. A. Blue, *Researcher looks at 'Digital Traces' to help students. Press release* (2018). URL: https://uanews.arizona.edu/story/researcher-looks-digital-traces-help-students. Accessed: 10.10.2019.

3. M. McGuirt, D. Gagnon, R. Meyer, *Embracing innovation: 2015-2016 higher education industry Outlook Survey* (KPMG, Amstelveen, The Netherlands, 2015).
4. B. Chechenova, The main features of the "mini-Islands": new ways of gaining knowledge, student tutors and the best digital footprint. *2035: In Progress* (2019). URL: https://ntinews.ru/in_progress/likbez/glavnye-fishki-miniostrova-novye-sposoby-polucheniya-znaniy-studenty-tyutory-i-luchshiy-tsifrovoy-sl.html. Accessed: 10.10.2019 [in Rus.].
5. N. Sokolov, K Gorshenin, On the digital footprint: An experiment will be launched in Russia to track artificial intelligence of student performance. *RT News* (2019). URL: https://russian.rt.com/russia/article/658196-studenty-ocenki-iskusstvennyi-intellekt. Accessed: 10.10.2019. [in Rus.].
6. TASS, *The analysis of social networks helped the Tomsk University to reduce the number of "C" students and expelled* (2019). URL: https://tass.ru/obrazovanie/6890954?fbclid=IwAR2n_oJUIZ_isEV_Vyfi2LL9LNcBjOKcezKy3wLzsYMpt56hD0Ql0AEzy. Accessed: 10.10.2019. [in Rus.].
7. G.V. Mozhaeva, A.V. Slobodskaja, A.V. Feshhenko, Information potential of social networks to identify the educational needs of senior pupils. *Open and Distance Education*, **3**(67), 25-31 (2017). [in Rus.].
8. J. Bleicher, P. Forrester, A Honal, E. Velinov, Virtual business projects in the classroom: Enhancing intercultural and business skills of students. In Gonzalez-Perez M.A., Lynden K., Taras V. (Eds.), *The Palgrave Handbook of Learning and Teaching International Business and Management* (pp. 747-768). Cham: Palgrave Macmillan (2019).
9. E. Velinov, E. Tiit, I. Denisov, Dealing with students' aversion in global virtual teams: Evidence from Czech Republic and Estonia. In A. Belousova (Ed.), *Proceedings of 2nd International Conference on Education Science and Social Development (ESSD 2019), Advances in Social Science, Education and Humanities Research,* **298**. Paris: Atlantis Press (2019). DOI: 10.2991/essd-19.2019.80.
10. A.G. Furin, I. I. Akhmatov, The concept of "educational product" of general education: The essence and features of formation in the educational cluster. *Modern Problems of Science and Education,* **2**. Retrieved from: http://science-education.ru/ru/article/view?id=8717. Accessed: 10.10.2019 (2013). [in Rus.].
11. M.V. Khokhlova, S.V. Lukashov, The implementation of the principles of quality monitoring of the basic professional educational programs in engineering and technology at the university. *Modern Problems of Science and Education*, **5**. URL: http://science-education.ru/en/article/view?id=27006. Accessed: 10.10.2019 (2017). [in Rus.].
12. M. Kurzweil, M. Stevens, Setting the table: Responsible use of student data in higher education. *EDUCAUSEREVIEW* (2018). URL: https://er.educause.edu/-/media/files/articles/2018/5/er183101.pdf. Accessed: 10.10.2019.

AUTHOR INDEX

AUTHOR INDEX